Smooth muscles wrap around blood vessels and many internal organs and, in general, are involved in moving fluids and slurry around the body; smooth muscle is also an important component of the uterus. These muscles are controlled by the action of hormones and by nervous stimulation, and can be influenced by drugs. Controlling factors act through a complex system of signals involving receptors, enzymes, and inorganic ions. Calcium, in particular, plays a major regulatory role.

This book provides an up-to-date review of our understanding of smooth muscles and integrates molecular, cellular, and physiological information with tissue and anatomical studies. Well-known researchers have written chapters giving detailed reviews of our current knowledge of the biochemistry, pharmacology, physiology, and anatomy of smooth muscles. These chapters cover the seven most important areas of smooth muscle function: morphology, electrophysiology, mechanisms of electromechanical and pharmacomechanical coupling, calcium homeostasis, signal transduction, mechanics of contraction, and the contractile proteins.

All those interested in muscular contraction will find this book worthwhile, be they biochemists, physiologists, pharmacologists, or cell biologists.

CELLULAR ASPECTS OF
SMOOTH MUSCLE FUNCTION

CELLULAR ASPECTS OF SMOOTH MUSCLE FUNCTION

Edited by

C. Y. Kao
State University of New York
Health Sciences Center, Brooklyn

Mary E. Carsten
University of California, Los Angeles
School of Medicine

CAMBRIDGE UNIVERSITY PRESS
Cambridge, New York, Melbourne, Madrid, Cape Town, Singapore,
São Paulo, Delhi, Dubai, Tokyo, Mexico City

Cambridge University Press
The Edinburgh Building, Cambridge CB2 8RU, UK

Published in the United States of America by Cambridge University Press, New York

www.cambridge.org
Information on this title: www.cambridge.org/9780521482103

First published 1997

A catalogue record for this publication is available from the British Library

Library of Congress Cataloguing in Publication Data
Cellular aspects of smooth muscle function / edited by C. Y. Kao, Mary
E. Carsten.
p. cm.
Includes index.
ISBN 0-521-48210-0 (hardback)
1. Smooth muscles - Physiology. 2. Muscle cells. 3. Smooth
muscles - Molecular aspects. 4. Muscle contraction. I. Kao, C. Y.
II. Carsten, Mary E.
[DNLM: 1. Muscle, Smooth - physiology. 2. Muscle Contraction-
physiology. WE 500 C393 1997]
QP321.5.C45 1997
573.7'536 - dc21
DNLM/DLC
for Library of Congress 96-37371

ISBN 978-0-521-48210-3 Hardback
ISBN 978-0-521-01860-9 Paperback

This book is dedicated to
R. Christine Kao
and
William M. Bowers

Contents

Contributors

Mary E. Carsten
Departments of Obstetrics/Gynecology and Anesthesiology, School of Medicine, University of California Los Angeles, Los Angeles, California 90095.

Joseph M. Chalovich
Department of Biochemistry, East Carolina University School of Medicine, Greenville, North Carolina 27858.

Giorgio Gabella
Department of Anatomy, University College London, London WC1E 6BT, U.K.

Luke J. Janssen
Department of Medicine, Division of Respirology 3UI, McMaster University Medical School, Hamilton, Ontario, Canada L8N 3Z5.

Kristine E. Kamm
Department of Physiology, University of Texas Southwestern Medical Center, Dallas, Texas 75235.

C. Y. Kao
Department of Pharmacology, State University of New York Health Science Center at Brooklyn, Brooklyn, New York 11203.

Richard A. Meiss
Departments of Physiology and Biophysics and Obstetrics/Gynecology, Indiana University School of Medicine, Indianapolis, Indiana 46202.

Jordan D. Miller
Department of Anesthesiology, School of Medicine, University of Calilfornia Los Angeles, Los Angeles, California 90095.

Gabriele Pfitzer
Institut für Physiologie, Medizinische Fakultat der Humboldt Universitat zu Berlin, D-10098 Berlin, Germany.

Stephen M. Sims
Department of Physiology, The University of Western Ontario, London, Ont. Canada, N6A 5C1.

R. Ann Word
Department of Obstetrics/Gynecology, University of Texas Southwestern Medical Center, Dallas, Texas 75235.

Preface

Smooth muscles occur in many parts of the body, usually in sheets enveloping hollow organs or tubes which are either occasionally or always filled. The ultimate function of smooth muscles is to move the fluid or slurry contents forward to some destination. Thus, they can receive and produce signals which then lead to contraction or relaxation. Smooth muscles have been studied since at least the founding of physiology, and have been indispensable assay systems for studying autonomic neural functions, actions of hormones and drugs, and mechanisms of second-messenger systems. They are named for the absence of striations by the early microscopists, and hence are also known as plain muscles. Plain as they might appear, there is nothing bland about them. They are ancient structures in which ionic channels, contractile machinery, and regulatory processes were laid down before the emergence of fast-twitch muscles. In malfunctioning, they afflict more sufferers than all dysfunctions of striated muscles put together. Possibly, in one form or another, ailing smooth muscles consume the largest share of health-care costs.

Yet, we do not know them very well. Their dispersion in relatively small amounts, the abundance of connective and other nonmuscle tissues amongst them, the diversity in their properties, and the small sizes of individual myocytes are all qualities which discourage and deter concerted studies. Perhaps for these reasons, they remain like strangers, recognized sometimes by incompatible transplanted information from other systems.

However, over the past decades, substantial advances have been made in our knowledge of smooth muscles – in such areas as their fine structures, the contractile systems and their regulations, ionic currents of single smooth myocytes, and neurotransmission mechanisms. In some areas, certain general principles are emerging as underlying the functional diversity. One purpose of this volume is to seek out such general principles without trivializing the cell-specific properties. In moving into these exciting realms, aided by new methodology, we are mindful that in their natural states smooth myocytes do not function as isolated cells. Rather, they always function in concert with innumerable other smooth myocytes, enmeshed neural terminals and their transmitter-bearing varicosities, connective tissues, and other cellular or noncellular matrices. By the old saw of physiology: we study the parts to better understand the whole.

In this volume we have brought together a collection of seven chapters on various cellular aspects of smooth muscle functions. Our aim is to bridge new information on general and specific cellular properties with functions of the parent smooth muscle tissue. The choice of topics was determined in part by the size of the book and in part by our desire to let contributors delve as deeply into a topic as necessary for appropriate presentation of their material. All of us will be gratified if any part of this book encourages the further study of smooth muscles.

C. Y. Kao
Mary E. Carsten

Abbreviations

[X]	concentration of X
$[X]_i$	intracellular concentration of (cytosolic-free) X
$[X]_o$	extracellular concentration of X
4-AP	4-aminopyridine
ACh	acetylcholine
AMP-PNP	5'-adenylyl imidodiphosphate
ATP	adenosine 5'-triphosphate
Ca^{2+} pump	Ca,Mg-ATPase
cADPR	cyclic adenosine diphosphate ribose
CaM	calmodulin
CaMK II	calcium-calmodulin-dependent protein kinase II
CK II	casein kinase II
CNBr	cyanogen bromide
CRAC	calcium release-activated current
CSA	cross-sectional area
DAG	1,2-diacylglycerol
diC_8	sn-1,2-dioctanoyl-glycerol
E	elastic modulus
E_{ion}	equilibrium potential for an ion
E_K	potassium equilibrium potential
ED_{50}	dose required to produce half-maximal effect
EGTA	ethylene glycol-bis-(β-aminoethyl ether) N,N,N',N' tetra acetic acid
ELC	essential light chain

γ	unitary channel conductance
G protein	guanine nucleotide-binding (regulatory) protein
GTP	guanosine triphosphate
HMM	heavy meromyosin, double-headed catalytic region of myosin
i	current (single-channel)
I	current (macroscopic, whole-cell)
IC_{50}	half-maximal inhibition
i_{ion}	single-channel current amplitude
I_{Ca}	calcium current
I_K	potassium current
I_{Na}	sodium current
I_{TO}	transient outward current
$InsP_2$	(d-myo-)inositol bisphosphate
$InsP_3$	(d-myo-)inositol 1,4,5-trisphosphate
$InsP_3R$	inositol trisphosphate receptor
$InsP_4$	(d-myo-)inositol tetrakisphosphate
K_{ATPase}	$1/K_m$, reciprocal of actin concentration required for half-maximal activity
$K_{binding}$	association constant of myosin to actin during ATP hydrolysis
K_{CaM}	concentration of Ca^{2+}-CaM required for half-maximal enzyme activity
k_{cat}	maximum rate (per second) of ATP hydrolysis extrapolated to an infinite actin concentration
K_d	equilibrium dissociation constant
K_m	concentration of substrate required for half-maximal enzyme activity
$K_v1.2$, $K_v1.5$	voltage-gated delayed rectifier K^+ channels of known amino acid sequences
kDa	kilodalton
L_i, L_o	initial, optimal tissue length
MAP	mitogen-activated protein
MLCK	myosin light chain kinase
MLCP	myosin light chain phosphatase
N	number of functional channels
NCDC	2-nitro-4-carboxyphenyl-N,N-diphenyl-carbamate
NMR	nuclear magnetic resonance (spectroscopy)
P	permeability
p_o	probability of channel opening
PKA	(cAMP-dependent) protein kinase A
PKC	protein kinase C
PKG	(cGMP-dependent) protein kinase G

PLC	phospholipase C
PM	plasma membrane
PMCA	plasma membrane Ca,Mg-ATPase
PP	protein phosphatase
$PP1_M$	myosin protein phosphatase type 1
ρPDM	N,N'-ρ-phenylenedimaleimide
RLC	regulatory light chain
RyR	ryanodine receptor
S	siemens (unit of conductance)
S1	subfragment 1, single-headed catalytic domain of myosin
S2	subfragment 2, region of myosin adjacent to S1
SERCA	sarco[endo]plasmic reticulum Ca,Mg-ATPase
SDS-PAGE	sodium dodecyl sulfate polyacrilamide gel electrophoresis
SITS	4-acetamido-4'-isothiocyanostilbene-2,2'-disulfonic acid
SM-1	204-kDa isoform of smooth muscle myosin heavy chain
SM-2	200-kDa isoform of smooth muscle myosin heavy chain
SMP	smooth muscle phosphatase
SR	sarcoplasmic reticulum
STIC	spontaneous transient inward current
STOC	spontaneous transient outward current
STX	saxitoxin
τ	time constant
τ_f	time constant of fast component
τ_m	time constant of intermediary component
τ_s	time constant of slow component
TEA	tetraethylammonium (ion)
TRP	transient receptor potential gene product
TTX	tetrodotoxin
U73122	1-[6[[17-3-Mesoxyestra-1,3,5(10)trien-17β-yl]amino]hexyl]-1H-pyrrole-2,5-dione
V_m	membrane potential
V_{max}	maximal velocity (of enzyme activity)
V_{rev}	reversal potential

1

Morphology of Smooth Muscle

GIORGIO GABELLA

1. Introduction

Smooth muscle has a wide distribution in the body, and its functional specializations are extraordinarily varied. Smooth muscle is chiefly located in the wall of hollow organs, where it occurs in broad and thin sheets, in arrays of bundles, or, in the case of the taeniae, as conspicuous long cords. Other cordlike smooth muscles connect the ovaries, duodenum, and rectum to the posterior abdominal wall, and link hair follicles to the surrounding connective tissue. In mammals, a ring of smooth muscle lies close to the pupillary edge of the iris. Individual smooth muscle cells are scattered within connective tissues or close to epithelia in many organs. The total amount of musculature in the body is difficult to calculate. On the basis of the estimates shown in Table 1, smooth muscle may represent 2% of human body weight.

Similarities and dissimilarities with striated muscles are obvious. Smooth muscles are made of small, elongated, uninucleated cells, embedded in abundant extracellular material with a large fibrous component (Figure 1): intercellular cooperativity and role of extracellular material are greater than in striated muscles. A smooth muscle cell has no transverse striations in spite of its actin and myosin filaments, has no T tubules, has numerous caveolae, and has an extensive insertion of the contractile apparatus on its cell membrane over the entire cell length. Its contractile machinery, based on actin and myosin filaments and on a dominant role of calcium (as in striated muscle), has an architecture and consequent mechanical properties that are substantially different from those of striated muscles. Although capable of producing an amount of force comparable to striated muscles, smooth muscles have a much slower and more variable speed of contraction, and their utilization of energy is markedly different. Save for the occurrence of external constraints, the extent of shortening can be exceedingly large.

Table 1. *Rough estimate of total amount of smooth muscle in the human body*

Bladder	30–60 grams
Ureter, vesicles, vas	30–60 grams
Uterus	200–400 grams
Gut	700–1,000 grams
Airways	50–100 grams
Blood Vessels	150–300 grams
Skin	10–30 grams
diffuse	50–100 grams

Note: These give a total of 1,020–1,600 grams, or between 1.5% and 2.2% of body weight, or 4–5 times the amount of cardiac muscle.

2. Structure of smooth muscle

2.1. Size and shape of muscle cells

Smooth muscle cells are spindle-shaped, the wider portion usually measuring 2–4 μm in width (Figures 1 and 2). Their length ranges up to 1,000 μm in visceral muscle cells, but is considerably less in vascular muscle cells. The characteristic shape seems dictated by the insertion of the contractile apparatus over the entire cell surface (rather than exclusively at the ends, as is the case in striated muscle fibers, which have a distinctive cylindrical shape) and is essential for the mechanical properties of these cells. Rare exceptions include the myo-epithelial cells of glands, which are cup-shaped but break up at the edge into several tapering processes, as well as disc-shaped myoid cells of the testis, whose actin filament bundles run in two directions in each cell, parallel and perpendicular to the axis of the seminiferous tubule (Vogl et al. 1985; Maekawa et al. 1991).

Visceral muscle cells have a volume of 2,500–3,500 cubic microns (approximately the volume of a monocyte), with modest differences between different muscles and modest differences in the same muscle in different species. The cell surface is relatively large (about 5,000 square microns, not taking into account the caveolae), giving a surface-to-volume ratio of about 1.5 μm^{-1}(approximately the same as in a human erythrocyte; Linderkamp and Meiselman 1982). Muscle cells of elastic arteries are much smaller and have a more irregular surface (see Figure 8), with a correspondingly higher surface-to-volume ratio—for example, 2.7 μm^{-1} (or 2.7 μm^2 of cell surface per μm^3 of cell volume) in rat

Figure 1. (*facing page*) A. Transverse section of the circular muscle layer of the rat ileum, showing muscle cell profiles (one nucleated) with mitochondria and myofilaments. To the top right and bottom left are two intramuscular nerve bundles. In the intercellular space there are collagen fibrils, mainly in transverse section. [16,000×] B. Longitudinal section of the adjacent longitudinal muscle layer. The muscle profiles (one nucleated) show elongated mitochondria and packed filaments (mainly myosin filaments are visible). [16,000×]

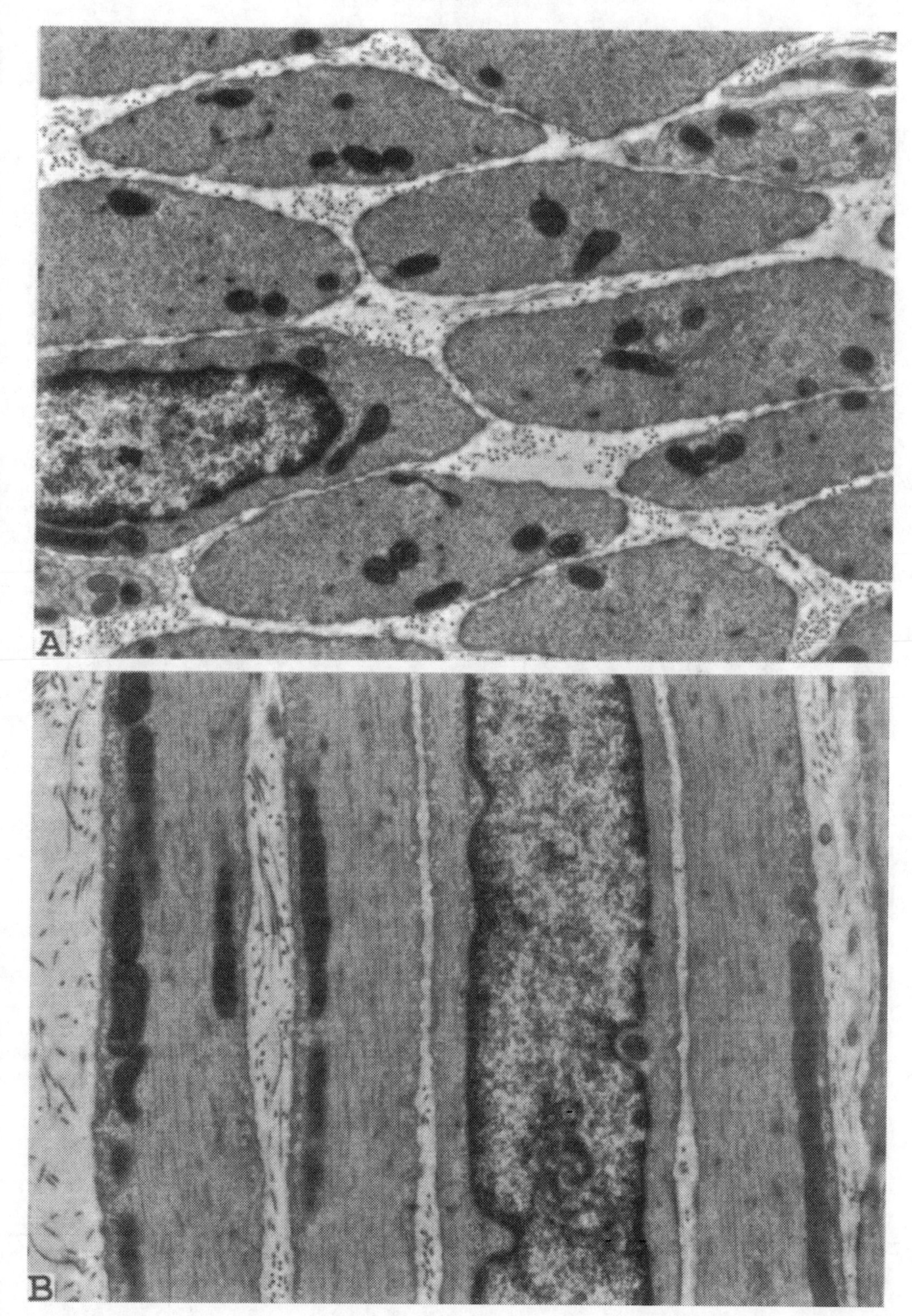
A
B

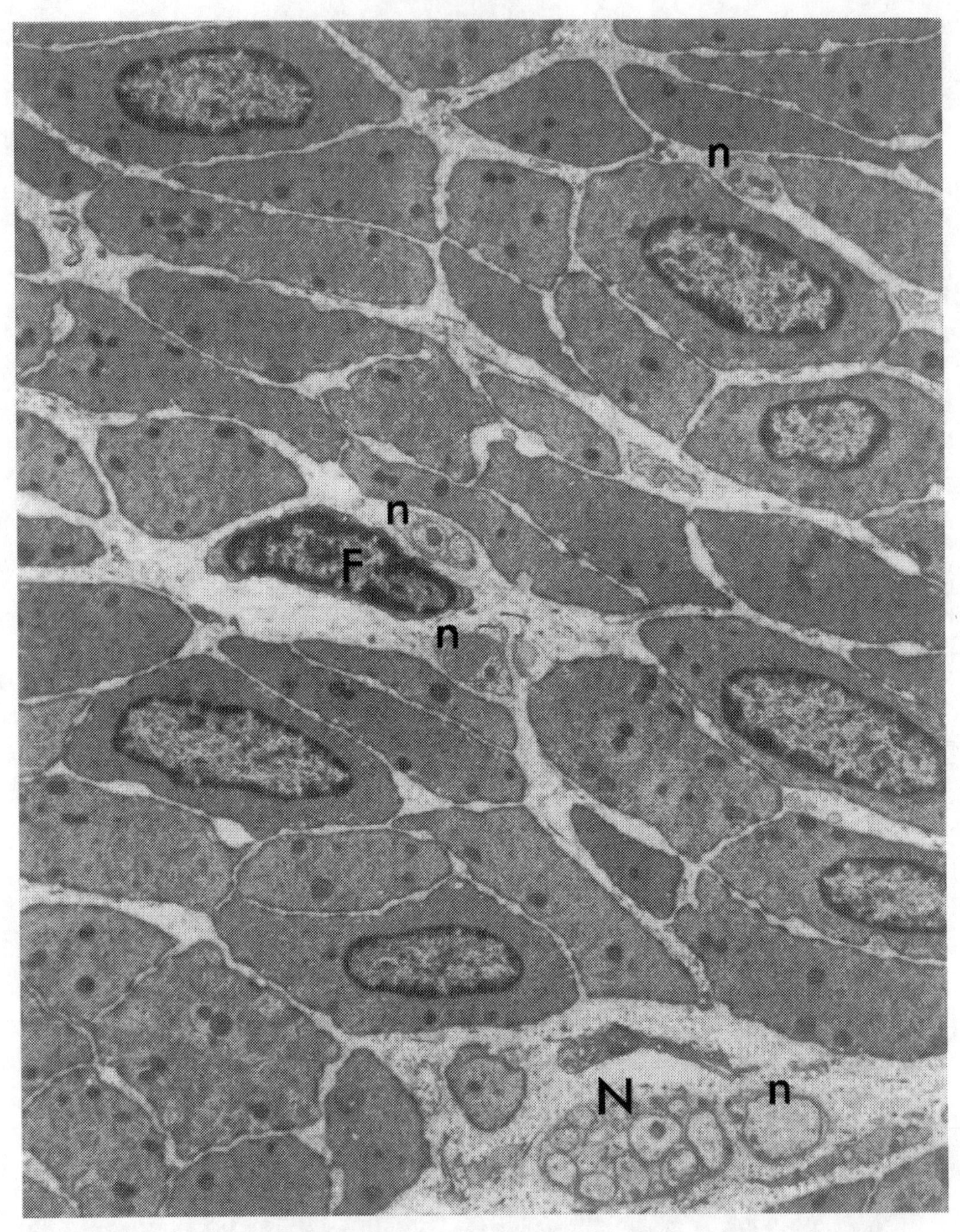

Figure 2. Musculature of the bladder of a rat, in transverse section. Among the many muscle cell profiles are several nerve fibres (n), a nerve bundle (N) and a fibroblast (F). [9,000×]

aorta muscle cells (Osborne-Pellegrin 1978; for changes in cell volume with contraction, see p. 22).

It has been calculated that in the guinea-pig taenia coli, where the extracellular space amounts to about 30% of the muscle volume, there are approximately 190,000 muscle cells per cubic millimeter; the amount of cell membrane is therefore roughly one square meter per gram of tissue (and 70% more if the caveolae are taken into account) (Gabella 1981). This estimate can be used for other visceral muscles, but it should be noted that preparations for pharmacological or biochemical experiments tend to include a substantial amount of extramuscular connective tissue and therefore give a smaller yield of muscle cells.

2.2. Cell membrane

Located between sarcoplasm, cytoskeleton, and contractile apparatus on the one side, and amidst basal lamina, stroma, extracellular space, and adjacent cells on the other side, the cell membrane is at the center of all the mechanical, metabolic, and signaling processes of the tissue. Its structural specializations include caveolae, various cell-to-cell junctions, cell-to-stroma junctions, and incrustations forming the dense bands (Figures 1 and 3).

Different regions, or domains, can be distinguished in the cell membrane. The most obvious distinction is between regions occupied by caveolae and those occupied by dense bands (Figure 4). Intramembrane particles, as seen in freeze-fracture preparations (about 450 per square μm on the P face and 300 on the E face; Devine and Rayns 1975) are mainly found in the regions occupied by the caveolae. Chemically, the domain of the caveolae is characterized by the presence of dystrophin (Byers et al. 1991, North et al. 1993) and of a specific calcium pump (Fujimoto 1993), whereas the regions occupied by the dense bands are associated with vinculin and other cytoskeletal molecules.

2.3. Caveolae

The invaginations of the muscle cell membrane known as caveolae measure 70 nm across and 120 nm in length, their long axis orthogonal to the surface of the cell at rest. In visceral muscle cells they are arranged in longitudinal rows interposed between dense bands.

The significance of caveolae is still obscure. They are regular in shape and size (although they occasionally form composites), they are stable structures, and their number is relatively constant for a given muscle and is unaffected by muscle contraction or relaxation. Caveolae are not expression of micropinocytotic activity (Devine et al. 1972) and are not seen to separate from the cell surface and turn into internalized vesicles. In the taenia coli of the guinea pig, there are 32–35 caveolae per square micron of cell surface or about 170,000 caveolae in one muscle cell (Gabella and Blundell 1978). In vascular muscle cells, caveolae are more numerous on the luminal than on the adluminal side (Lever et al. 1965, Forbes et al. 1979). Immunocytochemistry has shown that the plasmalemmal calcium pump ATPase is localized in the membrane of caveolae in smooth muscle and other tissues (Fujimoto 1993), in agreement with earlier immunohisto-

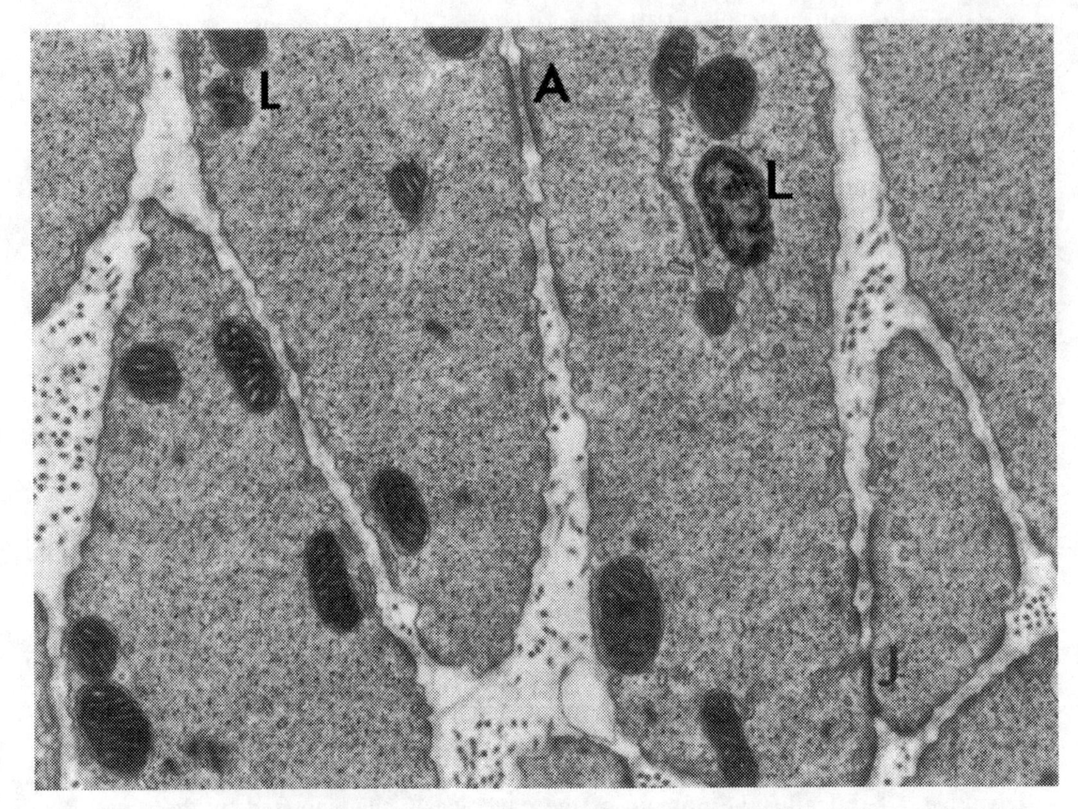

Figure 3. Circular muscle layer of the rat ileum in transverse section. The muscle cell profiles show several organelles: mitochondria, lysosomes (L), rough sarcoplasmic reticulum, vesicles, and cisternae of smooth sarcoplasmic reticulum. An intermediate junction (A) and a gap junction (J) are visible. Dense bodies encrust the cell membrane. The rest of the cytoplasm is packed with myofilaments and dense bodies. Collagen fibrils are visible in the extracellular space. [31,000×]

chemical observations (Ogawa et al. 1986, Nasu and Inomata 1990) and with electron-probe studies showing a high concentration of calcium inside caveolae (Popescu and Diculescu 1975). A protein probably involved in calcium entry through the plasma membrane, an inositol 1,4,5-trisphosphate receptor, is localized in the caveolae (Fujimoto et al. 1992).

Similar caveolae are found in cardiac muscle cells, in skeletal muscle fibers, in endothelial cells, and in pneumocytes, all of which are exposed to unusual mechanical stresses. It is possible that they serve a mechanical role. Other equally hypothetical suggestions include a role for caveolae in calcium transport across the cell membrane (Popescu and Diculescu 1975; Crone 1986; Fujimoto 1993), in the control of cell volume (Garfield and Daniel 1976), as miniature stretch receptors (Prescott and Brightman 1976) or as organelles creating a specialized compartment of the extracellular space. In muscle cells of *Aplysia*, a longitudinal stretch flattens the caveolae and distributes their membrane at the cell surface proper (Prescott and Brightman 1976). However, this role does not seem to extend to vertebrate smooth muscle cells, because their caveolae do not normally

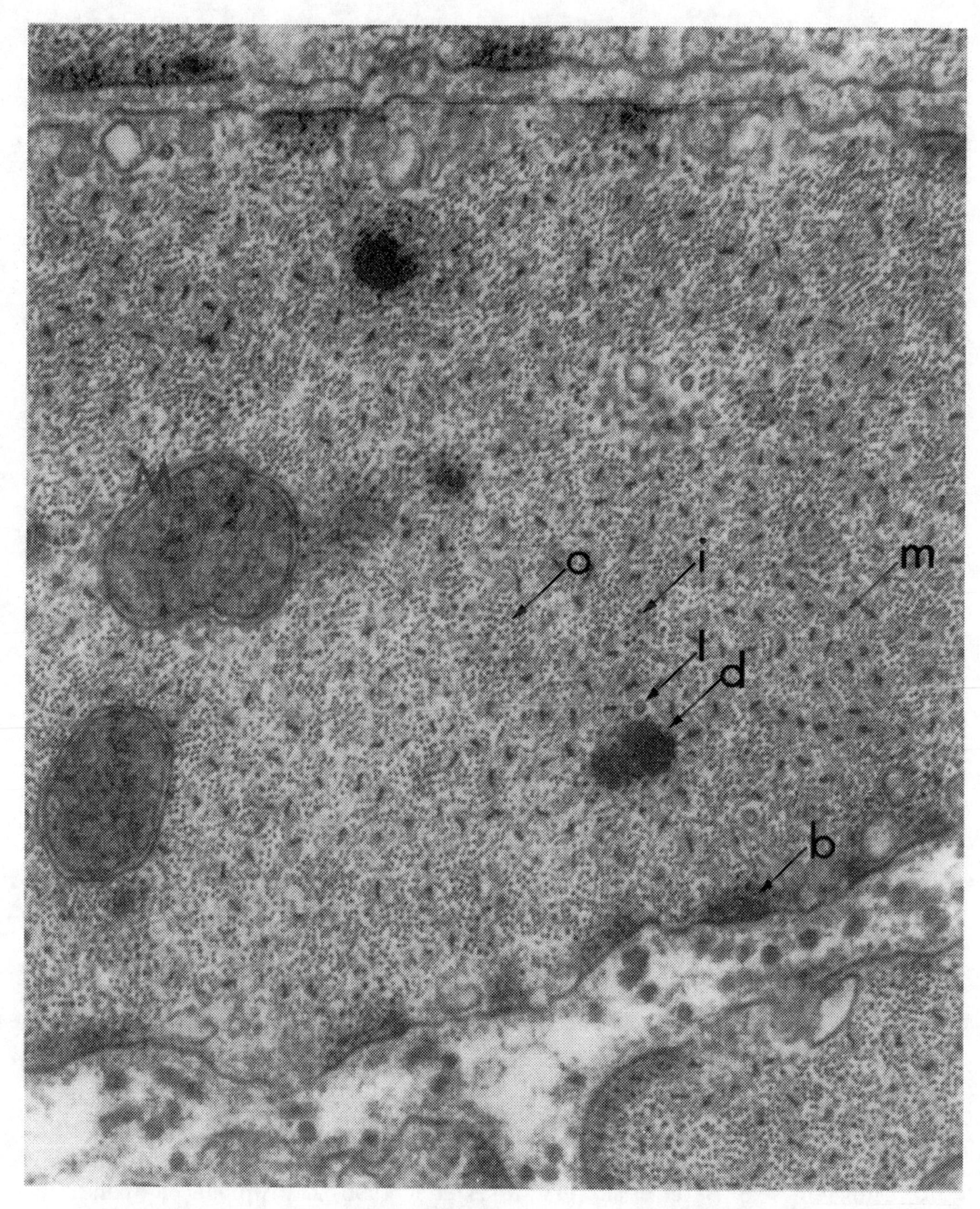

Figure 4. A muscle cell from the circular muscle layer of the chicken ileum, in transverse section. It is possible to recognize dense bodies (d), mitochondria (M), microtubules (t), caveolae, sarcoplasmic reticulum, dense bands (b), thin filaments (a), thick filaments (m) and intermediate filaments (i). [86,000×]

stretch out when the muscle is distended. An analogy between caveolae and T tubules of striated muscles has been mentioned in the literature (e.g. Forbes et al. 1979), but there is in fact no genuine similarity.

The possibility that caveolae are artefacts (i.e., invaginations arising from cytoplasmic vesicles or from a smooth cell membrane during chemical fixation) is discussed by Severs (1988), mainly in the context of striated muscle and endothelial cells. Although caveolae are seen also in muscles prepared without chemical fixation and by rapid freezing (e.g. Poulos et al. 1986), doubts about the true extent of this organelle remain (Severs 1988).

2.4. Cell junctions

Cell-to-cell junctions are membrane specializations that provide mechanical coupling (transmission of force between cells) and ionic coupling (transmission of excitation from cell to cell).

The mechanical coupling is supported by junctions of the adherens type, usually referred to as *intermediate* junctions (Figure 3). They correspond to two paired dense bands (see page 20) from adjacent cells, separated by a membrane-to-membrane gap of 40–60 nm, through the middle of which appears an ill-defined layer of electron-dense material. In addition, and particularly important in supporting transmission of force along the muscle, are sites of anchorage of a dense band to fibrous components of the extracellular matrix called cell-to-stroma junctions (see p. 12).

Ionic or electrical coupling is provided by gap junctions, whose structure conforms to that observed in great detail in other tissues (Bennett et al. 1991); see Figures 3 and 5. The membranes of the two cells lie strictly parallel to each other at the level of the junction. The reduction of the intercellular gap to ~2 nm makes possible the pairing of the transmembrane channels (connexons), so as to form single channels connecting the two cytoplasms. The connexons, as seen in freeze-fracture preparations, are about 10 nm wide, are packed to a density of 7,000 per μm^2 and form patches up to 0.2 μm^2 in size (larger gap junctions are found in other tissues). Therefore, in smooth muscles a gap junction may contain as many as 1,400 connexons from each of the two membranes (Gabella 1981). The smallest junctions are made of only 3–6 connexons; if individual connexons exist in the cell membrane, and if they are linked to connexons in the adjacent cell, they would work like the connexons of a junctional patch but they would not be recognized under the microscope. Six molecules of the protein connexin form a connexon. The connexin isolated from smooth muscles of the uterus and the colon is a connexin-43 that is immunologically identical to that of cardiac muscle (Beyer et al. 1989; Risek et al. 1990) and is produced by the same gene.

The number of gap junctions per cell varies in different smooth muscles. A very high density is found in the sphincter pupillae of the guinea-pig, which has scores and probably hundreds of junctions per cell; at the opposite end of the range, some muscles are devoid of gap junctions. In the circular muscle of the guinea-pig duodenum and ileum, 0.5% and 0.2% of the cell surface is occupied by gap junctions, which represent about 175,000 and 77,000 connexons per cell, respectively (Gabella and Blundell, 1978). In the adjacent longitudinal muscle, in contrast, gap junctions are either absent or rare and very small. Surprisingly, these muscle cells in vitro are dye-coupled (Zamir and Hanani 1990) and are well-coupled electrically, evidence suggesting that in those experimental conditions the cells are connected by gap junctions. By immunohistochemistry, the gap junction protein connexin-43 is localized in the circular muscle of the intestine of the mouse, dog and human, but absent in the longitudinal muscle and in the inner portion of the circular muscle (Mikkelsen et al. 1993). The distribution of connexin-43 immunofluorescence is, therefore, dramatically different in the circular and longitudinal muscle layers. This is confirmed by electron microscopy: in all the species and intestinal segments studied, the longitudinal muscle shows no gap junctions or an insignificant number of very small junctions, whereas the circular muscle is well endowed. In the rat bladder muscle, the absence of gap junctions has been reported by Gabella and Uvelius (1990); electrophysiological studies on guinea-pig bladder muscle show that its electrical coupling is different from and much less extensive than in other muscles (Bramich and Brading 1996). In the coronary artery of the pig, muscle cells are electrically and dye-coupled, and show discrete immunofluorescent spots for connexin-43 but no gap junctions by electron microscopy (Bény and Connat 1992), prompting the authors to suggest that dye coupling could occur via isolated connexons.

The poor correlation between electrical coupling and occurrence of gap junctions (detected by electron microscopy or by immunofluorescence) is noted by several authors (Daniel et al. 1976; Bény and Connat 1992; Garfield et al. 1992). The difficulty of making exact comparisons and the limitations of each technique are obvious factors to discuss when trying to solve this puzzle.

Although they are structurally stable during muscle contraction or elongation, gap junctions assemble and disassemble rapidly *in vitro* – for example, in the tracheal muscle treated with potassium channel blockers (Kannan and Daniel 1978). Gap junction formation in the myometrium is controlled by oestrogen and progesterone, and many large gap junctions are formed anew in the muscle cells a few hours before parturition (Garfield et al. 1977; Dahl and Berger 1978; Mackenzie and Garfield 1985).

The junctions of the adherens type (intermediate junctions) provide strong adhesion between the patches of cell membranes in adjacent cells (Figure 3). These patches are usually elongated, lie between rows of caveolae, and correspond to some of the dense bands. The intercellular gap is variable, most commonly 60 nm wide, and is occupied by a single layer of electron-dense material that splits at the edges and continues with the basal lamina of the two cells. Other dense bands are not paired and face the extracellular space and the stroma, and in most cases are mechanically linked with the latter (cell-to-stroma junctions) (Figures 4 and 5).

2.5. Organelles

Smooth and rough sarcoplasmic reticulum is present in variable amounts in vascular and visceral muscle cells (Figures 3 and 4). Cisternae and tubules of retic-

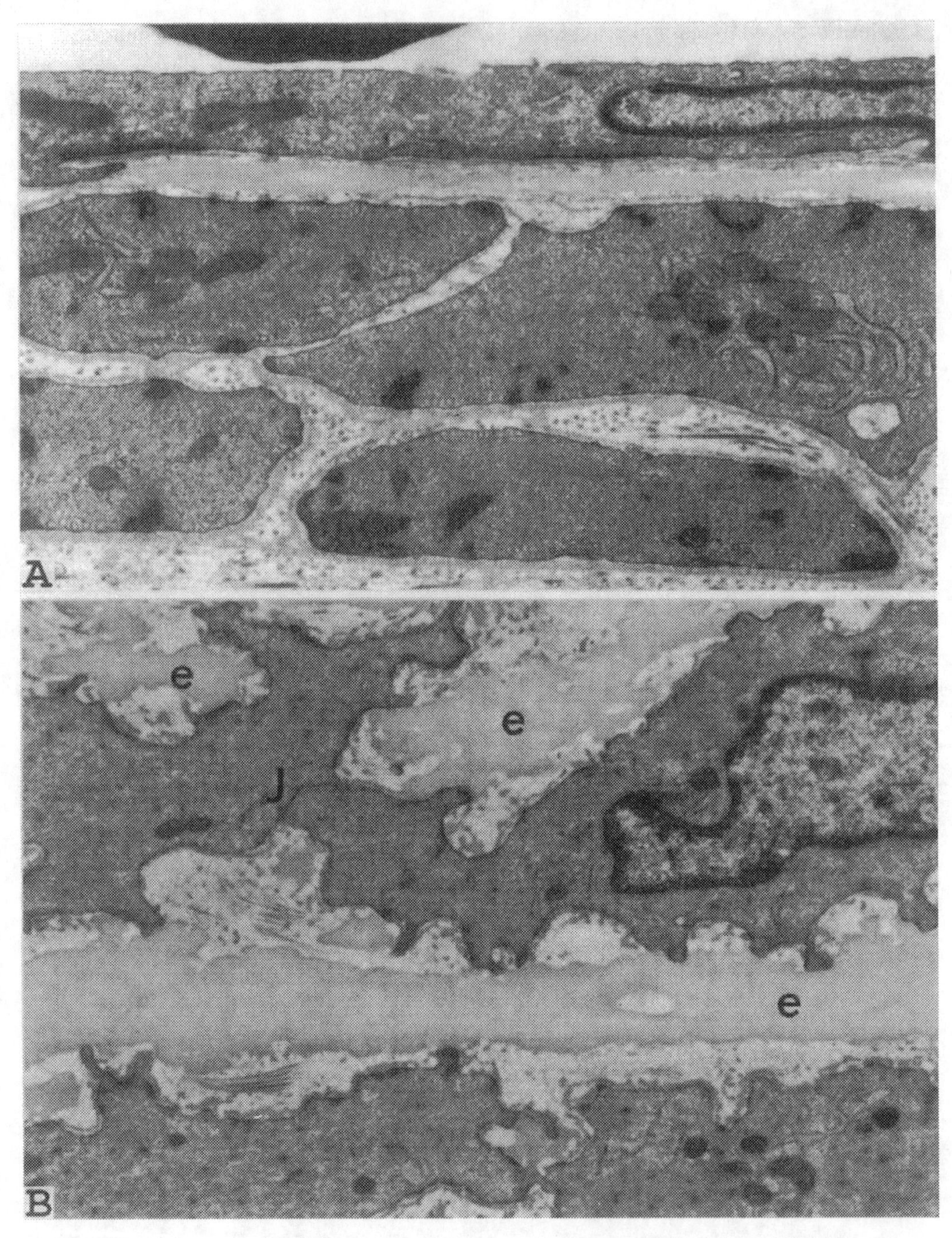

Figure 5. A. Longitudinal section of a small branch (arteriole) of the mesenteric artery of a rat. The lumen is at top, showing part of a red blood cell. The endothelial cell is flattened, displays the nucleus, and lies above the inner elastic lamina. The muscle cell profiles of the media are relatively smooth-surfaced, show a basal lamina, and are separated from each other only by a small amount of collagen. Prominent in the muscle

ulum, with or without attached ribosomes, are spread beneath the cell membrane, throughout the cytoplasm, and near the poles of the nucleus. They create an intracellular compartment and probably serve several roles.

The reticulum can only be seen by electron microscopy, in fractured or thin-sectioned preparations; its elaborate architecture is best appreciated in muscles impregnated *in toto* with osmium ferrocyanide (Forbes et al. 1979). Some of the cisternae of smooth reticulum lie parallel and close to the cell membrane, and the intervening gap is bridged by ill-defined electron-dense elements with a periodic distribution (Devine et al. 1972). These appositions of sarcoplasmic reticulum and cell membrane resemble the peripheral coupling or surface coupling of cardiac muscle cells (Sommer and Johnson 1968; Fawcett and McNutt 1969). Tubules and cisternae of reticulum also occur beneath or between rows of caveolae, but bridging elements are not seen in this case.

The smooth sarcoplasmic reticulum can sequester calcium ions from the cytoplasm and is regarded as a major storage site for calcium. Calcium concentration in the sarcoplasmic reticulum (28 mmol/Kg dry weight, as determined by electron-probe analysis in muscle cells of the guinea-pig portal vein) is sufficiently high to activate, when released, a full contraction of the muscle in the absence of extracellular calcium (Bond et al. 1985). The amount of sarcoplasmic reticulum and (less clearly) its distribution vary in different muscles. Values range from a total volume of over 5% of the cytoplasm in aorta and pulmonary artery muscles to about 2% in the rabbit taenia coli (Devine et al. 1972) or to about 1.5% in the guinea-pig taenia coli and vas deferens (McGuffee and Bagby 1976).

Mitochondria are most abundant in two conical regions near the poles of the nucleus, and are also often found near the cell surface (Figures 1 and 2). They are usually elongated, up to 1–2 μm long and 0.1–0.3 μm wide, and are clearly deformable during contraction. Mitochondria occupy 3–10% of the cytoplasm volume; the percentage volume, within the same muscle, is higher in animal species of small body size than in those of large size (see Figure 13). There are also consistent variations in mitochondrial percentage volume between different muscles. In the guinea-pig stomach, mitochondria occupy about 7.7% of the cytoplasm in the circular muscle cells and 4.6% in the longitudinal muscle cells (Moriya and Miyazaki 1980); a value of 3.5–4.0% was found in the guinea-pig taenia coli, as opposed to 7–9% in the circular muscle of the ileum (Gabella 1981).

The rough sarcoplasmic reticulum of muscle cells must be involved in protein and glycoprotein synthesis, and it accounts for the secretory activity of the cells, especially as regards elements of the intercellular stroma.

Other organelles found in smooth muscle cells include lysosomes (Figure 3), Golgi apparatus, and ribosomes. Occasionally there are large dense-cored vesicles, multivescicular bodies, lipid vacuoles, and clusters of glycogen granules.

(Figure 5 cont.)
cells are mitochondria, rough sarcoplasmic reticulum, dense bodies, and dense bands. [19,000×] B. Longitudinal section of a rat mesenteric artery, showing the abundance of elastic material (e) in the media. There are several points of contact of the muscle cell with the elastic lamella and the elastic fibers. A gap junction is visible (J). [14,000×]

Glycogen granules can become very abundant in muscles of aging subjects or in certain diseases, and form glycogen "lakes" up to 1 μm in diameter scattered through the cytoplasm (Figure 6B). It is not uncommon for visceral muscle cells to display a short cilium, often included in an invagination of the cell membrane beside the nucleus.

2.6. *Basal lamina*

All muscle cells are fully coated by a basal lamina, except at certain discrete junctional sites (at gap junctions; at some desmosome-like junctions; at the junctions between cell membrane and elastic fibers in vascular muscle cells; at some rare sites of direct neuro-muscular apposition). A basal lamina is lacking over large patches of the membrane of myometrial muscle cells and in muscle cells in vitro. In the latter case, the lack of basal lamina is related to the isolation procedure involving the use of proteolytic enzymes, which wipe away the original basal lamina; according to Gimbrone and Cottran (1975), it takes about two weeks of culture for a basal lamina to reappear. The term "basal lamina" is used here in preference to "basement membrane" since it is shorter and simpler.

The lamina conforms in appearance to the structure observed in other cell types (Leblond and Inoue 1989; Martinez-Hernandez and Amenta 1984) (Figure 5). An innermost component (*lamina lucida*) is electron-lucent in standard electron micrographs, and is coextensive with the glycocalyx of the cell membrane; it has great affinity for stains such as ruthenium red. The middle and most prominent component (the basal lamina in the strict sense) is a layer of electron-dense material (*lamina densa*) with fuzzy outline, mainly composed of collagen IV, entactin, and laminin. At the outer surface, the outer component of the basal lamina contains microfibrils (10-nm fibrils, many of which are made of fibronectin) and other fibrous materials, including collagen, which blur the outer limit of the lamina and complete the mechanical link between cell membrane and tissue stroma.

Specific features of the basal lamina exist in different tissues (Martinez-Hernandez and Amenta 1984; Sanes et al. 1990) that are variations on the basic theme, but little is known of this in the case of smooth muscle. In vascular muscle cells, the basal lamina is scanty at birth (Clark and Glagow 1979; Guyton et al. 1983). It is more prominent in muscle cells of muscular arteries than in those of elastic arteries (Clark and Glagow 1979), and contains mainly S laminin and laminin B2 rather than laminin B1 (Walker-Caprioglio et al. 1995). Other variations include an increase in the thickness of the basal lamina with age and the appearance of duplications (Gabella 1991) (Figure 6B). In the human aorta, the basal lamina around muscle cells is often multilayered (Dingemans et al. 1981).

2.7. *Extracellular materials*

The extracellular material in smooth muscles can be very abundant. The extracellular space is rarely less than 20% of the muscle volume, and it can be more than 60% in large elastic arteries (Figure 7). The major components of the extracellular material, collectively known as *stroma* of the muscle or as the *inter-*

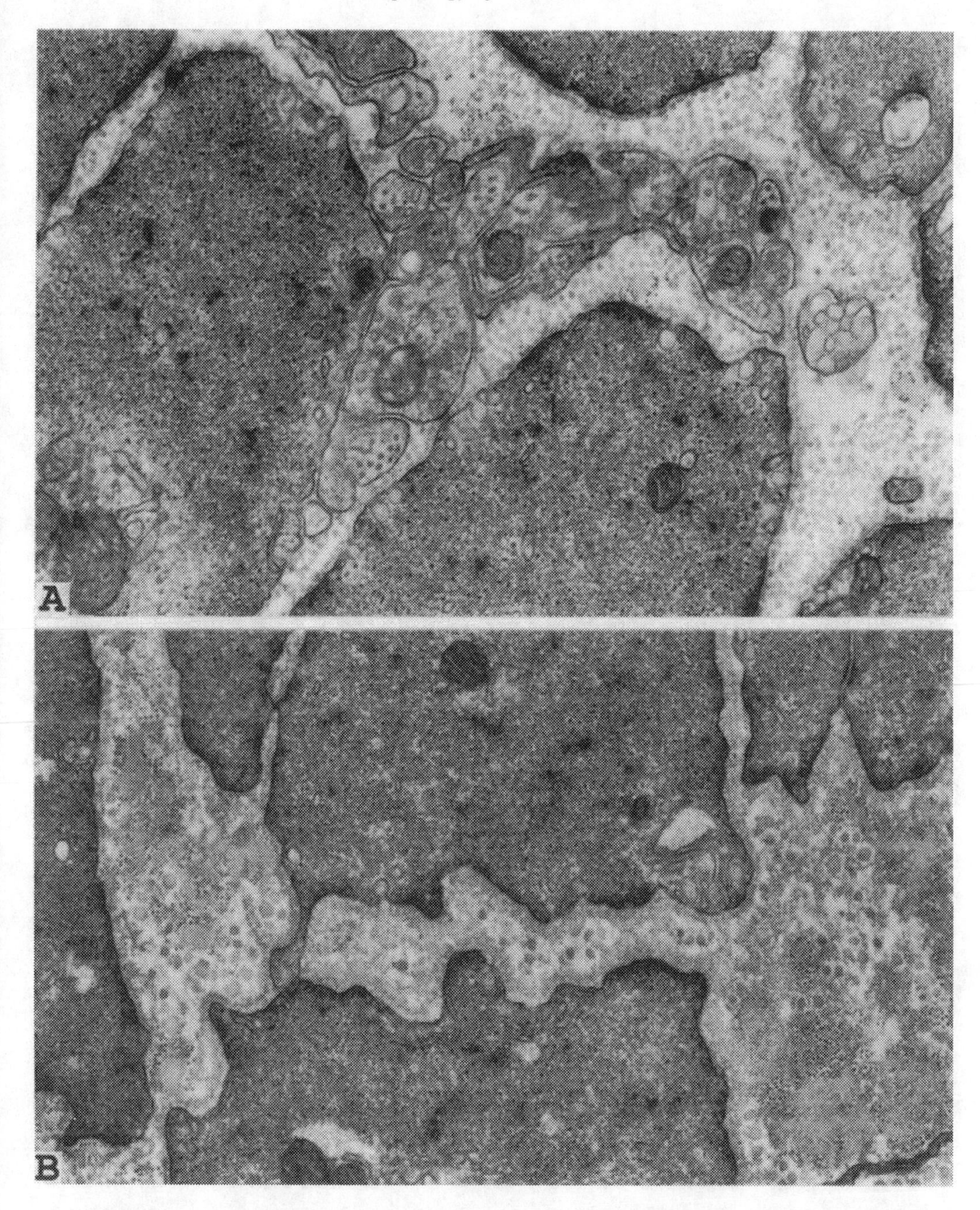

Figure 6. A. Taenia coli of a newborn guinea-pig, showing muscle cell profiles and a nerve bundle. The muscle cells are fully differentiated. The nerve bundle contains several axons, some of which make an extensive contact with a muscle cell; at the junction the gap is reduced to less than 20 nm. [33,000×] B. Tracheal muscle of an aging guinea-pig, showing muscle cell profiles and abundant extracellular material. Note the considerable thickness of the basal lamina and the special features of the extracellular fibrils. [30,000×]

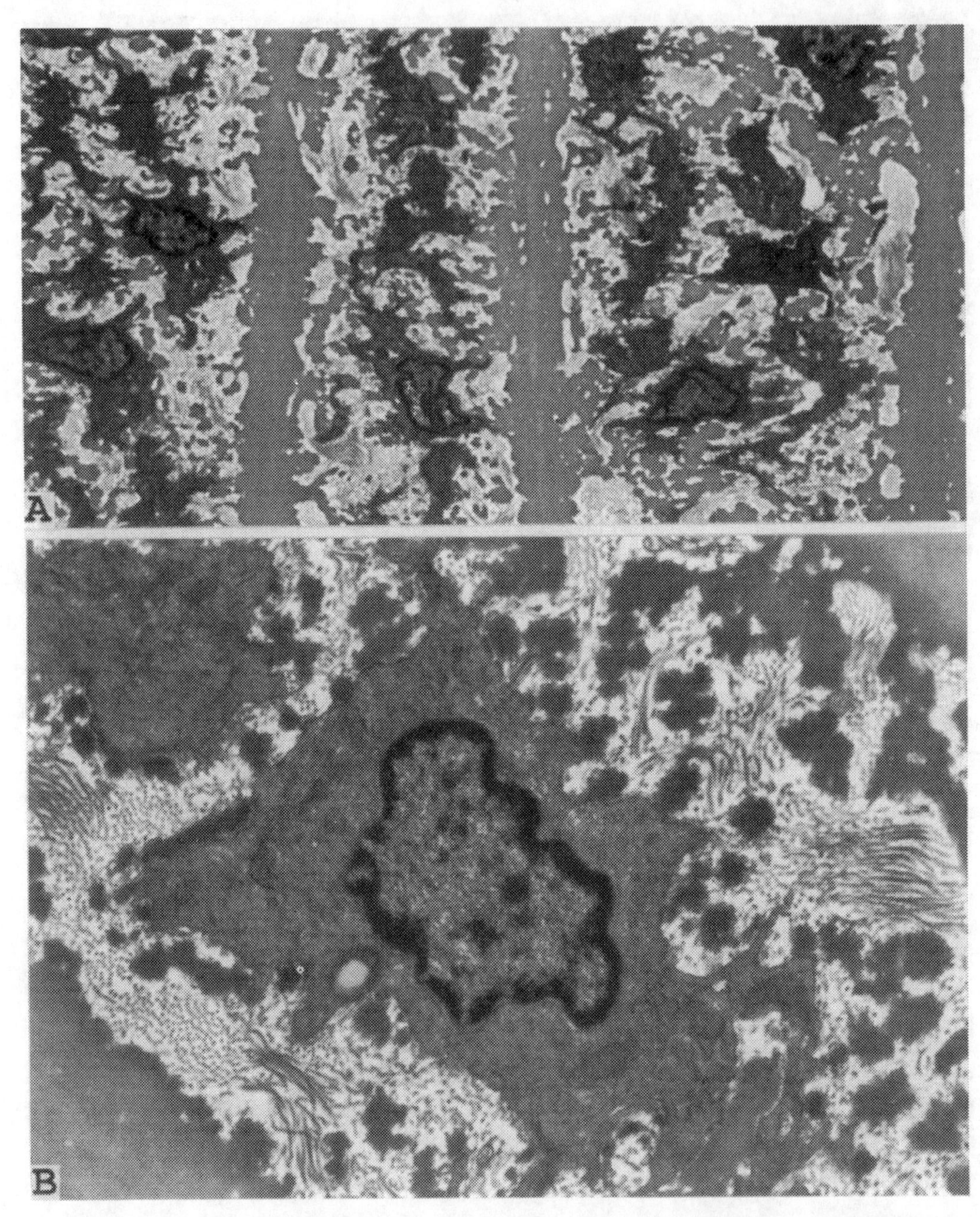

Figure 7. A. Longitudinal section of a rat aorta, showing part of the media. Arrays of muscle cells, with a small sectional area and a very convoluted profile, are separated by lamellae of elastic tissue. Note the large extent of the extracellular space. [4,000×] B. A muscle cell of the media of a rat aorta, showing the nucleus, mitochondria, and abundant sarcoplasmic reticulum and myofilaments. The cell is surrounded by electron-dense elastic fibers and by collagen fibrils. To the bottom left and top right is part of two elastic lanellae. [16,000×]

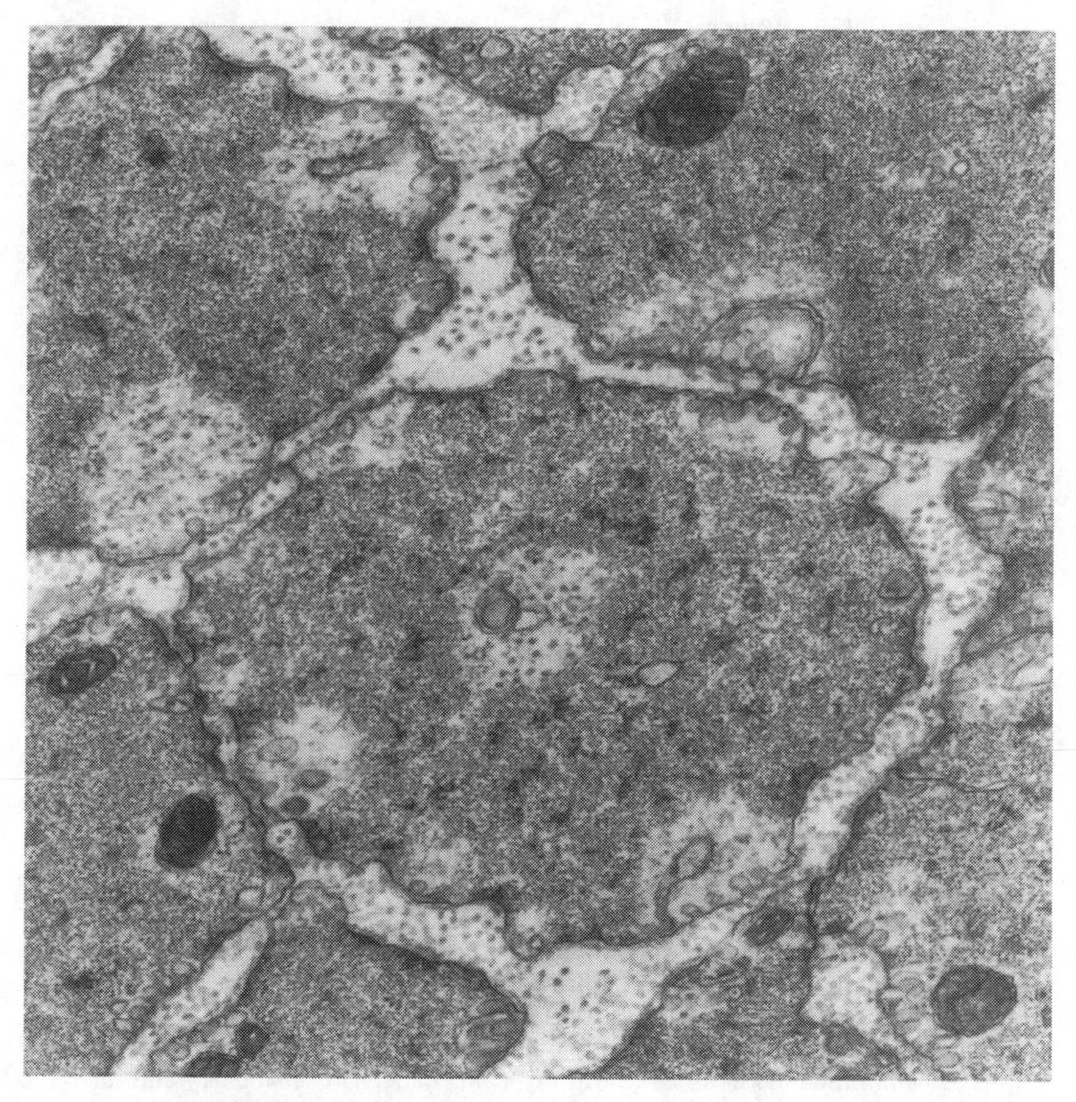

Figure 8. Transverse section of a guinea-pig taenia coli, isometrically contracted. In each muscle cell profile there are well-defined domains that have low electron density and are devoid of myofilaments. [33,000×]

cellular matrix, are collagen fibrils, elastic fibers, elastic lamellae, and the so-called ground substance. All these components are produced to a great extent by the muscle cells themselves.

Collagen concentration in smooth muscles, measured via the hydroxyproline content, is invariably greater than in striated muscles. In the guinea-pig taenia coli (Figure 8) there are 32 mg of collagen per gram of tissue, in the myometrium 36, in the aorta 78, in the myocardium 6, in skeletal muscle 9, and in tendon 197 (Gabella 1984). Most of the collagen content is accounted for by the collagen fibrils, cablelike structures 30–35 nm in diameter and of indeterminate but extremely long length. However, some of the collagen – for example, that of the basal lamina, which is type IV collagen – is in nonfibrillar form. Elastic fibers and lamellae are very abundant in elastic arteries, where the extracellular space amounts to more than 50% of the volume of the media (Figures 7 and 9). The elastic material is progressively reduced along the arterial tree (Figures 5 and 9).

The arrangement of this material – the continuity between large tubular lamellae and discrete elastic fibers, the links between elastic structures and collagen fibrils, the adhesion between muscle cell membrane and fibrous extracellular material – appears highly characteristic and is likely to have a complex mechanical significance. In spite of many studies on sections and by scanning microscopy, the details of these arrangements remain unknown.

3. Cytoskeleton and contractile apparatus

3.1. *Intermediate filaments*

More abundant in smooth muscles than in other muscles, intermediate filaments constitute at least 5% of the total protein content in the chicken gizzard muscle (Huiatt et al. 1980). They measure 10 nm in diameter, have a smooth and sharp outline, and are of indeterminate length (Figure 4). They are insoluble in high–ionic strength solutions (e.g., 0.25–0.6 M KCl, which extracts actin and myosin) and are more resistant than myofilaments to preparative procedures for microscopy (Cooke and Chase 1971).

The main component of intermediate filaments in visceral muscles is desmin, a 55-kDa protein, which has been sequenced (Geisler and Weber 1983) and from which intermediate filaments can be reconstituted in vitro (Ip et al. 1985). In vascular muscle cells, intermediate filaments are usually made of vimentin, with little or no desmin (Gabbiani et al. 1981). In the aortic tunica media the muscle cells contain vimentin, and only a minority of muscle cells contain both desmin and vimentin (Schmid et al. 1982). In myoepithelial cells of mammary and salivary glands, the intermediate filaments are of the keratin type and do not contain either desmin or vimentin (Franke et al. 1980). Expression of cytokeratins, both in vivo and in vitro, is found in muscle cells of human myometrium, in some fetal smooth muscle cells, and in leiomyosarcomas (Gown et al. 1988; Turley et al. 1988).

Intermediate filaments are associated with dense bodies (Cooke and Chase 1971; Tsukita et al. 1983; Somlyo and Franzini-Armstrong 1985), surrounding them rather than penetrating into them, and forming side-to-side links (Tsukita et al. 1983). They are also linked with some of the dense bands, and form a network through the cell with the true function of a cytoskeleton. Other elements of the cytoskeleton in smooth muscle cells are microtubules (always abundant) and cytoplasmic actin. It may be useful to distinguish contractile apparatus from cytoskeleton, but it should not be ignored that the two components (or "domains") are structurally and functionally linked.

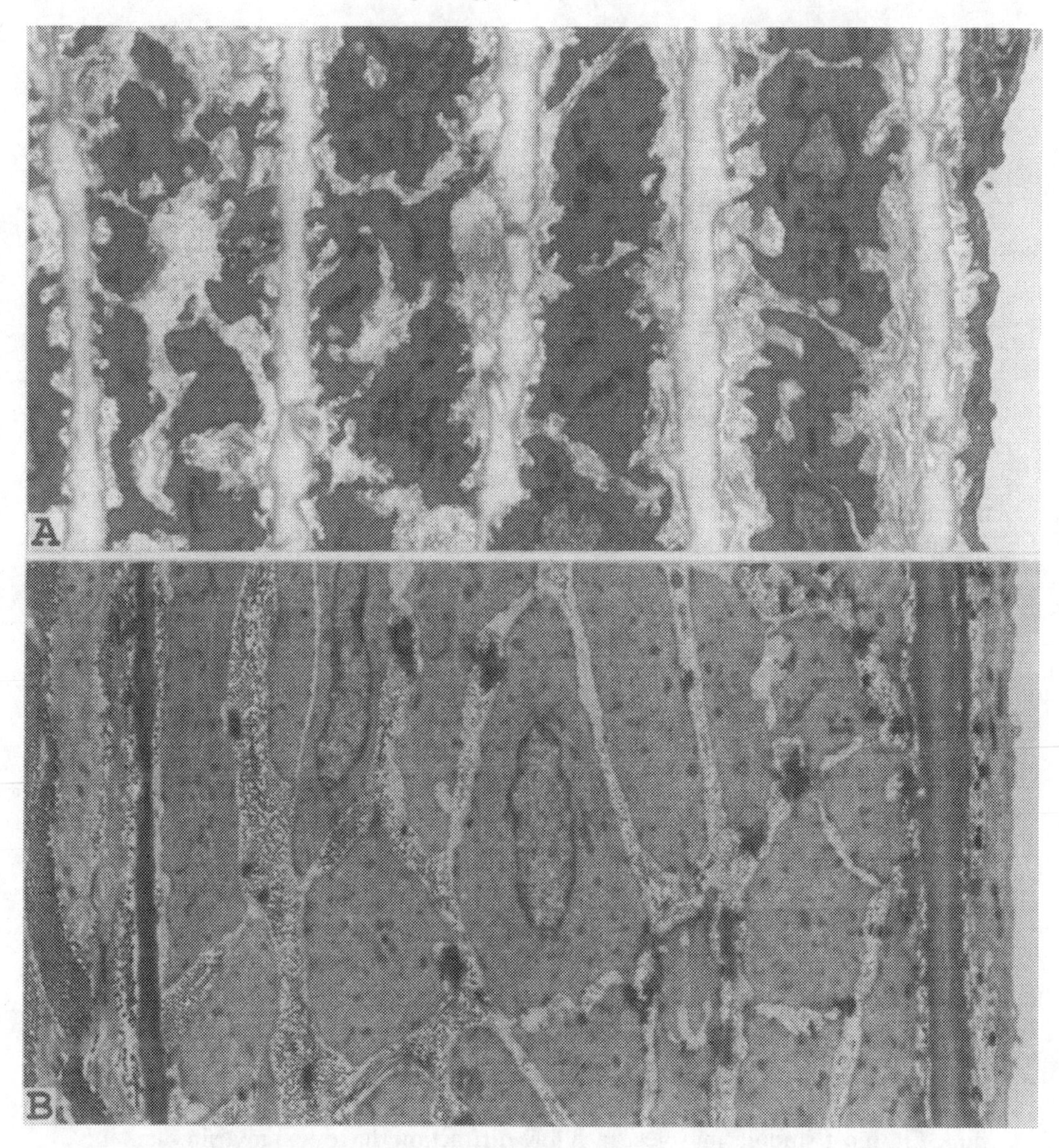

Figure 9. A. Longitudinal section of the pulmonary artery of a rat, showing bands of muscle cells regularly alternating with elastic lamellae. To the right are the lumen, the endothelium, and the inner elastic lamina. Between the muscle cells there is abundant collagen. [6,000×] B. Longitudinal section of a rat mesenteric artery, showing virtually the full thickness of the wall. To the right are the lumen, the endothelium, and the inner elastic lamina. To the left are the adventitia, in which a nerve bundle is visible, and the outer elastic lamina. The muscle cell profiles of the media are separated from each other by electron-dense elastic fibers and by collagen fibrils. [6,000×]

3.2. Actin filaments

Actin filaments, or thin filaments, measure 7–8 nm in diameter in stained sections of muscle cells (Figure 4). In some preparations they are arranged in rosettes around myosin filaments; more commonly, they are gathered into bundles or cables, which are many microns long and often merge or branch and terminate by penetrating into a dense body or a dense band.

The numerical ratio of actin to myosin filaments is difficult to estimate, but it is markedly higher in smooth than in striated muscles. Likewise, the weight ratio of actin to myosin is much higher in smooth muscles than in skeletal muscles, and it is higher in vascular than in visceral muscles (Murphy et al. 1977). The high frequency of actin filaments by comparison with myosin filaments and their arrangement in cables indicate that many actin filaments do not interact with a myosin filament, at least over part of their length.

Tsukita et al. (1982) have noted that the distance of myosin filaments from the nearest actin filaments is 10–20 nm in the cell at rest, but considerably less (figure not given) in contracted cells. A change in interfilament distance during contraction is also implied by the observations of Gillis et al. (1988).

After exposure of permeabilized muscle cells to heavy meromyosin subfragment S1, arrowheads are formed on actin filaments revealing the polarity of these filaments, with one end free and one end attached. The arrowheads point away from the site of insertion of the filament to the dense body or the dense band (Bond and Somlyo 1982; Tsukita et al. 1983). In deep-etched smooth muscle cells, decoration of actin filaments with myosin S1 produces rope-like structures, where the actin filament is surrounded by a double helix of S1 particles with a half-repeat rate of 38 nm (Somlyo and Franzini-Armstrong 1985).

3.3. Myosin filaments

Thick filaments are large aggregates of myosin molecules, and are approximately cylindrical, measuring 14-18 nm in diameter (Figure 4); they show an irregular contour because of the projection of the myosin heads, and taper at both ends. The filament length, estimated at 2.2 μm by serial section and high-voltage electron microscopy in the rabbit portal vein (Ashton et al. 1975), remains undetermined in most smooth muscles. In X-ray diffraction diagrams, myosin filaments produce a single, axial reflection of 14.3 nm, which represents the principal periodicity of the filament, owing to the myosin heads (Lowy et al. 1973).

The composition and distribution of thick filaments is still poorly known. Several forceful descriptions of these filaments have been produced in the past, and have often turned out to be untenable. Notwithstanding earlier observations, myosin filaments are observed both in contracted and in relaxed muscle cells (Somlyo et al. 1981; Tsukita et al. 1982; Gillis et al. 1988). However, according to Gillis et al. (1988), the filaments may vary in length and in number during contraction, raising again the possibility that part of the myosin may exist in monomeric rather than filamentous form.

The myosin molecules may be arranged as in the thick filaments of striated muscles, with bipolar symmetry and a bare area in the middle; of the latter, however, there is no evidence in electron micrographs (Small and Squire 1972; Ashton et al. 1975). An alternative, and more convincing, molecular plan envisages a "side polarity" arrangement in which the myosin heads are oriented all in the same direction on one side of the filament, and in the opposite direction on the other side (Craig and Megerman 1977; Cooke et al. 1989). Relatively similar is the "mixed polarity" model of Small and collaborators (Small 1977; Hinssen et al. 1978).

The lability of myosin filaments in cells subjected to current preparation procedures is at the root of the uncertainties about structure and arrangement of these filaments. It could be argued that the same lability exists when cells are isolated from a living tissue with collagenase (typically for 30–60 minutes) at 37°C (which would produce an extreme contracture), followed by centrifugation and storage, before the fine structure is examined.

In muscle cells isolated from the guinea-pig taenia coli and stimulated to contract with $CaCl_2$ and ATP in vitro, Bennett et al. (1988) have shown that myofilaments are organized in contractile zones where myosin is concentrated; these zones have not been seen in the cells at rest or by electron microscopy. Although myosin filaments are clearly not in register, some regularity in their transverse pattern (Rice et al. 1971) and some order in their distribution along the length of the cell (Ashton et al. 1975) have been noted.

Two isoforms of myosin heavy chain are detected by gel electrophoresis in visceral and vascular muscles of mammals (Cavaillé et al. 1986; Rovner et al. 1986) including humans (Sartore et al. 1989), generated from a single gene through alternative RNA splicing (Babij and Periasamy 1989; Hamada et al. 1990). A third isoform (MHC-3, 190 kDa) occurs in the human pulmonary artery (Sartore et al. 1989). The two main myosin heavy chain isoforms occur in different proportions in different muscles and are differentially regulated during development (Mohammad and Sparrow 1988; Borrione et al. 1989; Kuro-o et al. 1989). In development, vascular and visceral muscle cells express also a nonmuscle isoform of myosin heavy chain (Giuriato et al. 1992; Paul et al. 1994).

3.4. Dense bodies

Dense bodies have the appearance that the term suggests, the density being due to their affinity for electron-dense substances used in fixation and staining for electron microscopy (Figure 4). Dense bodies are scattered throughout the cytoplasm and resemble the dense structures bound to the cell membrane, which for sake of clarity are referred to as dense *bands* (although continuity between the two sets of structures is evident in some vascular muscle cells; see Section 3.5).

Dense bodies are found in all parts of the cell, except in small processes and where the cell profile measures less than 1 μm^2 (Figures 3–5). They are elongated along the cell length and occasionally are Y-shaped. Dense bodies are rigid (they remain straight, however twisted or contracted the cell may be), and are associated with thin and intermediate filaments. In visceral muscle cells they are quite variable in size, up to 1.2 μm long and up to 0.3 μm wide, their transverse profile usually being round. The structural heterogeneity and irregularity of distribution are apparent also in vascular muscle cells when examined by serial section reconstructions (McGuffee et al. 1991).

Although firmly denied in some earlier studies, the association of dense bodies with thin filaments is distinct and well documented (Somlyo et al. 1973; Bond and Somlyo 1982; Tsukita et al. 1983). The attachment of actin filaments to dense bodies is reminiscent of that found at Z lines of striated muscle fibers, a notion confirmed by the presence of alpha actinin at both sites (Schollmeyer et al. 1976; Fay et al. 1983) and by the orientation of the filaments. The actin filaments have longitudinal polarity, and at both sites the arrowheads formed by decoration of the actin filaments with myosin S1 point away from the site of insertion (Bond and Somlyo 1982; Tsukita et al. 1983).

Dense bodies are more prominent (an expression of both size and stainability of these structures) in some muscles than in others. Reptiles and birds have muscles with prominent dense bodies; amphibian muscles have small and pale dense bodies. The problems related to electron density are circumvented by using immunolabeling. However, the latter approach by light microscopy has limited resolution, a difficulty only partly overcome by confocal microscopy and computer analysis (Kargacin et al. 1989).

Studies on dense bodies are based on the assumption that they are discrete entities, an image clearly suggested by many microscope images but perhaps not fully correct. Major questions that remain unresolved concern how the dense bodies are linked to each other, since it is hardly conceivable that they should be "free-floating," and how these links are altered or displaced during contraction.

A similarity of dense bodies and Z lines is indicated by the links that both have with thin filaments, as if the dense bodies were an equivalent but fragmented structure. Notional intermediate conditions between dense bodies and Z lines would be the oblique striations and the perforated Z lines of certain muscles. The ill-defined limits of dense bodies, their heterogeneity in size, the examples of continuity of dense bodies with one another, the many examples of continuity of dense bodies with dense bands (notably in vascular muscle cells), and the general irregularity of their distribution all suggest that the "fragmentation" is not so complete as to create discrete entities.

Observations of dense bodies in isolated and skinned muscle cells from the toad stomach by Kargacin et al. (1989) confirm this impression by showing that large groups of dense bodies (stained with antibodies against alpha-actinin) have a similar pattern of motion when the cell contracts: within these groups, the distance between dense bodies remains relatively constant during contraction. These groups are referred to as *semirigid* structures and are found at approximately 6 μm intervals along the long axis of the cell (Kargacin et al. 1989).

3.5. Dense bands

Dense bands (or membrane-associated dense bodies) are structures linking the contractile apparatus to the cell membrane and, beyond the cell membrane, to the stroma or to other muscle cells. In visceral muscle cells they measure 30–40 nm in thickness and 0.2–0.4 μm in width; their length is difficult to measure but can exceed 1 μm. They are distributed over the entire cell surface, occupying 30–50% of the cell profile and even more along the tapering ends of the cell. In muscle cells of the large arteries, dense bands have a wedge shape (Büssow and Wulfhekel 1972) and penetrate deep into the sarcoplasm, an arrangement confirming the similarity and closeness (and, in this case, fusion) of dense bands and dense bodies (Figure 5).

The structural and functional similarity between the two is based on their providing insertion to bundles of actin filaments. In both cases the filaments are polarized; when decorated with S1 fragments, the arrowheads point away from the attached end (see page 18). The protein alpha-actinin is present in both (and in Z lines of striated muscles), although vinculin is found only in dense bands. Talin is found in the dense bands linked to the stroma but not in those forming cell-to-cell junctions (Drenckhahn et al. 1988). In isolated muscle cells of the chicken gizzard, the dense bands, visualized with labeled antibodies against talin or vinculin, display a periodic longitudinal arrangement and are aligned across the cell (Draeger et al. 1989). This arrangement has not been observed in the cells in situ (Gabella 1985).

3.6. Tendinous insertions

The great majority of smooth muscles do not possess proper tendons and instead form continuous sheets or cords. Even such discrete muscles as the taeniae of the guinea-pig caecum have no points of insertion, but continue into each other at the apex of the caecum and spread out and continue with the longitudinal coat of the ileum and colon at the other end. Some smooth muscles, however, are anchored to rigid structures via myo-tendinous attachments. The anococcygeus, the rectococcygeus, and the suspensor duodeni are smooth muscles inserted to the spine; the tracheal muscle is inserted to cartilages; and proper tendons are present in the expansor secundariorum of birds, in the muscle of the nictitating membrane and in tarsal muscles. In the arrectores pilorum of the mammalian skin, short elastic tendons connect the muscle cells to the connective tissue (Rodrigo et al. 1975). In birds, short elastic tendons connect the feather muscles to the follicles and are also interspersed along the length of the muscles (Langley 1904; Drenckhahn and Jeikowski 1978). A muscle–tendon attachment is found in the avian stomach (gizzard), where the two tissues are intimately related (Watzka 1932). The tendon contains collagen fibrils ranging in diameter between 30 nm and 160 nm as well as fibroblasts; in addition, near the interface between muscle and tendon, there are cells packed with intermediate filaments (Gabella 1985). At the myo-tendinous junction, the muscle bundles and the collagen bundles form an angle of about 45 degrees with each other; the two tissues are firmly linked but do not actually interpenetrate. Near the junction, the terminal parts of the muscle cells break into many longitudinal and cylindrical processes that are very rich in dense bands and surrounded by a thick basal lamina (Gabella 1985). These muscle cells are described as asymmetric: on one side of the nucleus they are long and tapering like ordinary muscle cells, while on the opposite side they are short and cylindrical and end with terminal brushlike processes (Gabella 1985). In the feather muscles, too, the muscle cells that approach the elastic tendon have a richly folded surface and deep fingerlike or wedgelike invaginations, and the cell membrane is heavily encrusted with dense bands (Drenckhahn and Jeikowski 1978).

4. Functional specializations of smooth muscle cells

4.1. Muscle contraction

Contraction is the paramount specialization of smooth muscles. In the form of contraction most commonly investigated in vitro – the *isometric contraction* – the muscle increases its tension (it ''hardens'') without changing length; this occurs when the impediment against a change in length is greater than the force generated by the muscle. In practice, some microscopic shortening always takes place with contraction, at least to the extent needed to take up the slack of the muscle and to bring under tension the tendons or the tissue stroma.

In an *isotonic* contraction, which is the more common form of contraction occurring in situ, the muscle does change its length or (more accurately) its shape. This is an active, energy-dependent process that should not be confused with passive changes in muscle length or shape.

4.1.1. Isotonic contraction

The shortening of the muscle can be very pronounced and is mirrored by a comparable change in length of the individual cells. Single cells, dissociated and isolated from the stroma, can contract down to a spherical shape, although this process is irreversible and nonphysiologic. In practice, the shortening is limited by external constraints, mechanical factors such as the lumen contents or the adjacent layers of tissue. In tubular organs, muscle contraction reduces the volume of the lumen by reducing the diameter or the length of the tube or both. In addition, the wall thickens (since both its contractile and noncontractile tissues are incompressible), and this contributes substantially to the reduction of the lumen. In a vessel of the diameter of an arteriole, where the combined thickness of media and endothelium at rest may represent half the outer radius of the vessel, a 15% shortening of the muscle is sufficient to produce occlusion of the lumen, whereafter no further muscle shortening can take place.

The shortening of the muscle is accompanied by its lateral expansion (along one or both axes, depending on external constraints); thus, an increase in muscle cross-sectional area accompanies the decrease in length. At the microscopic level, the size of the muscle cell profiles is increased. In the taenia coli, about 90,000 muscle cell profiles are counted in a square millimeter of muscle at rest, but only about 18,000 in the muscle isotonically contracted. The long axis of the contracted muscle cells is no longer parallel to the length of the muscle but deviates appreciably, a change more apparent in muscles containing many intramuscular septa, where a herringbone appearance may become evident. Cells are straight and virtually parallel to each other in the muscle at rest, but are somewhat contorted in the contracted muscle; a torsion of their long axis gives them the shape of irregular and long-pitched helices. These changes in shape are imposed in part by the stroma and in part by adjacent muscle cells; however, Warshaw et al. (1987) have observed that isolated muscle cells acquire a corkscrew shape upon contraction, and have discussed this observation in the light of a possible helical arrangement of the cytoskeleton and of the contractile apparatus. The insertion of the contractile units to the cell membrane would be helical, as envisaged also in the models of Fisher and Bagby (1977), Small (1977), and Small and Sobieszek (1980).

As to the arrangement of the myofilaments in contracted muscle cells, there is almost universal consensus. Authors who have investigated the biochemical composition of the muscle and its morphology, those who have discussed the literature in review articles, and virtually all textbooks (including *Gray's Anatomy*; see Salmon 1995) state that myofilament orientation is substantially altered in contraction and that myofilament units then acquire a criss-cross pattern (see e.g. Rosenbluth 1965; Fay and Delise 1973; Fisher and Bagby 1977; Small 1977; Small and Sobieszek 1980; Bagby 1983). However, I have maintained that the myofilaments remain basically parallel to the long axis of the cell even when it is fully contracted (Gabella 1984). This unorthodox conclusion is based on observations of visceral muscles made of large bundles or sheets of parallel muscle cells of regular shape. The situation may be different in other muscles – for example, in the media of elastic arteries, where the architecture of the myofilaments is more complex and the filaments do not run parallel to each other through the entire cell even when the cell is at rest.

All authors who have presented full accounts of the organization of the contractile apparatus (in particular, the relation between the contractile "units" and their links with the cytoskeleton) have described an oblique arrangement of the contractile units (in particular, the thin filaments) in the cells at rest and an accentuation of this arrangement during contraction. The so-called Rosenbluth model (1965), Cooke and Fay model (1972), Fay and Delise model (1973), Fisher and Bagby model (1977), and Small model (Small 1977; Small and Sobieszek 1980) are all thoroughly presented and analyzed in Bagby's review articles (1983, 1986). However, the issue of the orientation of the myofilaments and their displacement during contraction remains unresolved, and the amount of experimental evidence available is disproportionately small.

The surface of the isotonically contracted muscle cells is thrown into myriad projections that are mainly laminar and run at an angle to the cell length; the projections bear mainly caveolae, whereas the dense bands are mainly located in the regions of the membrane that have expanded least and thus appear to be invaginating.

In the muscle at rest, the collagen fibrils surrounding the muscle cells run mainly longitudinally or at a small angle with the axis of the cells; in the shortened muscle, however, the fibrils are wound around the cell and run obliquely or almost transversely to its long axis. The experimental observations of visceral and vascular muscles support the theoretical model of Mullins and Guntheroth (1965), according to which collagen provides a mechanism for binding cells together and for transmitting the tension generated by the myofilaments in directions up to 90 degrees away from their orientation.

Similarly, the collagen fibrils of intramuscular septa run oblique to the length of the muscle; when the muscle shortens the septa also shorten, and their depth increases and the collagen fibrils increase the angle they form with the length of the muscle.

4.1.2. Contraction as change in shape

In smooth muscles, because of the arrangement of collagen and of cell junctions, a mechanical effect is obtained by both the cell shortening and the cell fattening; indeed, given the tightly knit three-dimensional mesh of the muscle stroma, any change in muscle cell shape probably contributes to the mechanical performance of the muscle. It may therefore be more accurate to describe smooth muscle contraction as a change in shape, of the muscle as well as of its cells, rather than simply as a change in length. The vast displacements of material associated with isotonic contractions are compatible with their relative slowness.

4.1.3. Isometric contraction

A distinctive structural feature of muscle in isometric contraction is a checkered appearance of the cell profiles owing to discrete cytoplasm areas that are devoid of myofilaments and are therefore of lighter appearance (Figure 8). These domains are mostly distributed beneath the plasma membrane, sometimes in a regular array around the cell perimeter, and often causing small protrusions of its profile. Light areas of similar appearance form a cone tens of micrometers long at both poles of the nucleus. These large conical domains of nonfilamentous cytoplasm are centrally placed (as is the nucleus itself), but often become eccentric and reach the cell membrane. The nonfilamentous domains are sharply outlined and contain mitochondria, ribosomes, glycogen-like granules, cisternae of smooth and rough endoplasmic reticulum, Golgi apparatus, a few filaments, and many microtubules. Gillis et al. (1988) observed in the rat anococcygeus muscle that the packing density of myofilaments is increased by a factor of two during isometric contraction, and in this preparation the contracted muscle cells are reported to be of smaller volume than when at rest. In either condition, the myofilaments appear perfectly longitudinally oriented (Gillis et al. 1988).

4.1.4. Passive contraction

Within a strip of muscle, especially one set up and stimulated in vitro, there are often some cells that fail to be activated and are shortened passively by the action of the adjacent contracting cells (Gabella 1984). The passively contracted cells appear regularly coiled, and their surface is devoid of evaginations. Cells have been found that are passively contracted over a portion of their length and actively contracted over the remaining part.

There are also changes in shape of the muscle cells that occur by the effect of contraction in adjacent muscles. A contraction of the longitudinal muscle of the intestine shortens the gut and compresses sideways the muscle cells of the circular layer, which thus acquire a highly elliptical profile (unlike the circular profile they have in the gut at rest); when the circular muscle contracts in its turn, the shortening and fattening of its cells produce a lateral expansion of the muscle so that the gut elongates and the longitudinal muscle cells grow thinner and longer. Clearly, the profile of muscle cells can change markedly, from approximately circular to very elliptical. The arrangement of the cytoskeleton and of the contractile material is such that these changes in shape are permitted, are fully reversible, and do not seem to affect the ability of the cells to contract.

4.2. *Other functional specializations*

In addition to contractile activity, other functional specializations characterize smooth muscle and are manifested to different extents in different organs. For example, muscle cells grow and undergo adaptive structural changes, can hypertrophy and atrophy, and are subject to metabolic turnover. Muscle cells also have some kind of sensory activity, in the sense of responding to stretch and in general being sensitive to mechanical deformation. Muscles exhibit immediate, short-term responses (i.e., a contraction) as well as such cumulative, long-term effects as the growth observed in hypertrophy (see page 34).

Muscle cells produce factors that play a role in the interaction between autonomic nerves and target tissues. These factors include both neurotrophic diffusible factors and fixed factors, located in the cell membrane, the basal lamina, and the stroma.

Muscle cells are the main source of extracellular materials within the muscle, and are therefore involved in an important secretory activity. There are also more specialized forms of secretion in certain types of muscle cells. Muscle cell secretory activity is especially intense *in vitro*. Amyloid (Alzheimer's β-protein) is produced by muscle cells of cerebral vessels both in situ and in vitro (Frackowiak et al. 1995).

5. The musculature in different organs

Each of the numerous smooth muscles of the body possesses such distinctive functional properties as speed of contraction, force developed, amount of shortening, fatigability, changes in stiffness, energy consumption, duration of contraction, and various mechanical effects. These differences derive from chemical properties of the contractile apparatus, from the distribution of channels and receptors in the cell membrane, and from the extent of the connections between the cells and between cells and stroma. The functional individuality of smooth muscles is apparent when one considers the vast range of organs and tissues in which they are found. There are various features, biochemical and anatomical, that account for this individuality. The existence of isoforms of heavy (Rovner et al. 1986; Babij and Pariasamy 1989) and light chains (Nabeshima et al. 1987) of myosin goes some way toward accounting for the differences between tonic and phasic smooth muscles; the former are found mainly in large arteries, the latter mainly in viscera (reviewed in Somlyo 1993).

There are many structural features and a general plan that are common to the cells of all muscles; however, in different smooth muscles, the cells also have specific structural characteristics. A prominent element of the structural individuality of each muscle is the architecture of its muscle cells, the three-dimensional arrangement of cells and stroma and allied components in the organ. Beautiful three-dimensional views of smooth muscles in whole organs can be obtained by scanning electron microscopy, after digestion of the tissue stroma and micro-dissection (Uehara et al. 1990).

A variety of configurations is evident. Compact, cordlike muscles, such as the taeniae of the colon or caecum or the muscles connecting abdominal viscera to the spine (e.g., the levator duodeni, retractor ovarii, anococcygeus muscles) can be contrasted with the sheetlike muscles that form an integral part of the wall of hollow organs (Figures 1 and 10). The sheets can be formed by a very regular array of muscle cells that run parallel to each other or, when there is more than one layer, orthogonal to each other in contiguous layers (Figure 11C). There can be as many as four layers of musculature with this ply arrangement, for example, in the avian small intestine. While remaining orthogonal to each other in adjacent layers the muscle layers can acquire a complex configuration, as is the case in the mammalian stomach. Other sheetlike muscles are made of small muscle bundles that run in all directions in an apparently random arrangement, splitting and merging with each other; this is typically observed in the gallbladder and the urinary bladder (Figure 11B). A special starlike pattern is observed in the chicken amnion, where muscle cells are radially arranged converging and inserting onto focal points (Figure 11A).

Distinct muscle layers, circular and longitudinal, exist in some blood vessels, such as the avian mesenteric artery, the mammalian portal vein (Figure 10C), and the sheep carotid artery. More commonly, in arteries the musculature runs predominantly circumferentially, although there are large deviations from this pattern (Figure 10A). The exact orientation of the long axis of the cells with respect to the axis of the vessel may be of less consequence than the fact that the muscle cells are fully embedded in the stroma of the media and are tightly linked with it. Extensive variation in muscle cell size, cell shape, amount of extracellular material, and type of fibrous material is apparent in the media of arteries of different caliber (Figures 5, 7, and 9). In small arterioles the musculature is one-cell thick and each muscle cell makes several turns around the lumen, acquiring a corkscrew shape.

Smooth muscles contain other types of cells in addition to muscle cells, a further source of specific features. A feature common to all smooth muscles is the absence of a perimysium or equivalent structure. Among the intramuscular nonmuscle cells are vascular cells (mainly endothelial cells of capillaries and pericytes), interstitial cells, fibroblasts, mast cells, Schwann cells, and axons. Interstitial cells of Cajal are present in all the smooth muscles of the alimentary tract (Figure 12B). Mast cells are common in smooth muscles of the airways (Gabella 1991). Fibroblasts are found in many smooth muscles, commonly occupying intramuscular septa or lying around muscle bundles. Fibroblasts are very rare in the media of mammalian blood vessels; in contrast, the large arteries of birds have a media made of several layers of muscle cells alternating with layers of fibroblasts and elastic material (Figure 13).

6. Innervation of smooth muscle

A proper account of innervation would be very long and is beyond the scope of this chapter. Only a few aspects of innervation will be presented here.

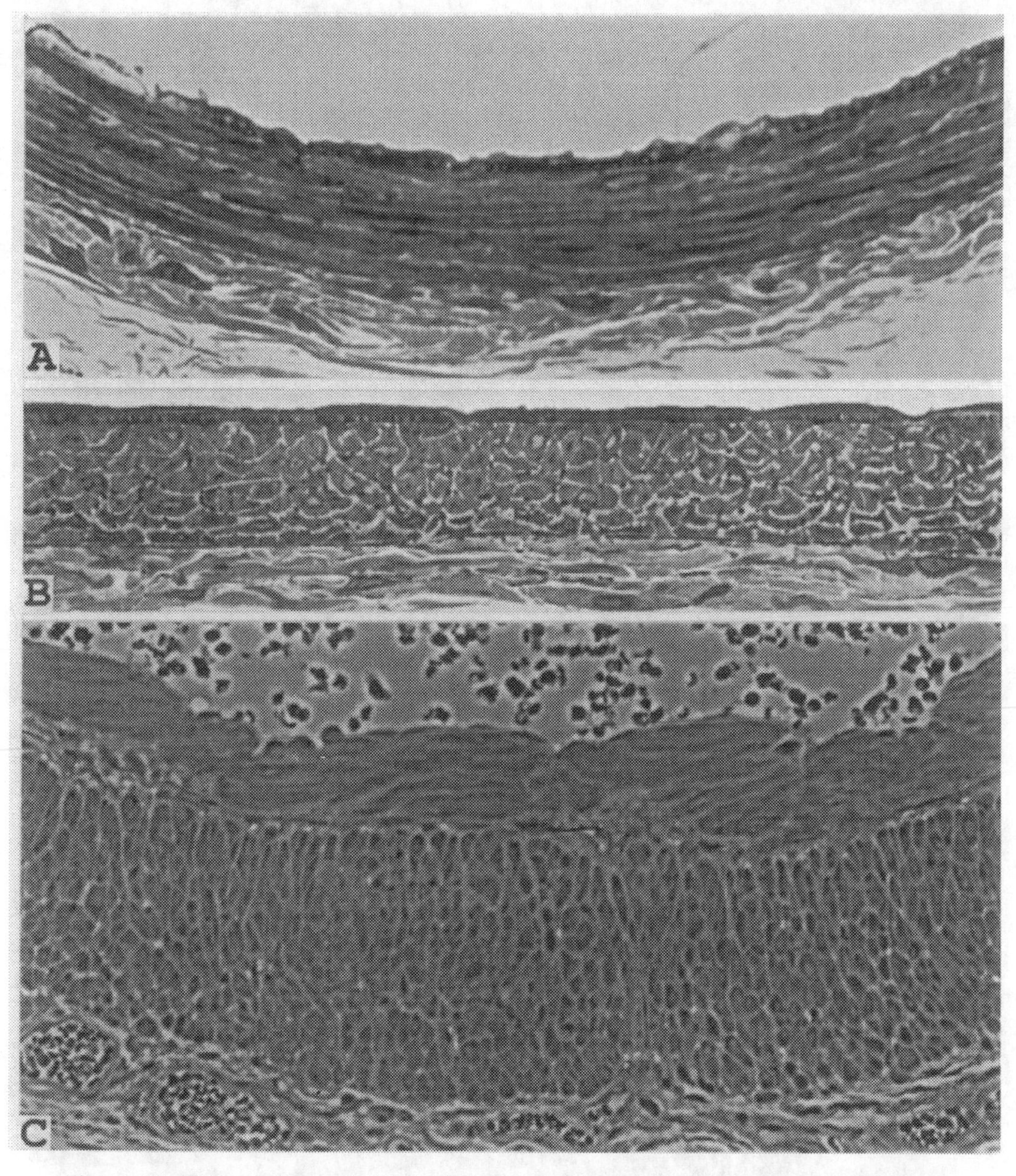

Figure 10. Thin sections of blood vessels seen in the light microscope. A. Transverse section of the mesenteric artery of a rat. The lumen is at the top and the medial musculature is in longitudinal section. [800×] B. Longitudinal section of the same vessel through its center. The medial musculature is in transverse section. [800×] C. Transverse section of the portal vein of a rat. The lumen is at the top. The musculature consists of an inner circular layer and an outer thicker longitudinal layer. [800×]

The role of nerves in smooth muscles is somewhat special, because many muscles have myogenic activity and can contract independently of nervous influences (some muscles, such as the chick amnion or certain large arteries, are devoid of nerves). Muscle cells are often electrically coupled, so excitation can

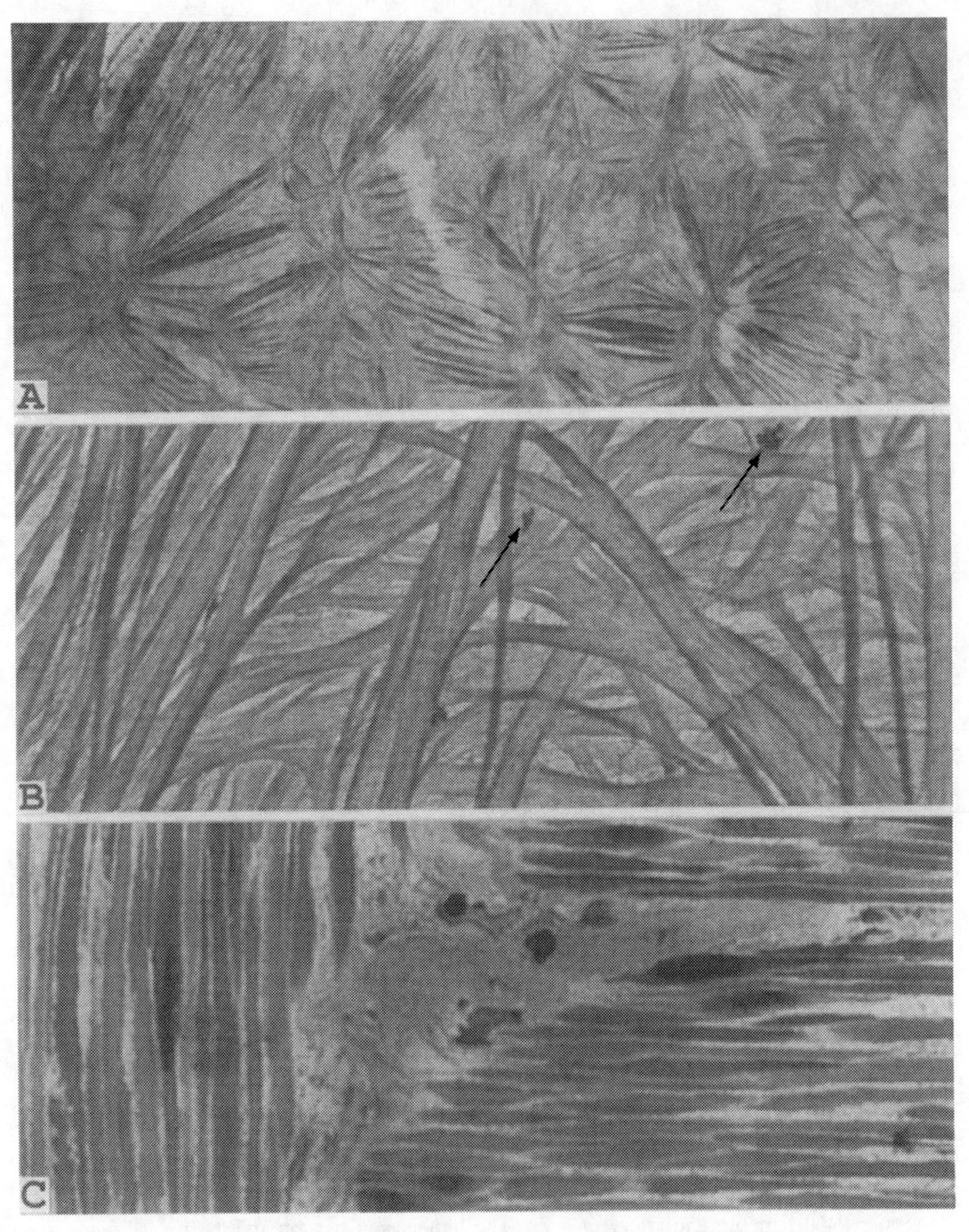

Figure 11. A. Whole-mount preparation of a chick amnion, approximately at day 11 *in ovo*. The muscle cells are arranged radially, diverging from focal points of insertion. B. Whole-mount preparation of the urinary bladder of a guinea pig, showing muscle bundles of different sizes and orientations. Some intramuscular nerve ganglia are visible (arrows). [12.7×] C. Small intestine of a guinea pig sectioned parallel to the serosal surface, showing the circular and the longitudinal muscle, both in longitudinal section. The two muscle layers run orthogonal to each other, and between them is visible a strand of the myenteric plexus. [800×]

spread quickly from innervated cells to cells that are not directly innervated. Muscle cells can also be stimulated to contract by such physical stimuli as stretch, light, and changes in temperature. Moreover, there are not only several types of excitatory stimuli but also many inhibitory stimuli, including some that are mediated by nerves.

The efferent innervation of smooth muscles is provided by postganglionic fibers that branch within the muscle and acquire a varicose shape along their terminal branches. Varicosities are regarded as functional nerve endings, insofar as they are possible points of release of transmitters. It should be realized, however, that varicosities support both neurosecretion (of transmitters) and conduction (of nerve impulses). With this pattern, hundreds of varicosities can be produced by a single axon with a limited amount of branching. Only about 1 in 50 varicosities is a true ending, the proper anatomical termination of an axon.

Axons that are destined for a smooth muscle branch repeatedly, but usually not before they have penetrated, or are in close proximity with, the target tissue. Two contrasting patterns of innervation are observed in smooth muscles. An innervation by nerve bundles is found typically in the intestine (Figures 12B and 14A): the intramuscular nerve bundles branch and merge repeatedly, but they rarely end by issuing individual axons. The axons, even when they are varicose and form neuromuscular junctions, remain grouped together within a bundle. In contrast, in muscles such as the sphincter pupillae, the vas deferens, and the urinary bladder, the intramuscular bundles branch repeatedly until most of them end by issuing individual axons (Figure 12A).

The bladder muscle is densely innervated by efferent fibers whose detailed relations with the muscle cells are best seen by serial section electron microscopy (Gabella 1995). Upon penetrating into the muscle, nerve bundles in the rat branch repeatedly and almost all become single varicose fibers. Being varicose is a feature expressed by degrees, and is not an all-or-none state. Some varicosities are greater than 1 μm in diameter, and some intervaricose segments are as little as 0.05 μm in diameter. The Schwann cell sheath around a varicosity is often incomplete; the axolemma thus exposed (a "window") is covered directly by the basal lamina (Figure 14B). Some varicosities have a window only 200–400 nm wide; others have more than one window, and some are devoid of sheath altogether, the entire axolemma being in contact with basal lamina (Figure 14B). The Schwann cell never extends beyond the axon, whereas very often the axon extends beyond the Schwann cell.

Neuromuscular junctions are identified by four features: the axon is a varicosity packed with vesicles; the axolemma has a window; the intermembrane distance is 10–100 nm, (usually 30–50 nm); and the intercellular gap contains a single basal lamina but excludes fibrils such as collagen. These parameters, however, can also occur uncoupled – for example, a window on intervaricose segments, varicosities without a window, or exposed axolemma far from muscle cell. On average there is more than one neuromuscular junction per muscle cell, and examples of muscle cells receiving multiple nerve endings from one or from two axons are picked up by serial sections. A striking feature is the variability of these ultrastructural parameters. The bladder innervation does not appear to be built on a rigid plan, and the notion of "loose-patterned" innervation is suggested.

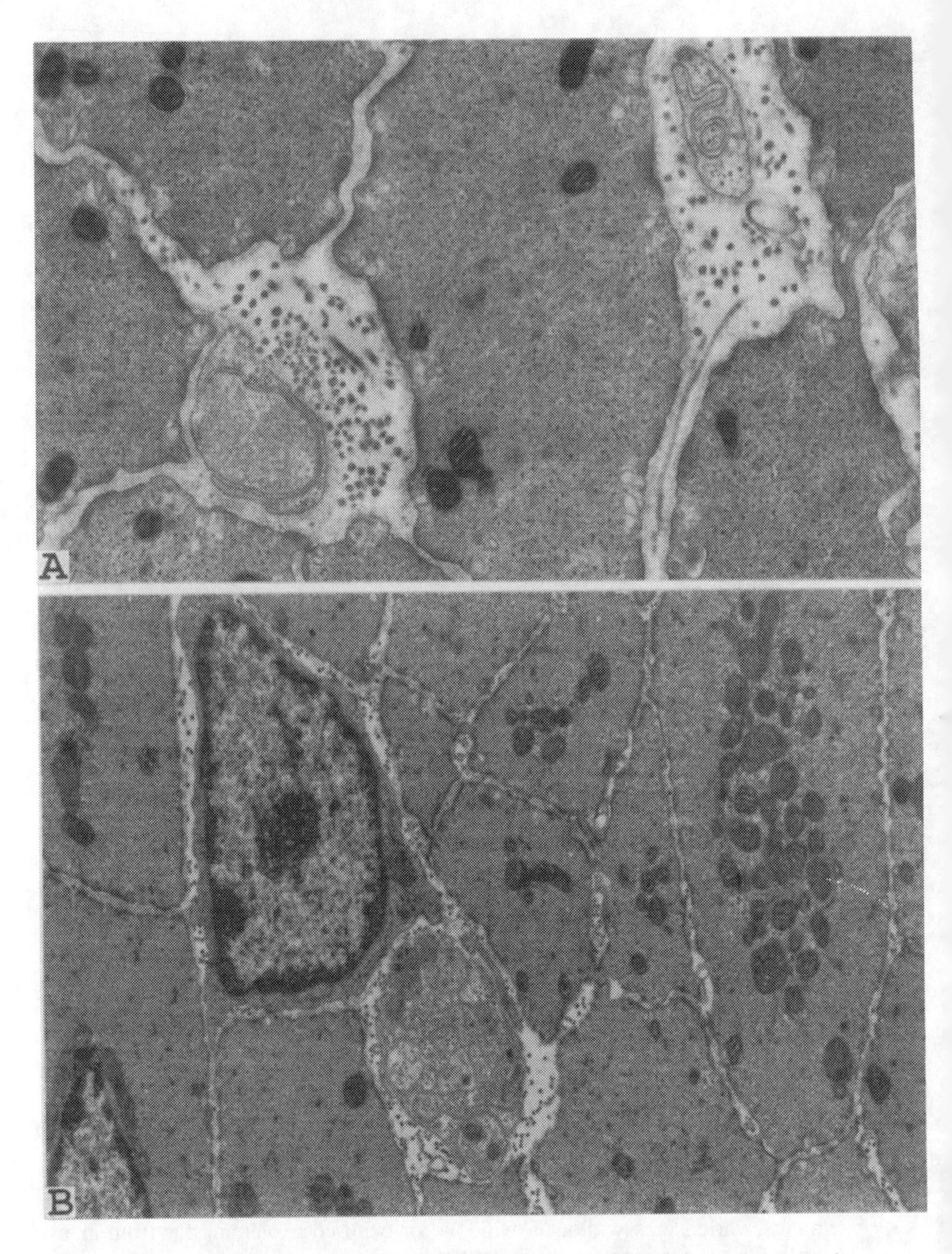

Figure 12. A. Musculature of the rat bladder, showing two nerve fibers among muscle cells. The fiber at bottom left is expanded into a varicosity, packed with vesicles and only partly sheathed by a Schwann cell process; the exposed part of the axolemma (''window'') lies very close to a muscle cell and forms a neuromuscular junction. The fiber at top right is an intervaricose segment, containing only microtubules and neurofilaments, fully wrapped by a Schwann cell process. [25,000×] B. Circular muscle of the

Even when the axons remain grouped in a bundle, they not only develop varicosities but can also form close junctions with muscle cells (Figures 12B and 14A). In certain blood vessels (e.g., the arterioles of the intestinal submucosa), the perivascular fibers have a well-developed varicose pattern and form neuromuscular junctions (Luff et al. 1987). By using serial section electron microscopy, Luff et al. (1987) have shown how frequent these junctions are.

7. Smooth muscle development

The muscles of viscera and vessels derive from mesenchymal cells; in the cranial and cervical region these are of neural crest origin (Le Douarin 1982), while elsewhere they are of mesodermal origin. The undifferentiated precursor cells become slightly elongated (in the direction of the future muscle cells of that layer) and aggregate by means of numerous points of cell-to-cell adhesion and a marked reduction of the intercellular spaces. Synthesis of proteins of the contractile apparatus and the cytoskeleton ensues.

Morphological and histochemical aspects of visceral muscle development have been investigated mainly in the gut of the chick embryo. According to Paul et al. (1994), the differentiation of muscle cells in the chicken gizzard proceeds from the serosal to the mucosal side, while capillaries invade the growing muscle in a mucosal-to-serosal direction. Such a gradient is not apparent in the intestine; however, the circular muscle appears some days earlier than the longitudinal muscle. Within a given muscle, all the muscle cells appear to develop synchronously.

In the chicken gizzard, the myo-fibrillar proteins myosin (heavy chain) and actin are detected from the end of the first week *in ovo*, first biochemically and then immunohistochemically (Stuewer and Gröschel-Stewart 1985; Hirai and Hirabayashi 1983, 1986). Actin filaments are first observed at about the same time, whereas intermediate filaments and myosin filaments become visible a day or two later (Bennett and Cobb 1969; Gabella 1989), when the first signs of spontaneous contractile activity are also detected (Donahoe and Bowen 1972). Dense bands appear later than dense bodies. In the early stages, the small bundles of thin (actin) filaments are inserted mostly to dense bodies; the extensive insertion of the contractile apparatus to the cell membrane develops later. Talin and vinculin are initially found in the cytoplasm (Volberg et al. 1986). Talin becomes associated with the cell membrane at the time of formation of dense bands, around the 16-18th day in the chick embryo; vinculin does not bind to dense bands until after hatching (Volberg et al. 1986).

The sarcoplasmic reticulum is well developed from the earliest stages, when both the smooth and the rough type are observed. Mitochondria are more abundant than in mature muscle cells (Gabella 1989). Caveolae are absent or rare in the early stages of differentiation.

(Figure 12 cont.)
mouse small intestine, showing an interstitial cell of Cajal and several muscle cells in transverse section; the large number of mitochondria is characteristic of this species. Below the center is a nerve bundle with about 25 axons. Two of the axons, containing vesicles, lie very close to a muscle cell and to the interstitial cell. [14,000×]

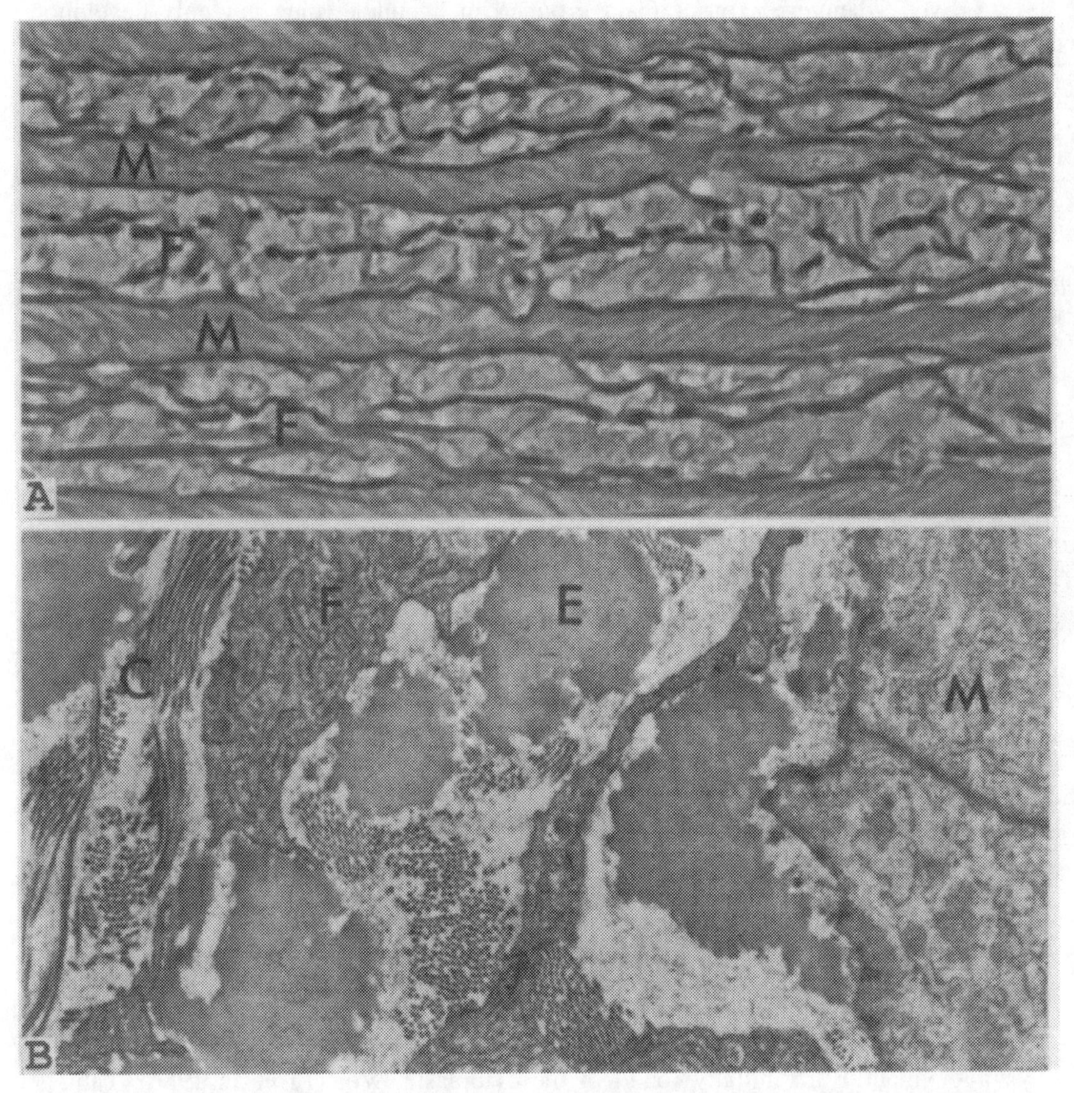

Figure 13. A. Partial view of the aorta of a 6-week-old chick. In the media, lamellae of muscle cells (M) alternate with lamellae of fibroblasts (F) and elastic material. [800×] B. Fibroblasts (F), elastic fibers (E), collagen fibrils (C), and muscle cells (M) are equally abundant in the media of the mesenteric artery of a chick. [14,000×]

Gap junctions are detected only from the late embryonic stages onward. In the chicken gizzard, gap junctions appear at around day 16 *in ovo* (La Mantia and Shafiq 1982; Gabella 1989) and in the guinea-pig ileum around the time of birth. These junctions consist of only a few intramembrane particles, and then grow by progressive addition of new connexons.

The large increase in the mass of a smooth muscle during development is partly due to an increase in cell size (a fivefold growth in cell volume is found in the chicken gizzard) and partly to an increase in cell number by mitotic division

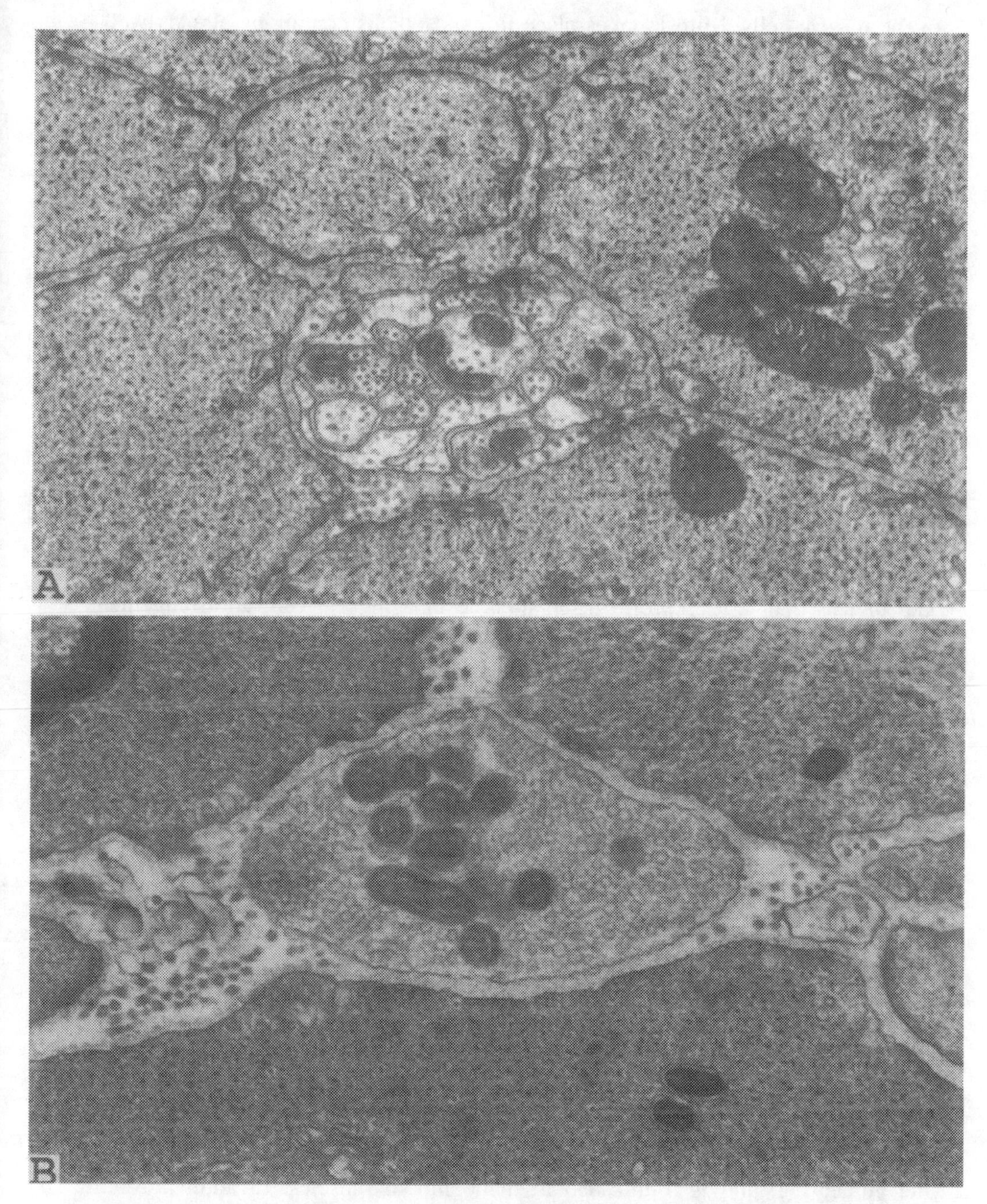

Figure 14. A. Circular muscle of the mouse ileum, showing muscle cell profiles and an intramuscular nerve bundle. The bundle is made of 11 axons and some Schwann cell processes. The axon to the right contains granular and agranular vesicles and lies very close to a muscle cell. [31,000×] B. Musculature of the rat bladder, showing muscle cell profiles and a large varicosity. The varicosity is packed with small agranular vesicles and mitochondria, and lies very close to three muscle cells. There is no Schwann cell sheath. Note that this varicosity is larger than the nerve bundle shown in A. [31,000×]

of the muscle cells. Mitosis takes place in muscle cells containing already well-formed membrane specialization and contractile apparatus: terminal differentiation remains compatible with ability to divide. The central part of the cell becomes bulbous and develops the mitotic apparatus, including paired cisternae of sarcoplasmic reticulum associated with the chromosomes and possibly deriving from the breakdown of the nuclear envelope (Kamio et al. 1977). The remaining parts of the cell are indistinguishable from the neighboring cells in terms of organelles, filaments, and membrane junctions. The cell invariably divides across its length, and the two daughter cells are initially lined up end-to-end; they remain in cytoplasmic communication with each other for some time after division via a cytoplasmic bridge occupied by a bundle of microtubules.

From an early stage, developing smooth muscles are accompanied by fibroblasts, usually in intramuscular septa and sometimes within muscle bundles. Developing muscle cells and fibroblasts are distinct cell types, and apparently there are no cells with shared features.

Nerve fibers do not penetrate into muscle bundles until the differentiation of muscle cells is advanced, that is, toward the end of the embryonic stage. In the intestine, innervation is provided by axons grouped into bundles even in their varicose portion: around the time of birth the intramuscular nerve bundles are large, and several of their axons (some packed with vesicles) are separated by a gap of less than 20 nm from the membrane of a muscle cell, an arrangement that becomes progressively less common with postnatal growth (Figure 6A). The embryonic development of avian intestinal smooth muscles appears to proceed normally even when, by an early transplant on the chorioallantoid membrane, the organ is devoid of nerves (Smith et al. 1977).

Extensive structural changes take place in smooth muscle in postnatal development. These processes are especially in evidence in blood vessels, and many are related to haemodynamic changes occurring around the time of birth. For example, programmed death of muscle cells is thought to occur in the abdominal aorta of the lamb, in connection with the decrease in blood-flow rate ensuing after loss of the placenta (Cho et al. 1995). Apoptotic cell death is well documented in muscle cells isolated from arterial media and grown *in vitro* (see e.g. Bennett et al. 1995), but less so in the vessels *in situ*. There is, however, extensive remodeling of the vessel wall in postnatal life (Langille 1993).

8. Hypertrophy in adult smooth muscles

Growth does continue in some animal species through adult life, albeit at a very slow rate. The process is noticeable in rodents (including the rat) and in the guinea-pig, and is found also in their smooth muscles. In adult muscles, mitotic muscle cells are rare but not absent altogether, both in viscera and in blood vessels (Figure 15). In aging guinea-pigs, visceral muscle cells are larger than in the same tissues at maturity, and there are other changes probably associated with aging. They include an increase in the amount of intramuscular collagen fibrils, a thickening or duplication of the basal lamina, and a reduction of the spatial density of nerve endings. These processes of protracted growth and aging should not be confused with hypertrophic growth.

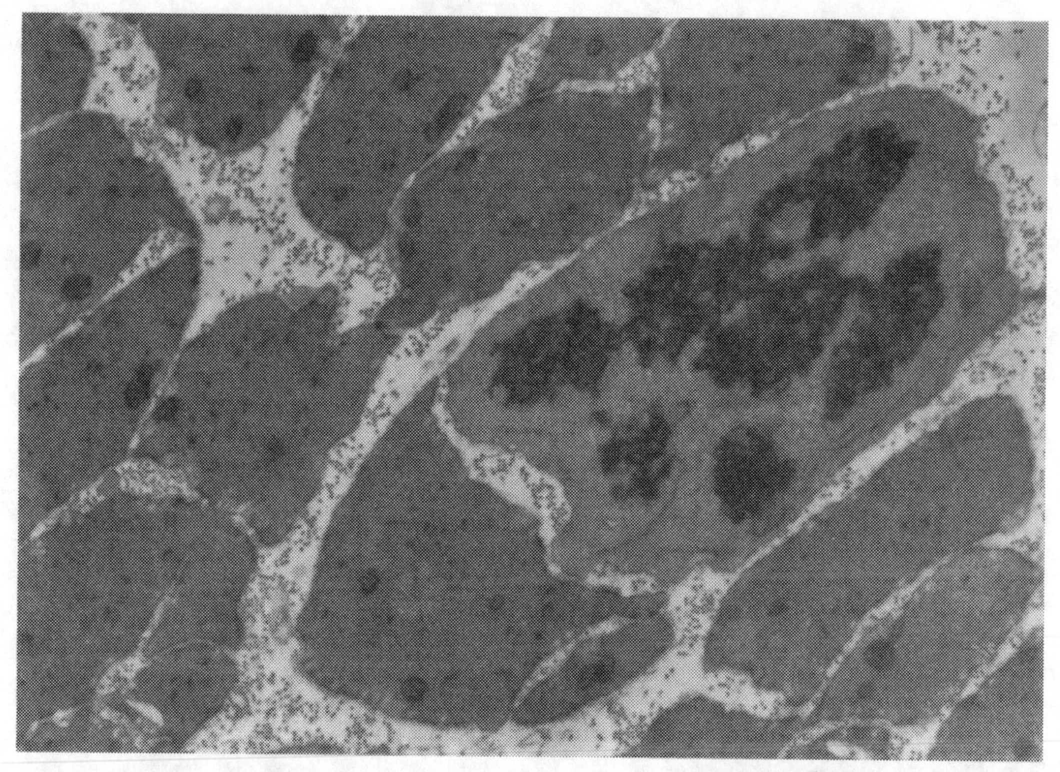

Figure 15. Transverse section of the taenia coli of a guinea pig, showing a muscle cell undergoing mitosis. The lumps of dense material are chromosomes. [14,000×]

Hypertrophy is the growth in excess to the physiologic size reached by an organ or a tissue, and it occurs in connection with increased functional demands. (Some authors refer to the postnatal growth of a muscle as "hypertrophy," a usage obviously at variance with the one proposed here.)

In the case of smooth muscle, an overdistension is often the stimulus to hypertrophic growth. Typically, a partial obstruction of the small intestine (by an ingrowing mass within the lumen, or by a surgical stenosis) causes accumulation of ingesta on the oral side and consequent distension: the muscle layers respond to this stimulus (whose main component is the stretch) and in time become considerably larger and thicker. In the urinary bladder, hypertrophy of the detrusor occurs when the urethral outlet is partially obstructed – for example, by an enlargement of the prostate gland. An entirely physiologic hypertrophy of smooth muscle takes place in the uterus during gestation. Other examples of smooth muscle hypertrophy are found in the caecum in rats fed a fiber-rich diet (Jacobs 1985), in blood vessels in certain forms of hypertension or in arterial stenosis (Folkow 1982; Owens 1989), and in pulmonary vessels in conditions of chronic hypoxia.

Fully hypertrophic visceral muscles have a volume 10 to 15 times larger than the controls. The volume increase is due to an increase in muscle cell number and to an enlargement of the cells, that is, to hyperplasia and to cell hypertrophy (Figure 16). Mitoses are commonly found in the circular and longitudinal muscle

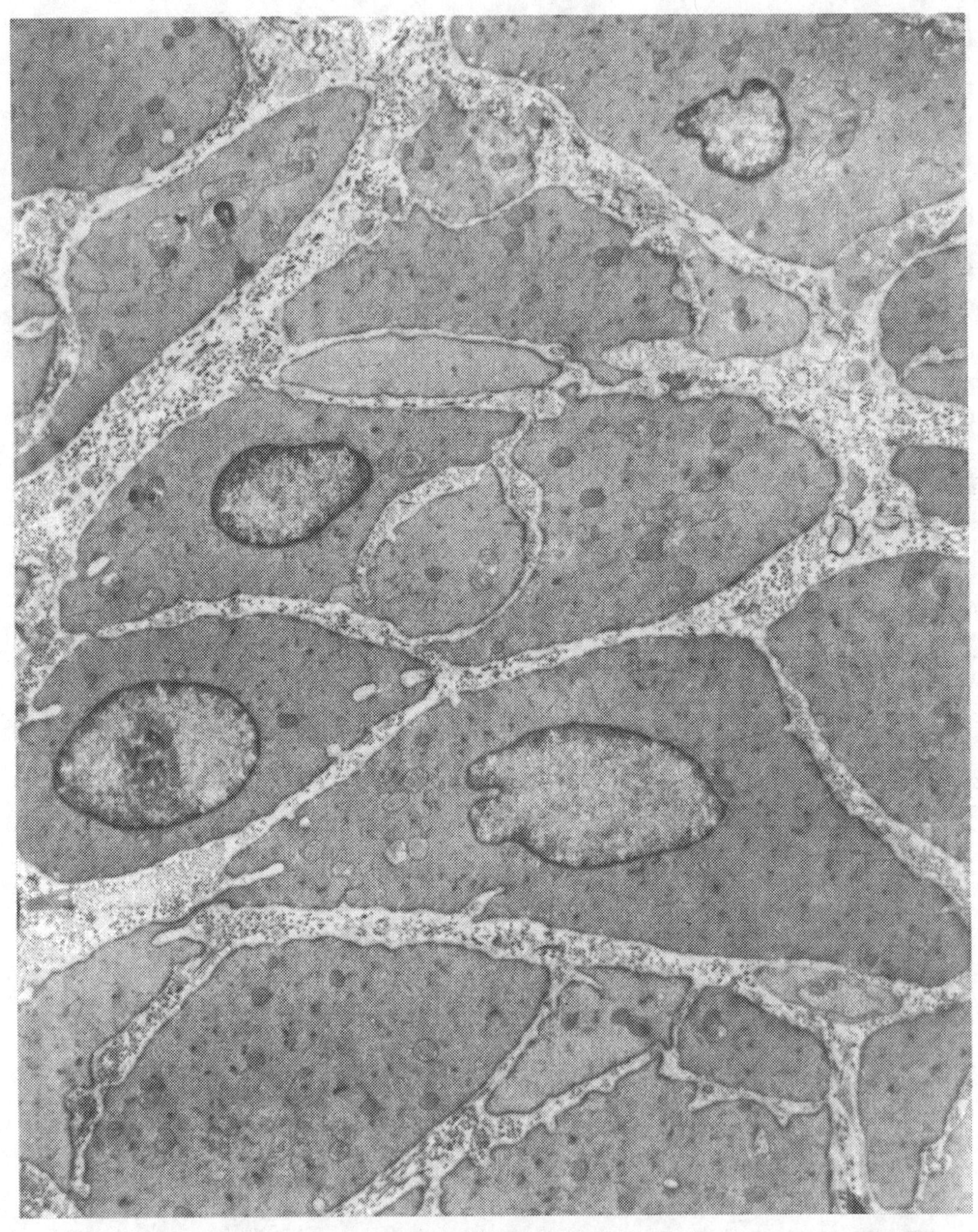

Figure 16. Bladder musculature hypertrophied with urethral obstruction; same magnification as in Figure 2 (control bladder). Note the large size and irregular shape of the muscle cell profiles, the invaginations of the cell membrane, the abundance of extracellular material, and the enlargement of the nuclei (less marked than that of the whole cells). [31,000×]

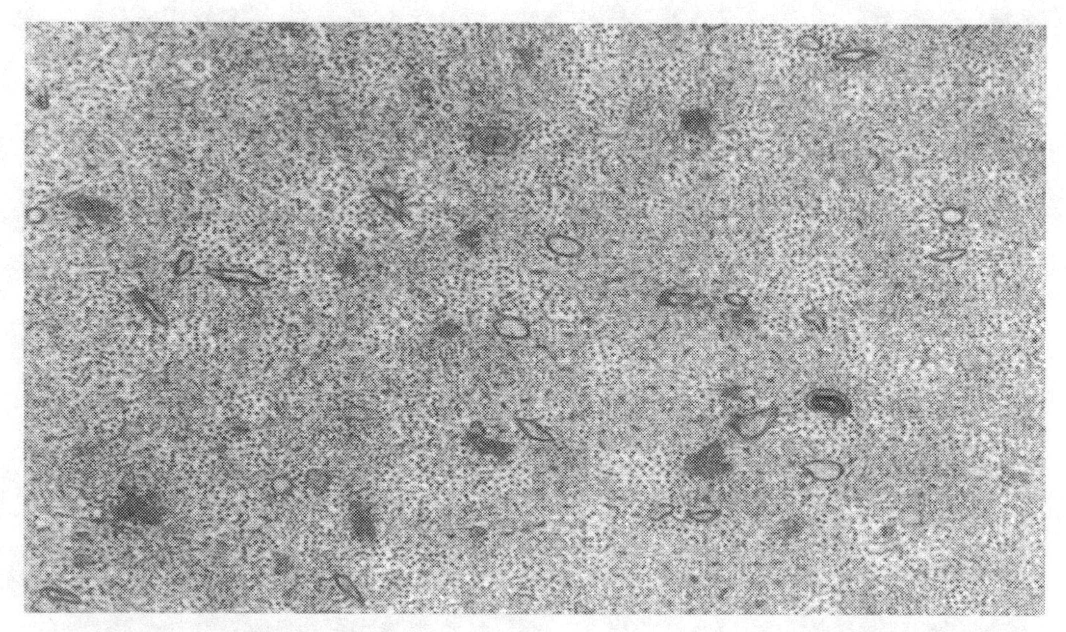

Figure 17. A hypertrophic muscle cell of the rat ileum, showing large groups of intermediate filaments and many tubules of sarcoplasmic reticulum. [61,000×]

layer of the hypertrophying ileum. Muscle cells do not de-differentiate prior to mitosis, but rather retain their full complement of myofilaments while dividing. The occurrence of mitosis in the hypertrophic bladder is much less apparent, when it occurs at all; there is, however, a significant increase in the total DNA content of the tissue, especially in the early stages of hypertrophy (Uvelius et al. 1984; Karim et al. 1992). In the muscle cells of the hypertrophic aorta of hypertensive rats, the increase in DNA content is accounted for by polyploidy and there is little or no cellular proliferation (Owens and Schwartz 1983). In contrast, in the rat hypertrophic mesenteric artery, the enlargement is mainly due to muscle cell division (Mulvany et al. 1985); the same is the case for rat and rabbit aorta in obstruction-induced hypertrophy (Bevan et al. 1976; Owens and Reidy 1985).

Most of the increase in muscle mass, however, is due to an enlargement of the muscle cells (Figure 16). Cell hypertrophy involves not only enlargement of the cells but also complex changes of all the structural parameters, notably in the bladder and intestine (Gabella 1990). The cell surface increases less than the cell volume, but develops deep invaginations that provide additional sites for myofilament insertion and metabolic exchanges and make less dramatic the fall in surface-to-volume ratio in these cells. Many new mitochondria are formed, but there is at the same time a decrease in the percentage mitochondrial volume.

During the second week after onset of bladder obstruction, there is a transient increase in expression of the protooncogenes c-*fos* and c-*myc* in muscle cells (Karim et al. 1992).

In the hypertrophic muscle cells of the intestine, all three types of filaments (thick, thin and intermediate) increase in number owing to the large increase in

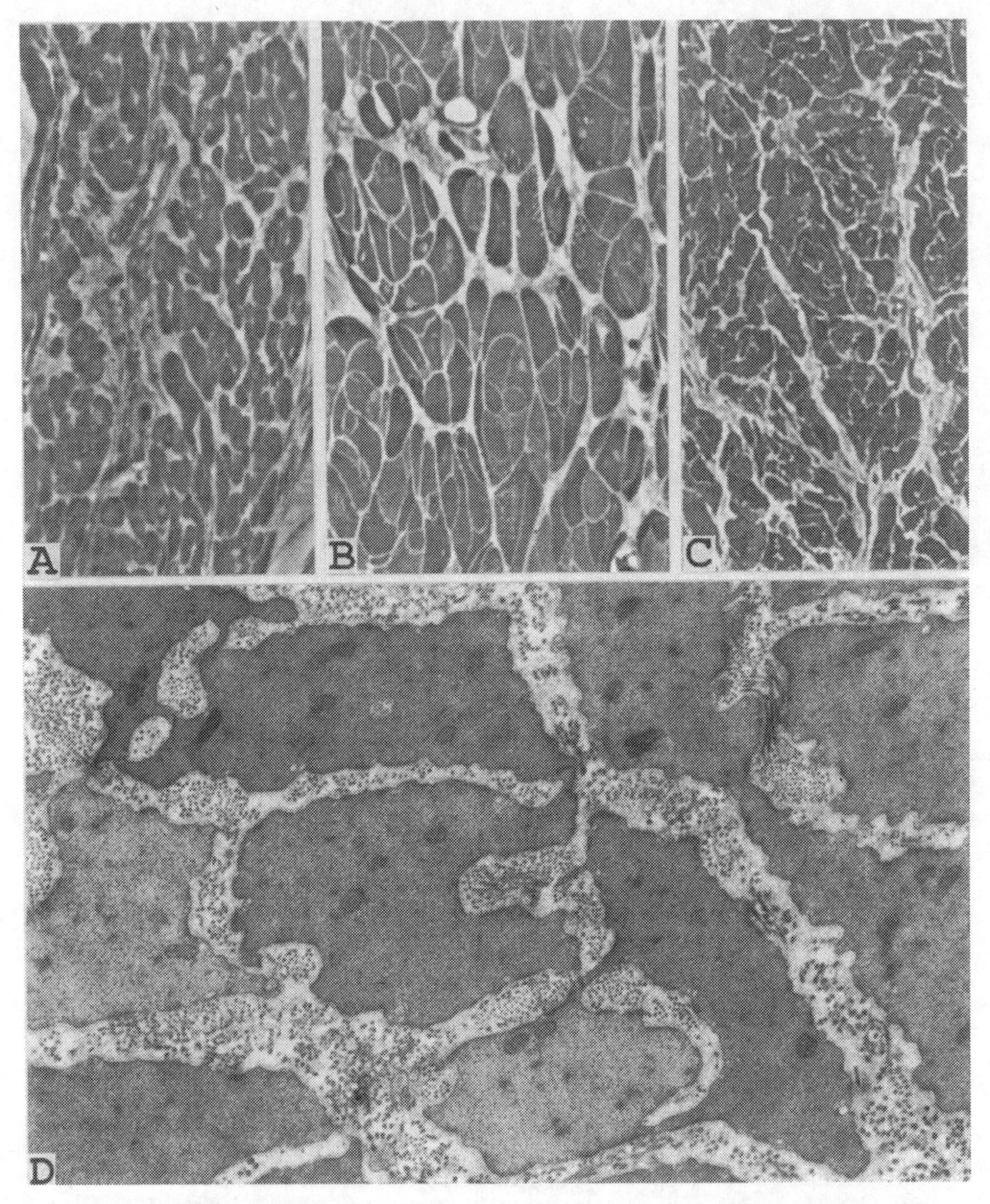

Figure 18. Urinary bladder of rats; transverse sections of the musculature photographed in the light microscope (A-C, 60×) and in the electron microscope (D, 14,000×). A. Control. B. Urethral obstruction for eight weeks and muscle hypertrophy. C and D. Urethral obstruction for eight weeks, followed by obstruction removal for six weeks (reversal of hypertrophy). The enlargement of the muscle cells after obstruction is visible in B. After obstruction and obstruction removal, the cells revert to their size in controls, but the shape of the profiles and the arrangement of the cells remain altered. Large amount of collagen, produced during hypertrophy, occupy the enlarged spaces between cells even six weeks after de-obstruction. [after Gabella and Uvelius 1994]

cell size. Counts of filaments are not available, but thin filaments seem to increase more than thick ones, and intermediate filaments increase substantially more than either type of myofilament (Figure 17). A large increase in intermediate filaments is also observed in the hypertrophic portal vein of the rabbit, accompanied by a less conspicuous increase in actin filaments and by no change in myosin filaments (Berner et al. 1981).

In hypertrophic muscles there is a large amount of smooth sarcoplasmic reticulum, especially the small tubules close to the myofilaments. The rough sarcoplasmic reticulum is also increased over the controls, probably resulting in the production of new stroma, an observation confirmed in the human bladder (Gilpin et al. 1985). The total amount of stroma and its percentage amount are both increased in hypertrophy, and the muscle supports the growth of many new capillaries and hence remains well vascularized.

Although nerves grow within the hypertrophic muscles, the density of innervation (i.e., the spatial frequency of nerve endings in the enlarged muscle) is reduced. Substantial growth takes place also in the neurons supplying the hypertrophic muscle, a process of neuronal hypertrophy observed in enteric ganglia, in the pelvic ganglion, and in dorsal root ganglia. In the case of the bladder, an important link in the neuronal hypertrophy is the production of Nerve Growth Factor and possibly other neurotrophic factors by the hypertrophied muscle cells of the bladder (Steers et al. 1991).

Hypertrophy is to a great extent reversible. In the urinary bladder after removal of the obstruction, the muscle volume returns to control values and the cell hypertrophy is reversed (Gabella and Uvelius 1994). However, the shape of the muscle cell profiles remains distorted, and large amounts of stroma produced during hypertrophy are not removed (Figure 18). In contrast, when the myometrial hypertrophy recedes after parturition, the excess of collagen – a tenfold increase in collagen content takes place during pregnancy (Morrione and Seifter 1962; Woessner and Brewer 1963) – is rapidly resorbed (Harkness and Moralee 1956) by collagenases produced by the myometrial muscle cells (Jeffrey et al. 1990).

Acknowledgments

Invaluable assistance was provided by Peter Trigg. This work is supported by the Wellcome Trust.

References

Ashton, F. T., Somlyo, A. V., and Somlyo, A. P. (1975). The contractile apparatus of vascular smooth muscle: Intermediate high voltage stereo electron microscopy. *J. Mol. Biol.* 98: 17–29.

Babij, P., and Periasamy, M. (1989). Myosin heavy chain isoform diversity in smooth muscle is produced by differential RNA processing. *J. Mol. Biol.* 210: 673–9.

Bagby, R. M. (1983). Organization of contractile/cytoskeletal elements. In: *Biochemistry of Smooth Muscle.* (N.L. Stephens, ed.) Boca Raton, FL: CRC Press, pp. 1–84.

Bagby, R. (1986) Toward a comprehensive three-dimensional model of a contractile system of vertebrate smooth muscle cells. *Int. Rev. Cytol.* 105: 67–128.

Bennett, J. P., Cross, R. A., Kendrick-Jones, J., and Weeds, A. G. (1988). Spatial pattern of myosin phosphorylation in contracting smooth muscle cells: evidence for contractile zones. *J. Cell Biol.* 107: 2623–29.

Bennett, M. V. L., Barrio, L. C., Bargiello, T. A., Spray, D. C., Hertzberg, E., and Sáez, J. C. (1991). Gap junctions: New tools, new answers, new questions. *Neuron* 6: 305–20.

Bennett, T., and Cobb, J. L. S. (1969). Studies of the avian gizzard: the development of the gizzard and its innervation. *Z. Zellforsch.* 98: 599–621.

Bény, J.-L., and Connat, J.-L. (1992). An electron-microscopic study of smooth muscle cells dye coupling in the pig coronary arteries. *Circ. Res.* 70: 49–55.

Berner, P. F., Somlyo, A. V., and Somlyo, A. P. (1981). Hypertrophy-induced increase of intermediate filaments in vascular smooth muscle. *J. Cell Biol.* 88: 96–101.

Bevan, R., Marthens, E., and Bevan, J. (1976). Hyperplasia of vascular smooth muscle in experimental hypertension in the rabbit. *Circ. Res.* 38 (suppl. II): 58–62.

Beyer, E. C., Kistler, J., Paul, D. L., and Goodenough, D. A. (1989). Antisera directed against connexin 43 peptides react with a 43-kD protein localized to gap junctions in myocardium and other tissues. *J. Cell Biol.* 108: 595–605.

Bond, M., Kitizawa, T., Somlyo, A.P., and Somlyo, A.V. (1985). Release and recycling of calcium by the sarcoplasmic reticulum in guinea pig portal vein smooth muscle. *J. Physiol.* (Lond.) 355: 677–95.

Bond, M., & Somlyo, A. V. (1982). Dense bodies and actin polarity in vertebrate smooth muscle. *J. Cell Biol.* 95: 403–413.

Borrione, A. C., Zanellato, A. M. C., Scannapieco, G., Pauletto, P., and Sartore, S. (1989). Myosin heavy chain isoforms in adult and developing rabbit vascular smooth muscle. *Eur. J. Biochem.* 183: 413–17.

Bramich, N. J., and Brading, A. F. (1996). Electrical properties of smooth muscle in the guinea-pig urinary bladder. *J. Physiol. (Lond.)* 492: 185–98.

Büssow, H., and Wulfhekel, U. (1972). Die Feinstruktur der glatten Muskelzellen in den grossen muskulären Arterien der Vögel. *Z. Zellforsch. Mikrosk. Anat.* 125: 339–52.

Byers, T. J., Kunkel, L. M., and Watkins, S. C. (1991). The subcellular distribution of dystrophin in mouse skeletal, cardiac, and smooth muscle. *J. Cell Biol.* 115: 411–21.

Cavaillé, F., Janmot, C., Ropert, S., and D'Albis, A. (1986). Isoforms of myosin and actin in human, monkey and rat myometrium: Comparison of pregnant and non-pregnant uterus proteins. *Eur. J. Biochem.* 160: 507–13.

Cho, A., Courtman, D. W., and Langille, B. L. (1995). Apoptosis (programmed cell death) in arteries of the neonatal lamb. *Circ. Res.* 76: 168–75.

Clark, J. M. and Glasgow, S. (1979). Structural integration of the arterial wall. I. Relationships and attachments of medial smooth muscle cells in normally distended and hyperdistended aorta. *Lab. Invest.* 40: 587–602.

Cooke, P. H., and Chase, R. H. (1971). Potassium chloride-insoluble myofilaments in vertebrate smooth muscle cells. *Exp. Cell Res.* 66: 417–25.

Cooke, P. H., and Fay, F. S. (1972). Correlation between fiber length, ultrastructure and the length-tension relationship of mammalian smooth muscle. *J. Cell Biol.* 52: 105–16.

Cooke, P. H., Fay, F. S., and Craig, R. (1989). Myosin filaments isolated from skinned amphibian smooth muscle cells are side-polar. *J. Muscle Res. Cell Motil.* 10: 206–20.

Craig, R., and Megerman, J. (1977). Assembly of smooth muscle myosin into side-polar filaments. *J. Cell Biol.* 75: 990–6.

Crone, C. (1986). Modulation of solute permeability in microvascular endothelium. *Fed. Proc.* 45: 77–83.

Dahl, G. P., and Berger, W. (1978). Nexus formation in the myometrium during parturition and induced by estrogen. *Cell Biol. Int. Rep.* 2: 381–7.
Daniel, E. E., Daniel, V. P., Duchon, G., Garfield, R. E., Nichols, M., Malhotra, S. K., and Oki, M. (1976). Is the nexus necessary for cell-to-cell coupling of smooth muscle. *J. Membr. Biol.* 28: 207–39.
Devine, C. E., and Rayns, D. G. (1975). Freeze-fracture studies of membrane system in vertebrate muscle. II. Smooth muscle. *J. Ultrastruct. Res.* 51: 293–306.
Devine, C. E., Somlyo, A. V., and Somlyo, A. P. (1972). Sarcoplasmic reticulum and excitation-contraction coupling in mammalian smooth muscles. *J. Cell Biol.* 52: 690–718.
Dingemans, K. P., Jansen, N., and Becker, A. E. (1981). Ultrastructure of the human aortic media. *Virchows Arch. Pathol. Anat.* 392: 199–219.
Donhahoe, J. R., and Bowen, J. M. (1972). Analysis of the spontaneous motility of the avian embryonic gizzard. *Am. J. Vet. Res.* 33: 1835–48.
Draeger, A., Steltser, E. H. K., Herzog, M., and Small, J. V. (1989). Unique geometry of actin membrane anchorage sites in avian gizzard smooth muscle cells. *J. Cell Sci.* 94: 703–11.
Drenckhahn, D., Beckerle, M., Burridge, K., and Otto, J. (1988). Identification and subcellular location of talin in various cell types and tissues by means of [^{125}I] vinculin overlay, immunoblotting and immunocytochemistry. *Eur. J. Cell Biol.* 46: 513–22.
Drenckhahn, D., and Jeikowski, H. (1978). The myotendinous junction of the smooth feather muscles (mm. pennati). *Cell Tissue Res.* 194: 151–62.
Fawcett, D. W., and McNutt, N. S. (1969). The ultrastructure of the cat myocardium. I. Ventricular papillary muscle. *J. Cell Biol.* 42: 1–45.
Fay, F. S., and Delise, C. M. (1973). Contraction of isolated smooth muscle cells – Structural changes. *Proc. Nat. Acad. Sci. USA* 70: 641–5.
Fay, F. S., Fujiwara K., Rees, D. D., and Fogarty, K. E. (1983). Distribution of α-actinin in single isolated smooth muscle cells. *J. Cell Biol.* 96: 783–95.
Fisher, B. A., and Bagby, R. M. (1977). Reorientation of myofilaments during contraction of a vertebrate smooth muscle. *Am. J. Physiol.* 232: C5–C14.
Folkow, B. (1982). Physiological aspects of primary hypertension. *Physiol. Rev.* 62: 347–504.
Forbes, M. S., Rennels, M. L., and Nelson, E. (1979). Caveolar systems and sarcoplasmic reticulum in coronary smooth muscle cells of the mouse. *J. Ultrastruct. Res.* 67: 325–39.
Frackowiak, J., Mazur-Kolecka, B., Wisniewski, H. M., Potempska, A., Carroll, R. T., Emmerling, M. R., and Kim, K. S. (1995). Secretion and accumulation of Alzheimer's β-protein by cultured vascular smooth muscle cells from old and young dogs. *Brain Res.* 676: 225–30.
Franke, W. W., Schmid E., Freudenstein, C., Appelhans, B., Osborn, M., Weber, K., and Keenan, T. W. (1980). Intermediate-sized filaments of the prekeratin type in myoepithelial cells. *J. Cell Biol.* 84: 633–54.
Fujimoto, T. (1993). Calcium pump of the plasma membrane is localized in caveolae. *J. Cell Biol.* 120: 1147–57.
Fujimoto, T., Nakade, S., Miyawaki, A., Mikoshiba, K., and Ogawa, K. (1992). Localization of inositol 1,4,5-trisphosphate receptor-like protein in plasmalemmal caveolae. *J. Cell Biol.* 119: 1507–13.
Gabbiani, G., Schmid, E., Winter, S., Chaponnier, C., de Chastonay, C., Vendekerckhove, J., Weber, K., and Franke, W. W. (1981). Vascular smooth muscle cells differ from other smooth muscle cells: Predominance of vimentin filaments and a specific a-type actin. *Proc. Nat. Acad. Sci. USA* 78: 298–302.

Gabella, G. (1981). Structure of smooth muscle. In: *Smooth Muscle: An Assessment of Current Knowledge*. (E. Bülbring, A. F. Brading, A. W. Jones, and T. Tomita, eds.). London: Arnold, pp. 1–46.

Gabella, G. (1984). Structural apparatus for force transmission in smooth muscles. *Physiol. Rev.* 64: 455–77.

Gabella, G. (1985). Chicken gizzard. The muscle, the tendon and their attachment. *Anat. Embryol.* 171: 151–62.

Gabella, G. (1989). Development of smooth muscle: ultrastructural study of the chick embryo gizzard. *Anat. Embryol.* 180: 213–26.

Gabella, G. (1990). Hypertrophy of visceral smooth muscle. *Anat. Embryol.* 182: 409–24.

Gabella, G. (1991). Ultrastructure of the tracheal muscle in developing, adult and ageing guinea-pigs. *Anat. Embryol.* 183: 71–9.

Gabella, G. (1995). The structural relations between nerve fibers and muscle cells in the urinary bladder of the rat. *J. Neurocytol.* 245: 159–87.

Gabella G., and Blundell, D. (1978). Effect of stretch and contraction on caveolae of smooth muscle cells. *Cell Tissue Res.* 190: 255–71.

Gabella, G., and Uvelius, B. (1990). Urinary bladder of the rat: Fine structure of normal and hypertrophic musculature. *Cell Tissue Res.* 262: 67–79.

Gabella, G., and Uvelius, B. (1994). Reversal of hypertrophy in rat urinary bladder after removal of urethral obstruction. *Cell Tissue Res.* 277: 333–9.

Garfield, R. E., and Daniel, E. E. (1976). Relation of membrane vesicles to volume control and Na^+-transport in smooth muscle: Studies in Na^+ rich tissues. *J. Mechanochem. Cell Motil.* 4: 157–76.

Garfield, R. E., Sims, S. M., and Daniel, E. E. (1977). Gap junctions: their presence and necessity in myometrium during parturition. *Science* 198: 958–9.

Garfield, R. E., Thilander, G., Blennerhassett, M. G., and Sakai, N. (1992). Are gap junctions necessary for cell-to-cell coupling of smooth muscle? An update. *Can. J. Physiol. Pharmacol.* 70: 481–90.

Geisler, N., and Weber, K. (1983). Amino acid sequence data on glial fibrillary acidic protein (GFA); Implications for the subdivision of intermediate filaments into epithelial and non-epithelial members. *EMBO J.* 2: 2059–63.

Gillis, J. M., Cao, M. L., and Godfraind-De Becker, A. (1988). Density of myosin filaments in the rat anococcygeus muscle, at rest and in contraction. II. *J. Muscle Res. Cell Motil.* 9: 18–28.

Gilpin, S. A., Gosling, J. A., and Barnard, R. J. (1985). Morphological and morphometric studies of the human obstructed, trabeculated urinary bladder. *Br. J. Urol.* 57: 525–9.

Gimbrone, M. A., and Cottran, R. S. (1975). Human vascular smooth muscle in culture. Growth and ultrastructure. *Lab. Invest.* 33: 16–27.

Giuriato, L., Scatena, M., Chiavegato, A., Tonello, M., Scannapieco, G., Pauletto, P., and Sartore, S. (1992). Nonmuscle myosin isoforms and cell heterogeneity in developing rabbit vascular smooth muscle. *J. Cell Sci.* 101: 233–46.

Gown, A. M., Boyd, H. C., Chang, Y., Ferguson, M., Reichler, B., and Tippens, D. (1988). Smooth muscle cells can express cytokeratins of ''simple'' epithelium. *Am. J. Pathol.* 132: 223–32.

Guyton, J. R., Lindsay, K. L., and Dao, D. T. (1983). Comparison of aortic intima and inner media in young adult versus aging rats. Stereology in a polarized system. *Am. J. Pathol.* 111: 234–46.

Hamada, Y., Yanagisawa, M., Katsuragawa, Y., Coleman, J. R., Nagata, S., Matsuda, G., and Masaki, T. (1990). Distinct vascular and intestinal smooth muscle myosin heavy chain mRNAs are encoded by a single-copy gene in the chicken. *Biochem. Biophys. Res. Commun.* 170: 53–8.

Harkness, R. D., and Moralee, B. E. (1956). The time course and route of loss of collagen from the rat's uterus during post-partum involution. *J. Physiol. (Lond.)* 132: 492–501.

Hinssen, H., D'Haese, J., Small, J. V., and Sobieszek, A. (1978). Mode of filament assembly of myosins from muscle and non-muscle cells. *J. Ultrastuct. Res.* 64: 282–302.

Hirai, S. I., and Hirabayashi, T. (1983). Developmental change in protein constituents in chicken gizzards. *Develop. Biol.* 97: 483–93.

Hirai, S. I., and Hirabayashi, T. (1986). Development of myofibrils in the gizzard of chicken embryos. Intracellular distribution of structural proteins and development of contractility. *Cell Tissue Res.* 243: 487–93.

Huiatt, T. W., Robson, R. M., Arakawa, N., and Stromer, M. H. (1980). Desmin from avain smooth muscle. Purification and partial characterization. *J. Biol. Chem.* 255: 6981–9.

Ip, W., Heuser, J. E., Pang, Y.-Y. S., Hartzer, M. K., and Robson, R. M. (1985). Subunit structure of desmin and vimentin and how they assemble into intermediate filaments. *Ann. N.Y. Acad. Sci.* 455: 185–99.

Jacobs, L. R. (1985). Differential effects of dietary fibers on rat intestinal circular muscle cell size. *Digest. Dis. Sci.* 30: 247–52.

Jeffrey, J. J., Roswit, W. T., and Ehlich, L. S. (1990). Regulation of collagenase production in uterine smooth muscle cells: An enzymatic and immunologic study. *J. Cell. Physiol.* 143: 396–403.

Kamio, A., Huang, W. Y., Imai, H., and Kummerow, F. A. (1977). Mitotic structures of aortic smooth muscle cells in swine and in culture: Paired cisternae. *J. Electron Micr.* 26: 29–40.

Kannan, M. S., and Daniel, E. E. (1978). Formation of gap junctions by treatment *in vitro* with potassium conductance blockers. *J. Cell Biol.* 78: 338–48.

Kargacin, G. J., Cooke, P. H., Abramson, S. B., and Fay, F. S. (1989). Periodic organization of the contractile apparatus in smooth muscle revealed by the motion of dense bodies in single cells. *J. Cell Biol.* 108: 1465–75.

Karim, O. M. A., Seki, N., and Mostwin, J. L. (1992). Detrusor hyperplasia and expression of ''immediate early'' genes with onset of abnormal urodynamic parameters. *Am. J. Physiol.* 263: R1284–R1290.

Kuro-o, M., Nagai, R., Tsuchimochi, H., Katoh, H., Yazaki, Y., Ohkubo, A., and Takaku, F. (1989). Developmentally regulated expression of vascular smooth muscle myosin heavy chain isoforms. *J. Biol. Chem.* 264: 18272–5.

La Mantia, J., and Shafiq, S. A. (1982). Developmental changes in the plasma membrane of gizzard smooth muscle of the chicken. A freeze-fracture study. *J. Anat.* 134: 243–53.

Langille, B. L. (1993). Remodelling of developing and mature arteries: Endothelium, smooth muscle and matrix. *J. Cardiovasc. Pharmacol.* 21: S11–S17.

Langley, J. N. (1904). On the sympathetic system of birds, and on the muscles which move the feathers. *J. Physiol. (Lond.)* 30: 221–52.

Leblond, C. P., and Inoue, S. (1989). Structure, composition, and assembly of basement membrane. *Am. J. Anat.* 185: 367–90.

Le Douarin, N. (1982). *The Neural Crest.* Cambridge, UK: Cambridge University Press.

Lever, J. D., Ahmed, M., and Irvine, G. (1965). Neuromuscular and intercellular relationships in the coronary arterioles. A morphological and quantitative study by light and electron microscopy. *J. Anat.* 99: 829–40.

Linderkamp, O., and Meiselman, H. J. (1982). Geometric, osmotic, and membrane mechanical properties of density-separated human red cells. *Blood* 59: 1121–7.

Lowy, J., Vibert, P., Haselgrove, J. C., and Poulsen, F. R. (1973). The structure of the myosin elements in vertebrate smooth muscles. *Phil. Trans. Roy. Soc. Lond. B* 265: 191–6.

Luff, S. E., McLachlan, E. M., and Hirst, G. D. S. (1987). An ultrastructural analysis of the sympathetic neuromuscular junctions on arterioles of the submucosa of the guinea pig ileum. *J. Comp. Neurol.* 257: 578–94.

Maekawa, M. T., Nagano, K., Kamimura, T., Murakami, T., Ishikawa, H., and Dezawa, M. (1991). Distribution of actin-filament bundles in myoid cells, Sertoli cells, and tunica albuginea of rat and mouse testes. *Cell Tissue Res.* 266: 295–300.

McGuffee, L. J., and Bagby, R. M. (1976). Ultrastructure, calcium accumulation, and contractile response in smooth muscle. *Am. J. Physiol.* 230: 1217–24.

McGuffee, L. J., Mercure, J., and Little, S. A. (1991). Three-dimensional structure of dense bodies in rabbit renal artery smooth muscle. *Anat. Rec.* 229: 499–504.

Mackenzie, L. W., and Garfield, R. E. (1985). Hormonal control of gap junctions in the myometrium. *Am. J. Physiol.* 248: C296–C302.

Martinez-Hernandez, A., and Amenta, P. S. (1984). The basement membranes in pathology. *Lab. Invest.* 48: 656–77.

Mikkelsen, H. B., Huizinga, J. D., Thuneberg, L., and Rumessen, J. J. (1993). Immunohistochemical localization of a gap junction protein (connexin43) in the muscularis externa of murine, canine, and human intestine. *Cell Tissue Res.* 274: 249–56.

Mohammad, M. A., and Sparrow, M. P. (1988). Changes in myosin heavy chain stoichiometry in pig tracheal smooth muscle during development. *FEBS Letts.* 228: 109–12.

Moriya, M., and Miyazaki, E. (1980). Ultrastructural differences between longitudinal and circular muscle cells of the guinea pig stomach. *Sapporo Med. J.* 49: 391–401.

Morrione, T. G., and Seifter, S. (1962). Alterations in the collagen content during pregnancy and post-partum involution. *J. Exp. Med.* 115: 357–65.

Mullins, G. L., and Guntheroth, W. G. (1965). A collagen net hypothesis for force transference of smooth muscle. *Nature* 206: 592–4.

Mulvany, M., Baandrup, U., and Gundersen, H. (1985). Evidence for hyperplasia in mesenteric resistance vessels of spontaneously hypertensive rats using a three-dimensional dissector. *Circ. Res.* 57: 794–800.

Murphy, R. A., Driska, S. P., and Cohen, D. M. (1977). Variations in actin to myosin ratios and cellular force generation in vertebrate smooth muscles. In: *Excitation–Contraction Coupling in Smooth Muscle.* (R. Casteels, ed.). Amsterdam: Elsevier, pp. 417–24.

Nabeshima, Y., Nabeshima, Y.-I., Nonomura, Y., and Fujii-Kuriyama, Y. (1987). Non-muscle and smooth muscle myosin light chain mRNAs are generated from a single gene by the tissue-specific alternative RNA splicing. *J. Biol. Chem.* 267: 10608–12.

Nasu, F., and Inomata, K. (1990). Ultracytochemical demonstration of Ca^{++}-ATPase activity in the rat saphenous artery and its innervated nerve terminal. *J. Electron Micr.* 39: 487–91.

North, A. J., Galazkiewicz, B., Byers, T. J., Glenney, J. R., and Small, J. V. (1993). Complementary distribution of vinculin and dystrophin define two distinct domains in smooth muscle. *J. Cell Biol.* 120: 1159–67.

North, A. J., Gimona, M., Lando, Z., and Small, J. V. (1994). Actin isoform compartment in chicken gizzard smooth muscle cells. *J. Cell. Sci.* 107: 445–55.

Ogawa, K. S., Fujimoto, K., and Ogawa, K. (1986). Ultracytochemical studies of adenosine nucleotidases in aortic endothelial and smooth muscle cells – Ca^{2+}-ATPase and Na^{+},K^{+}-ATPase. *Acta Histochem. Cytochem.* 19: 601–20.
Osborne-Pellegrin, M. J. (1978). Some ultrastructural characteristics of the renal artery and abdominal aorta of the rat. *J. Anat. (Lond.)* 125: 641–52.
Owens, G. K. (1989). Control of hypertrophic versus hyperplastic growth of vascular smooth muscle cells. *Am. J. Physiol.* 257: H1755–H1765.
Owens, G. K., and Reidy, M. (1985). Hyperplastic growth response of vascular smooth muscle cells following induction of acute hypertension in rats by aortic coarctation. *Circ. Res.* 57: 695–705.
Owens, G. K., and Schwartz, S. (1983). Vascular smooth muscle cell hypertrophy and hyperploidy in the Goldblatt hypertensive rat. *Circ. Res.* 53: 491–501.
Paul, E. R., Christian, A.-L., Franke, R., and Gröschel-Stewart, U. (1994). Embryonic chicken gizzard: Smooth muscle and non-muscle myosin. *Cell Tissue Res.* 276: 381–6.
Popescu, L. M., and Diculescu, I. (1975). Calcium in smooth muscle sarcoplasmic reticulum in situ. Conventional and X-ray analytical electron microscopy. *J. Cell Biol.* 67: 911–18.
Poulos, A. C., Rash, J. E., and Elmund, J. K. (1986). Ultrarapid freezing reveals that skeletal muscle caveolae are semipermanent structures. *J. Ultrastruct. Res.* 96: 114–24.
Prescott, L., and Brightman, M. W. (1976). The sarcolemma of Aplysia smooth muscle in freeze-fracture preparations. *Tissue & Cell* 8: 241–58.
Rice, R. V., McManus, G. M., Devine, C. E., and Somlyo, A. P. (1971). A regular organization of thick filaments in mammalian smooth muscle. *Nature New Biol.* 231: 242–3.
Risek, B., Guthrie, S., Kumar, N., and Gilula, N. J. (1990). Modulation of gap junction transcript and protein expression during pregnancy in the rat. *J. Cell Biol.* 110: 269–82.
Rodrigo, F. G., Lotta-Pereira, G., and David-Ferrera, J. F. (1975). The fine structure of the elastic tendons in the human arrector pili muscle. *Br. J. Dermatol.* 93: 631–8.
Rosenbluth, J. (1965). Smooth muscle: An ultrastructural basis for the dynamics of its contraction. *Science* 148: 1337–9.
Rovner, A. S., Thompson, M. M., and Murphy, R. A. (1986). Two different heavy chains are found in smooth muscle myosin. *Am. J. Physiol.* 250: C861–C870.
Salmon, S. (1995). Muscle. In: *Gray's Anatomy*. (P. L. Williams et al., eds.). Edinburgh: Churchill Livingstone, pp. 737–900.
Sanes, J. R., Engvall, E., Butkowski, R., and Hunter, D. D. (1990). Molecular heterogeneity of basal laminae: Isoforms of laminin and collagen IV at the neuromuscular junction and elsewhere. *J. Cell Biol.* 111: 1685–99.
Sartore, S., De Marzo, N., Borrione, A. C., Zanellato, M. C., Saggin, L., Fabbri, L., and Schiaffino, S. (1989). Myosin heavy-chain isoforms in human smooth muscle. *Eur. J. Biochem.* 179: 79–85.
Schmid, E., Osborn, M, Rungger-Brändle, E., Gabbiani, G., Weber, K., and Franke, K.K. (1982). Distribution of vimentin and desmin filaments in smooth muscle tissue of mammalian and avian aorta. *Exp. Cell Res.* 137: 329–40.
Schollmeyer, J. E., Furcht, L. J., Goll, D. E., Robson, R. M., and Stromer, M. H. (1976). Localization of contractile proteins in smooth muscle cells and in normal and transformed fibroblasts. In: *Cell Motility* (A. R. Goldman, T. Pollard and J. Rosenbaum, eds.) Cold Spring Harbor, New York: Cold Spring Harbor Laboratory, pp. 361–88.

Severs, N. J. (1988). Caveolae: static inpocketings of the plasma membrane, dynamic vesicles or plain artifacts? *J. Cell Sci.* 90: 341–8.
Small, J. V. (1977). Studies on isolated smooth muscle cells. The contractile apparatus. *J. Cell Sci.* 24: 327–49.
Small, J. V., and Sobieszek, A. (1980). The contractile apparatus of smooth muscle. *Int. Rev. Cytol.* 64: 241–306.
Small, J. V., and Squire, J. M. (1972). Structural basis of contraction in vertebrate smooth muscle. *J. Mol. Biol.* 67: 117–49.
Smith, J., Cochard, P., and Le Douarin, N. M. (1977). Development of choline acetyltransferase and cholinesterase activities in enteric ganglia derived from presumptive adrenergic and cholinergic levels of the neural crest. *Cell. Diff.* 6: 199–216.
Somlyo, A. P. (1993). Myosin isoforms in smooth muscle: How may they affect function and structure? *J. Muscle Res. Cell Motil.* 14: 557–63.
Somlyo, A. P., Devine, C. E., Somlyo, A. V., and Rice, R. V. (1973). Filament organization in vertebrate smooth muscle. *Phil. Trans. Roy. Soc. Lond. B* 265: 223–9.
Somlyo, A. V., Butler, T. M., Bond, M., and Somlyo, A. P. (1981). Myosin filaments have nonphosphorylated light chains in relaxed smooth muscle. *Nature* 294: 567–9.
Somlyo, A. V., and Franzini-Armstrong, C. (1985). New views of smooth muscle structure using freezing, deep-etching and rotary shadowing. *Experientia* 41: 841–56.
Sommer, J. R., and Johnson, E. A. (1968). Cardiac muscle. A comparative study of Purkinje fibers and ventricular fibers. *J. Cell Biol.* 36: 497–526.
Steers, D., Kolbeck, S., Creedon, D., and Tuttle, J. B. (1991). Nerve growth in the urinary bladder of the adult regulates neuronal form and function. *J. Clin. Invest.* 88: 1709–15.
Steuwer, D., and Gröschel-Stewart, U. (1985). Expression of immunoreactive myosin and myoglobin in the developing chicken gizzard. *Roux's Arch. Dev. Biol.* 194: 417–24.
Tsukita, S., Tsukita, S., and Ishikawa, H. (1983). Association of actin and 10 nm filaments with the dense body in smooth muscle cells of the chicken gizzard. *Cell Tissue Res.* 229: 233–42.
Tsukita, S., Tsukita, S., Usukura, J., and Ishikawa, H. (1982). Myosin filaments in smooth muscle cells of the guinea pig taenia coli: A freeze-substitution study. *Eur. J. Cell Biol.* 28: 195–201.
Turley, H., Pulford, K. A. F., Gatter, K. C., and Mason, D. Y. (1988). Biochemical evidence that cytokeratins are present in smooth muscle. *Br. J. Exp. Pathol.* 69: 433–40.
Uehara, Y., Fujiwara, T., Nakashiro, S., and De Shan, Z. (1990). Morphology of smooth muscle and its diversity as studied with scanning electron microscopy. In: *Ultrastructure of Smooth Muscle.* (P. M. Motta, ed.). Boston: Kluwer, pp. 119–36.
Uvelius, B., Persson, L., and Mattiasson, A. (1984). Smooth muscle cell hypertrophy and hyperplasia in the rat detrusor after short-term infravesical outflow obstruction. *J. Urol.* 131: 173–6.
Vogl, A. W., Soucy, L. J., and Lew, G. J. (1985). Distribution of actin in isolated seminiferous epithelia and denuded tubule walls of the rat. *Anat. Rec.* 213: 63–71.
Volberg, T., Sabanay, H., and Geiger, B. (1986). Spatial and temporal relationships between vinculin and talin in the developing chicken gizzrd smooth muscle. *Differentiation* 32: 34–43.
Walker-Caprioglio, H. M., Hunter, D. D., McGuire, P. G., Little, S. A., and McGuffee, L. J. (1995). Composition in situ and in vitro of vascular smooth muscle laminin in the rat. *Cell Tissue Res.* 281: 187–96.

Warshaw, D. M., McBride, W. J., and Work, S. S. (1987). Corkscrew-like shortening in single smooth muscle cells. *Science* 236: 1457–9.

Watzka, M. (1932). Sehnen glatter Muskelfasern. *Z. Mikr.-Anat. Forsch.* 30: 23–8.

Woessner, J. F., and Brewer, T. H. (1963). Formation and breakdown of collagen and elastin in the human uterus during pregnancy and post-partum involution. *Biochem. J.* 89: 75–82.

Zamir, O., and Hanani, M. (1990). Intercellular dye-coupling in intestinal smooth muscle. Are gap junctions required for intercellular coupling? *Experientia* 46: 1002–5.

2

Calcium Homeostasis in Smooth Muscles

JORDAN D. MILLER AND MARY E. CARSTEN

1. Introduction

The obligatory role of Ca^{2+} in muscle contraction was first described by Ringer (1882), and subsequently Ca^{2+} was shown to stimulate a host of regulatory enzymes. The change in intracellular free Ca^{2+} necessary for contraction is very small, on the order of 1 μM. This chapter deals with the variety of organelles, regulatory factors, and mechanisms that work together to maintain this tight control of the intracellular free Ca^{2+} concentration in smooth muscles.

2. Overview

The control of intracellular free Ca^{2+} concentration was first explored in skeletal and cardiac muscles. Now, knowledge of events in smooth muscles is also progressing rapidly. Major quantitative and qualitative differences between skeletal and smooth muscles are evident. Whereas in skeletal muscle Ca^{2+} flux and contraction are initiated by Na^+ influx and membrane depolarization, in smooth muscles Ca^{2+} translocation and contraction can occur independently of changes in membrane potential (Edman and Schild 1963). This phenomenon, called *pharmacomechanical coupling* (Somlyo and Somlyo 1968), can be initiated by agonist receptor binding, release of second messengers, release of Ca^{2+} from internal stores, or opening of receptor-operated Ca^{2+} channels. Influx of Ca^{2+} through Ca^{2+} channels is very important for tension development in smooth muscles, but not in skeletal muscle.

Although smooth muscles resemble each other more than they do cardiac or skeletal muscles, major differences are evident in the structures and functions among different smooth muscles. These differences may be determined by the number and specificity of receptors, the second messengers generated, the Ca^{2+} pathway involved, and the activity of Ca^{2+}-sensitive enzymes including the protein kinases. Whether these differences among smooth muscles are based on variation in the primary structure of receptors, receptor density, or the G protein is not known. On the other hand, the observed diversity in structure may not be related to a diversity in genes but may rather be the consequence of alternative splicing. Clearly, a fuller understanding than we now have of the mechanisms involved in maintenance of Ca^{2+} homeostasis will clarify many problems concerning smooth muscle function.

The focus of most investigators over the past few decades has been on the contractile aspect of muscle cells. It is not surprising that when one uses the term "Ca^{2+} homeostasis" it is generally meant to refer to the availability of Ca^{2+} for the contractile apparatus. Thus, control of Ca^{2+} has been attributed to mechanisms located in the plasma membrane (PM) and the sarcoplasmic reticulum (SR), whereas mitochondria and nuclei have been found not to play a role in the regulation of Ca^{2+}. This inference is based on the timing of changes in Ca^{2+} concentration ($[Ca^{2+}]$) in these organelles. If cytoplasmic and organellar $[Ca^{2+}]$ increase at the same time, the organelle is said not to participate in Ca^{2+} homeostasis. Such a simplistic picture has been replaced by a much more complex description that includes locally high Ca^{2+} concentrations within the cytoplasm. Changes in $[Ca^{2+}]$ in organelles buffer the cytoplasmic concentration and also have effects on the organelle. Transient changes in $[Ca^{2+}]$ in mitochondria may play a role in providing the energy necessary for contractile activity and the later re-uptake of Ca^{2+} by SR. From the perspective of the actomyosin the mitochondria play no role in Ca^{2+} homeostasis, but changes in $[Ca^{2+}]$ in this organelle clearly are important in cellular function (Rizzuto et al. 1994).

3. Maintenance of cellular $[Ca^{2+}]$

The classical description of Ca^{2+} homeostasis in muscle cells focuses on the cytoplasm and the control of contraction. Proper functioning of the smooth muscle cell requires maintenance of a low concentration of free Ca^{2+} in the cytoplasm, while a high concentration of Ca^{2+} is present in the extracellular fluid and in various intracellular compartments. Protection is afforded by the low Ca^{2+} permeability of the PM and the SR membranes. Upon excitation, Ca^{2+} channels in the PM open and allow Ca^{2+} to move down its concentration gradient, and second messengers release Ca^{2+} from the SR. Both processes evoke contraction. Ca^{2+} binding proteins may act as storage sites and thus buffer changes in free Ca^{2+}. Active movement of Ca^{2+} out of the cytoplasm across these membranes initiates relaxation. More recent work has also described the Ca^{2+} homeostatic mechanisms in other intracellular compartments, including mitochondria and nuclei. Although there is little evidence that $[Ca^{2+}]$ in these other compartments plays a direct role in contraction and relaxation, the maintenance of Ca^{2+} levels is critical for proper functioning of these organelles and thus ultimately of the cell as a whole.

4. Source of activator Ca^{2+}

4.1. Plasma membrane

The PM of smooth myocytes maintains a 10,000-fold electrochemical Ca^{2+} gradient, because the intracellular [Ca^{2+}] is approximately 10^{-7} M when the cell is relaxed and about 10^{-6} M when contracted, whereas the extracellular [Ca^{2+}] is 10^{-3} M. The notion that PM cannot be a source of activator Ca^{2+} is strongly influenced by Hill's (1948) arguments against diffusion of activator Ca^{2+} in the skeletal muscle fiber. These led to the concept of a major internal source of Ca^{2+}, now known to be the SR. However, considerations for smooth muscles are different. In skeletal muscle a single fiber is 80 μm in diameter, 10,000 μm in length, and 48,000 × 10^3 μm^3 in volume, with a surface-to-volume ratio of 1: 200 (assuming a cylindrical shape). A smooth myocyte is 5–10 μm in diameter, 300–600 μm in length, and 2,300–3,500 μm^3 in volume. Cell surface irregularities lead to a high surface-to-volume ratio, with estimates ranging from 1.5 to 2.7 μm^{-1} (Broderick and Broderick 1990). In view of the high surface-to-volume ratio and slow contraction of the smooth muscle cell, arguments against diffusion of a substance into the cell do not apply to smooth muscle contraction. One can therefore consider alternative hypotheses for contraction, such as the diffusion of second messengers to release Ca^{2+} from the SR and movement of Ca^{2+} ions across the PM (Yamamoto et al. 1989). However, the high surface-to-volume ratio does not prove that the PM is the source of the activator Ca^{2+}.

4.2. Sarcoplasmic reticulum

Electron microscopy has revealed the presence of SR in smooth muscles (Gabella 1971; Devine et al. 1972). The SR in the smooth muscle cell is less abundant than in skeletal muscle, occupying 2.0–7.5% of the cell volume, with the pregnant myometrium, main pulmonary artery, and aorta in the higher range. This compares with 9% in frog twitch muscle (Devine et al. 1972). The [Ca^{2+}] in the terminal cisternae of SR from skeletal muscle (120 mmol/kg dry weight) is higher than [Ca^{2+}] in the smooth myocyte (see Broderick and Broderick 1990). The Ca content of rabbit portal vein (Bond et al. 1984b) and pulmonary artery (Kowarski et al. 1985) were measured by electron-probe analysis. In the central SR, the calcium content was 42–49 mmol/kg dry weight in the relaxed state; when contracted, the content decreased to 19–32 mmol/kg dry weight (Kowarski et al. 1985). This loss of Ca^{2+} from the SR is thought to be more than enough to account for the measured rise in Ca^{2+} in the cytoplasm during a contraction, and suggests the SR as a source of Ca^{2+} for contraction.

Further evidence for the importance of an internal Ca^{2+} store is inferred from the poor response to contractile agents when the internal store is depleted, even in the presence of normal extracellular Ca^{2+}. Similarly, the rate of relaxation is markedly slowed when Ca^{2+} cannot be stored in the SR (Kanmura et al. 1988). Although the Ca^{2+} stored in the SR is sufficient to produce one contraction, only under special conditions using lanthanum can repeated contractions be demonstrated in the absence of extracellular Ca^{2+} (Bond et al. 1984a). Under physiologic conditions, the Ca^{2+} in the SR is at least partially replenished from extracellular sources. In summary, the relative importance of the PM and SR as sources of Ca^{2+} for activation of contraction is still not settled.

4.3. Mitochondria

Ca^{2+} is also stored in mitochondria, which occupy 5% of the smooth muscle cell volume. Unlike SR and PM, which cause increase in cytoplasmic [Ca^{2+}] during a contraction, mitochondria limit the increase by taking up Ca^{2+}. When cytoplasmic [Ca^{2+}] rises, so does mitochondrial [Ca^{2+}]. The main function of mitochondrial Ca^{2+} appears to be the regulation of Ca^{2+}-dependent enzymes in them (for a review see Broderick and Broderick 1990). Recent work seems to suggest that release of Ca^{2+} from SR by inositol trisphosphate ($InsP_3$) causes a large transient increase in mitochondrial free Ca^{2+} followed by an increase in energy production. Localized high perimitochondrial Ca^{2+} concentrations may allow the increase in Ca^{2+} in spite of the low K_m of the mitochondrial Ca^{2+} uniporter (Rizzuto et al. 1994). Ca^{2+} overload of mitochondria may be intimately involved in cell death.

4.4. Nuclei

Ten percent of the smooth muscle cell volume is occupied by the nucleus. Even though Ca^{2+} is not freely mobile between the nucleus and cytoplasm, changes in nuclear [Ca^{2+}] have been reported to follow the direction of change in the cytoplasm. Thus nuclear [Ca^{2+}] is low during relaxation and high during contraction, showing properties opposite to those shown by the SR. There is evidence that some nuclear Ca^{2+} may come from a subcompartment within the nucleus. Control of nuclear [Ca^{2+}] is complex, and varies with the tissue and with the type and concentration of the agonist. Mechanisms such as phosphorylation or dephosphorylation of proteins and $InsP_3$ binding have been proposed to account for the changes in nuclear [Ca^{2+}] (Gilchrist et al. 1994, Himpens et al. 1994).

5. Experimental preparations

Various model systems are used to study intracellular [Ca^{2+}]. From these, one can estimate the concentration of free Ca^{2+}, the movement of Ca^{2+} across the membranes, mechanisms of action, and concentration-dependent effects. The most common types of preparation employed include muscle strips, single cells, cultured cells, skinned or permeabilized muscle fibers or cells, and isolated cellular components such as microsomes. Each type has advantages and shortcomings for the understanding of the intact system.

5.1. Muscle strips and single cells

Intact muscle strips are infrequently used now because of problems with cellular inhomogeneity and diffusion gradients. These problems have been largely eliminated when using single intact smooth muscle cells obtained by enzymatic digestion of surrounding connective tissue, but variability may still remain. Damage to cells during the preparative procedure must also be considered.

5.2. Cultured cells

Many investigators have resorted to using cultured cells as a model. This model often does not mimic physiological conditions, since many properties of cells change in culture. Frequently the cells derived from smooth muscle no longer contract. Although this may be advantageous for studying the cells (Himpens et al. 1994), it certainly indicates gross changes in the cell: Ca^{2+} channels seem to be easily affected (Missiaen et al. 1991), and voltage-operated Ca^{2+} channels disappear (Reynolds and Dubyak 1986; Nelemans et al. 1990). (For a review of changes in cation channels, see Chapter 3 in this volume.) Long-term cell culture decreases the number of hormone receptors and lowers the amount of $InsP_3$ generated upon hormone stimulation (Bouscarel et al. 1990).

5.3. Permeabilized fibers and cells

Early estimates of the Ca^{2+} required for contraction were made in skinned fibers. The threshold for contraction was about 0.2 μM Ca^{2+} in vascular smooth muscle (Filo et al. 1965). The same threshold was observed in skinned fibers from pregnant rat myometrium, with a peak contraction at 6 μM (Savineau et al. 1988). One must consider that values obtained from these studies may not accurately reflect *in vivo* conditions; they depend on how well *in vitro* conditions mimic the normal intracellular environment.

Cells are permeabilized by chemical skinning, and various degrees of permeabilization can be produced. The aim is to make the PM permeable to the test agent without changing the integrity of the contractile apparatus or the SR. Equally desirable is to avoid leakage of compounds that may play a role in the contractile process. Skinned cells can be used for studying the intracellular effects of agents that cannot cross the PM, compounds normally originating inside the cell, and agonists, independent of their effects on permeability and membrane potential.

The techniques used for skinning may be summarized as follows.

1. Immersion in glycerol at low temperature, a technique long used for skeletal muscle, removes practically everything from the cell except the actomyosin contractile element.

2. Exposure to the detergent Triton X disrupts intracellular membranes in a time-dependent and concentration-dependent manner, but does not allow large proteins to cross the PM.

3. Chemical skinning with saponin preferentially attacks the PM because of its high cholesterol content, while supposedly leaving the SR intact. However,

other components of the cell may be affected depending on concentration and duration of exposure. Various enzyme activities may also be altered in saponin-skinned cells (Kwan and Lee 1984). Therefore, optimal skinning conditions must be experimentally determined for each tissue studied (Stout and Diecke 1983). The size of the holes in saponin-treated cells was measured to be 7–8 nm under precisely controlled conditions, and 30% of the protein including calmodulin (MW 17,000) was lost (Kargacin and Fay 1987).

4. A commercially available ester of saponin, β-escin, makes the membrane permeable to compounds with a molecular weight up to approximately 17,000 without uncoupling the receptor from the G protein and $InsP_3$. Muscle strips so loaded are responsive to depolarization with KCl and contract normally with an agonist (Kobayashi et al. 1989).

5. An α-toxin from staphylococcus *aureus* achieves low permeability and does not interfere with the function of SR or mitochondria. By inserting into the PM it forms 2–3 nm pores (Cassidy et al. 1979). The pores allow passage of inorganic ions and small molecules of molecular weight $< 1{,}000$ Da (e.g., ATP, EGTA), but not of proteins (Kitazawa et al. 1989). Unlike saponin treatment, β-escin and α-toxin treatment have the advantage of leaving PM receptors and G proteins intact, and therefore are valuable in studying pharmacomechanical coupling.

When used under controlled conditions, permeabilized cells or tissues allow correlation of contractile activity and changes in Ca^{2+}. However, the relationship of Ca^{2+} to force varies greatly in different studies and depends on the presence or loss of ions and cofactors. A common problem with these systems is the lack of stability with time, and fragility of the preparation. Thus, reliable conclusions for the physiological system often cannot be drawn.

5.4. Microsomal preparations

Studies of biochemical reactions, mechanisms, and enzyme kinetics require the isolation of the Ca^{2+} transport ATPases. As both lipid and protein components of the cellular membranes interact with the enzymes, the use of microsomes allows for a better understanding of enzyme activity in a more native environment. Isolation and purification of SR from smooth muscle are made difficult by the small amount of SR present and the large amount of collagen that is difficult to remove even by treatment with collagenase. Additional problems are that, unlike in skeletal muscle, there is relatively little SR in the cell and there is more PM than SR, since the smooth muscle cell is small and has a large surface area.

Preparative procedures for both SR and PM are based on methods used for isolation of skeletal SR. The tissue is first minced and homogenized. This is followed by stepwise centrifugation with increasing force to remove heavier particles, saving the supernatants and finally separating components on a density gradient. Identification of the source of the membranes is based on known or assumed differences, such as specific density, marker enzyme activities, specific receptor binding, oxalate or ouabain sensitivity, and so on. (Carsten and Miller 1980).

In the preparation of fractions of either SR or PM, some degree of cross-contamination often occurs because of the small difference in densities of the two fractions. Methods have been developed to change the density of one fraction or to use affinity chromatography to achieve better separation (for a review see Carsten and Miller 1990).

6. Assessment of activity

6.1. Measurement of Ca,Mg-ATPase activity

Often the activity of the Ca,Mg-ATPase is measured in parallel with ATP-dependent Ca^{2+} uptake. Usually inorganic phosphate release or ADP production is assayed in the presence and absence (zero Ca^{2+} plus added EGTA) of Ca^{2+}. The difficulties encountered are related to the high levels of intrinsic Mg-ATPase activity and the small increase owing to the Ca-ATPase activity. Linking Ca^{2+} transport and enzyme activity has the added problem that leaky and inside-out vesicles transport Ca^{2+} that is not accumulated (Carsten and Miller 1990).

6.2. Measurement of Ca^{2+} translocation

Methods used to measure ATP-dependent Ca^{2+} uptake or Ca^{2+} release are essentially the same for SR and PM preparations. ATP-dependent Ca^{2+} uptake is the difference in Ca^{2+} content in the presence and absence of ATP. Methods used include (1) centrifugation and atomic absorption spectroscopy and (2) filtration and counting of tracer ^{45}Ca.

In the centrifugation method, Ca^{2+} associated with the SR preparation is determined directly after incubation, centrifuging, washing, and resuspending the pellets and recentrifuging. Atomic absorption spectroscopy is then carried out on the pellets (Carsten and Miller 1977b). Inasmuch as the Ca^{2+} associated with the SR vesicles is directly measured, unequivocal results are obtained, but large amounts of material are needed and the method is time-consuming.

In the filtration method, ^{45}Ca is added to the incubation medium. Samples are filtered through a microporous filter at specified times, and ^{45}Ca is counted either in aliquots of the filtrate or on the filters after twice washing the filters. In the former variation, large amounts of protein (0.3–0.5 mg/ml) are needed, and EGTA buffers with high total Ca^{2+} concentrations cannot be used (Carsten and Miller 1977b). In the latter variation, protein concentration can be lower, and EGTA buffers can be used (Godfraind et al. 1976).

Although the filtration method lends itself to kinetic measurements, the centrifugation method is superior. It gives highly accurate results and avoids some pitfalls in the filtration method. One such error is caused by the exchange of ^{45}Ca with $^{40}Ca^{2+}$ intrinsically present in the microsomal preparation and filters. This problem is unique for smooth muscle, which has a low ATP-dependent Ca^{2+} uptake and a large intrinsic exchangeable Ca^{2+} pool. Further contributing to the problem is a low free Ca^{2+} concentration in the incubation medium and possible dependence of the size of the exchangeable Ca^{2+} pool on added agents such as ATP (Carsten and Miller 1977b). Additional technical problems are encountered with commercial filters. They contain large and variable amounts of Ca^{2+}, and must therefore be washed thoroughly before use. The releasable Ca^{2+}

per filter averaged 650 nmol and varied from batch to batch of filters. The time of contact of solution with the filter is another variable (Carsten and Miller 1977a).

6.3. Spectrophotometry

Spectrophotometric methods depend on complex-formation of a dye with Ca^{2+} in solution, the concomitant change in molar absorbance, and a shift in the wavelength of maximum absorbance or fluorescence. Much information has been generated from work with dyes such as murexide (requiring 25–250 μM [Ca^{2+}]) and arsenazo III (requiring 1–25 μM [Ca^{2+}]); however, obtaining information at intracellular concentrations of Ca^{2+} requires the use of fluorescent dyes that can be used in the [Ca^{2+}] <100-nM range. The great attraction of these methods is the ease of kinetic measurements and the small amount of protein needed. Because the amount of Ca^{2+} taken up is calculated from the reduction in the free [Ca^{2+}], difficulties include changes in dye sensitivity (Kargacin et al. 1988) and changes in free Ca^{2+} that occur independently of Ca^{2+} uptake. In all methods that measure the extravesicular free [Ca^{2+}] one must take into account the *increase* in free [Ca^{2+}] as ATP is hydrolyzed to ADP since ADP has a lower binding affinity for Ca^{2+} and Mg^{2+} than ATP (Miller and Carsten 1984). Various programs for calculating the free [Ca^{2+}] in buffered solutions containing multiple cations are available (see Bers et al. 1994).

6.4. Ca^{2+} measurement in intact cells

In living cells, Ca^{2+}-selective microelectrodes can be used for direct measurement of intracellular Ca^{2+} provided that one works with large, nonmotile cells. Measurement of the Ca^{2+} concentration in single smooth muscle cells is difficult because of the small size and motility of the cells. Microelectrodes allow quantitative measurement of Ca^{2+} activity in single cells larger than 10–20 μm. The detection limits are 10^{-8} M Ca^{2+}, with response times of one second or more. Resting levels and slow changes in Ca^{2+} levels can be measured accurately, but fast transients cannot be detected (Baudet et al. 1994).

New techiques based on the use of bioluminescent and fluorescent indicator dyes have contributed much to the determination of intracellular [Ca^{2+}]. Aequorin (21 kDa), a bioluminescent protein from the jellyfish *Aequorea forskalea*, emits a light signal as a function of the free [Ca^{2+}]. It can be injected or chemically loaded into reversibly permeabilized cells without affecting the cellular sensitivity to contractile agents. Muscle strips can be observed both at rest and during a contraction. In relaxed vascular smooth muscle, intracellular free [Ca^{2+}] was 0.1 μM, and during contraction with potassium it rose to 0.4 μM (Defeo and Morgan 1986). However, in the aequorin method the signal is not linearly related to the free Ca^{2+}; thus a larger signal will originate from an area containing localized regions of both low and high Ca^{2+} concentration than from a similar area with an average of the Ca^{2+} concentrations. Aequorin is degraded during the emission of the light, so repeated measurements are limited (for a review see Cobbold and Rink 1987). A further refinement in the technique is the use of recombinant aequorin targeted to mitochondria or nuclei to measure concentrations of Ca^{2+} in these organelles in intact cells (Rizzuto et al. 1994).

Several fluorescent dyes have been used for measuring free Ca^{2+} concentration (Cobbold and Rink 1987). These dyes can be loaded into single cells as permeable esters that are hydrolyzed by cellular esterases, trapping the dye inside the cell. The earliest results were obtained with Quin-2. However, disadvantages – such as buffering of Ca^{2+}, the requirement of a high dye concentration (>0.5 mM), and difficulty in obtaining absolute values of $[Ca^{2+}]$ – led to the use of other agents. Furthermore, the Ca^{2+} affinity of Quin-2 is affected by variations in the magnesium concentration (Grynkiewicz et al. 1985). In contrast to Quin-2, the dyes Fura-2 and Indo-1 when bound to Ca^{2+} not only change the intensity of fluorescence, but shift the excitation or emission wavelength. This allows calculation of Ca^{2+} to be based on the ratio at dual wavelengths independently of dye concentration and specimen path length, thereby reducing artefacts. These dyes also show greater sensitivity, produce less buffering, and have greater ionic selectivity for Ca^{2+} (Grynkiewicz et al. 1985). A problem with Fura-2 is leakage of dye from cell suspensions into extracellular fluid. The leakage increases with time and can complicate data analysis (Mitsui et al. 1993). Moreover, the Fura-2 signal is influenced by buffer composition, pH (Ganz et al. 1990), temperature (Shuttleworth and Thompson 1991) and viscosity changes (Moore et al. 1990).

Rhod-1, Rhod-2, Fluo-1, Fluo-2 and Fluo-3 have more recently been employed. Although they yield better resolution at high Ca^{2+} concentrations, the increase in fluorescence upon binding of Ca^{2+} is not accompanied by a shift in the wavelength. Hence, alternatives to ratio methods must be used to obtain quantitative data on single cells, a disadvantage in many situations. A decided advantage of these indicators is the visible excitation wavelength (Minta et al. 1989), especially important for experiments using microscopy and UV-triggered photolysis (see Section 10.2). Though each dye has advantages in specific applications, at present the most commonly used dye is Fura-2.

Questions remain as to the amount of dye that penetrates into the SR, mitochondria, and nucleus. Penetration would produce an artefact that would increase the measured free Ca^{2+} concentration in the cytoplasm in resting cells and decrease the change seen with stimulation. It is thought that aequorin stays in the cytoplasm whereas Fura-2 enters all compartments, but the distribution depends on time and is tissue-specific (for a review see Cobbold and Rink 1987 or Moore et al. 1990). The rate of diffusion of the indicator dyes Fura-2 and Indo-1 within the cardiac cell is lower than predicted when measured after photobleaching of the indicator in half the cell. This suggests that the dye is not free in the cytoplasm but is bound to or located inside structures such as SR (Blatter and Wier 1990). Digital imaging and microfluorometry allow a better assessment of differences in $[Ca^{2+}]$ in different regions of the cell. With these improvements, localized changes in $[Ca^{2+}]$ have recently been observed (Moore et al. 1990; Johnson et al. 1991; Hayashi and Miyata 1994). Nevertheless, it appears that the fluorescent methods still need to be improved; at present, careful and frequent calibration is advisable. Care must be exercised when hypotheses are based solely on indicator dye measurements.

7. Molecular genetics

Recent advances in molecular genetics have rapidly expanded our knowledge of the amino acid sequence of many proteins involved in Ca^{2+} homeostasis. Much of the reported data come from studies with whole organs; thus it is not clear whether such findings pertain to the smooth muscle, glandular epithelium, blood vessels, or some combination thereof. Furthermore, one must be careful not to overread the information available. For instance, the molecular structural information presented in this chapter discusses the presence of phosphorylation sites in specific proteins. It must be remembered that the presence of a site that may be phosphorylatable does not prove that the site is phosphorylated *in vivo,* nor that phosphorylation of the site will lead to a functional change. Tertiary structure and the environment of the protein *in vivo* may either make the site of no physiological significance, or make it physically unavailable to be phosphorylated. On the other hand, a protein that has no phosphorylatable site may have its function modified by interacting with adjacent proteins and lipids that are phosphorylated (see Section 8.5). Although molecular structure is a possible area for investigation, the actual proof of its importance requires showing a physiologic effect. Similarly, the presence of products of multiple genes and mRNA transcripts does not indicate the relative importance or functionality at the protein level in a given tissue.

The exact molecular weight of a protein can be calculated once the amino acid sequence is known, but transcript and species differences can substantially alter composition. Thus, although we give a specific number to the amino acid composition and the weight of a protein, the *in vivo* protein may be only an approximation of this value. The identification of a specific number is only meant to imply that the amino acid sequence has been determined, not that all species and transcripts will have the same value. As work continues on the identification of gene products, greater variety is being identified. Several mRNA transcripts for a protein may be from one gene or from a modified copy of the gene. Modifications to the original structure appear to be expressed in a tissue-specific manner. When a protein such as a channel consists of multiple subunits, the permutations and combinations become overwhelming. Further work at the protein level is required to decide whether these are only theoretical combinations or whether they actually exist.

8. Decrease of cytoplasmic Ca^{2+} – sarcoplasmic reticulum

8.1. Ca^{2+} translocation into SR

The function of the SR was discovered in the search for a ''relaxing factor'' in skeletal muscle. A vesicular fraction was isolated that rapidly took up Ca^{2+} against a concentration gradient. Ca^{2+} uptake was concomitant with ATP hydrolysis, and was dependent on the ATP and Mg^{2+} concentrations. Uptake of Ca^{2+} was considerably enhanced in the presence of oxalate, with the Ca^{2+} accumulated rising to 8 μmol per milligram vesicular protein. The effect of oxalate rests on the permeability of the SR membrane to potassium oxalate and the low solubility of calcium oxalate formed inside the SR vesicles (solubility product 2.6×10^{-9} M^2) (Hasselbach and Makinose 1961; Ebashi and Lipmann 1962). A similar Ca^{2+} pump was found in uterine smooth muscle. In smooth muscles, as in other muscles, ATP-dependent Ca^{2+} uptake requires free sulfhydryl groups (Carsten 1969) and varies with temperature and Ca^{2+} concentration (Carsten and Miller 1977a, 1977b). Ca^{2+} uptake in smooth muscle SR is small, and depends on the tissues and preparation. ATP-dependent Ca^{2+} uptake of approximately 35 nmol/mg protein/min has been observed in bovine pulmonary artery (Raeymaekers et al. 1990) and of 5 nmol/mg protein/min in myometrial preparations from EGTA-buffered solution without oxalate (Carsten and Miller 1977a). The maximum uptake of Ca^{2+} by SR *in vivo* is probably much higher, since the *in vitro* preparations contain only a small fraction of competent vesicles among inside-out and leaky vesicles, PM, and debris.

Proof that the Ca^{2+} is really inside the vesicles comes from the electron microscopic demonstration of calcium crystals (as oxalate or phosphate) inside smooth muscle SR vesicles (Raeymaekers et al. 1981). Further proof is the observation of increased rate of Ca^{2+} release from SR vesicles in the presence of a Ca^{2+} ionophore (Miller and Carsten 1984). Whether phosphate functions as a Ca^{2+} precipitating agent under physiological conditions is not known.

8.2. *Ca,Mg-ATPase of the SR*

To establish the mechanism of Ca^{2+} translocation into the SR, it is useful to look at partial reactions. In skeletal muscle the Ca^{2+} transport cycle consists of the reactions outlined in Table 1 (Inesi et al. 1990). It is thought that two Ca^{2+} ions, bound at the Ca^{2+}-specific transport sites on the cytosolic exposure of the enzyme, are translocated to the luminal side and then released. The energy for this reaction is derived from phosphorylation of the enzyme (for a review see Inesi and Kirtley 1992). Recent evidence suggests that the two high-affinity Ca^{2+} binding sites do not traverse the membrane to become low-affinity sites, but rather that the low-affinity lumenal sites are independent of the high-affinity cytoplasmic sites (Myung and Jencks 1994). The catalytic site of the enzyme requires one mole of a divalent cation, which physiologically is Mg^{2+}. Lanthanides inhibit the enzyme not by binding to the transport site but by binding to the catalytic site (Fujimori and Jencks 1990).

In smooth muscles the entire cycle has not been fully established, but some partial reactions have been demonstrated, notably reactions 1 and 3 or 4 of Table 1. Reaction 1, the binding of Ca^{2+} to the enzyme in the absence of ATP, has been shown in SR preparations. Two Ca^{2+} binding sites with different affinities have been shown. The association constants were $K_1 = 8.4 \pm 0.4 \times 10^6$ M^{-1} and $K_2 = 0.11 \pm 0.02 \times 10^6$ M^{-1}. Magnesium (5 mM) did not affect the affinity or the number of high-affinity Ca^{2+} binding sites, whereas it halved the number of low-affinity binding sites (Miller and Carsten 1987). The high-affinity site is in the correct range for the Ca,Mg-ATPase, suggesting that the binding sites are on the Ca,Mg-ATPase (although these experiments were not carried out with purified Ca,Mg-ATPase). The presence of high-affinity Ca^{2+} binding in the ab-

Table 1. *Ca^{2+} transport cycle (skeletal muscle)*

	Partial reactions			Equilibrium constant
1	E + 2 Ca^{2+}_{out}	⇌	E-Ca_2	3×10^{12} M^{-2}
2	E-Ca_2 + ATP	⇌	ATP-E-Ca_2	1×10^{-5} M^{-1}
3	ATP-E-Ca_2	⇌	ADP-P~E-Ca_2	0.3
4	ADP-P~E-Ca_2	⇌	P~E-Ca_2 + ADP	7×10^{-4} M
5	P~E-Ca_2	⇌	P~E + 2 Ca^{2+}_{in}	3×10^{-6} M^2
6	P~E	⇌	E-P_i	1
7	E-P_i	⇌	E + P_i	1×10^{-2} M

Notes: E is the enzyme, Ca^{2+}_{out} is external Ca^{2+}, and Ca^{2+}_{in} is internal Ca^{2+}.
Source: Inesi et al. (1990); reprinted with permission.

sence of ATP suggests that Ca^{2+} binding does not require any modification of the enzyme.

Reactions 3 and 4 have been demonstrated in bovine uterine muscle. SR preparations were phosphorylated using [^{32}P]ATP. The yield was proportionate to the ATP or Ca^{2+} concentration, respectively, and magnesium was required. A phosphorylated intermediate with a molecular weight of 100,000 to 110,000 was shown (Carsten and Miller 1984). Sensitivity to hydroxylamine confirmed that it was an acylphosphate intermediate, similar to findings in skeletal muscle SR where an aspartyl residue (Asp^{351}) has been identified as the site of phosphorylation (for a review of skeletal muscle see Inesi and Kirtley 1992). Phosphoenzyme intermediates have been described in smooth muscles from aorta (Chiesi et al. 1984) and stomach (Raeymaekers and Hasselbach 1981).

Although reactions 3 and 4 require Mg^{2+}, magnesium ion is not the counter-ion for Ca^{2+}, as the Mg^{2+} concentration of the vesicles stays constant during ATP-dependent Ca^{2+} uptake. Mg^{2+} is bound specifically to bovine uterus SR with an association constant of $K_{Mg} = 2.9 \pm 0.3 \times 10^3$ M^{-1} (Miller and Carsten 1984). This association constant for Mg^{2+} is three to four orders of magnitude lower than that for Ca^{2+} in accordance with the physiological intracellular concentration of these ions. Membrane electroneutrality may still be maintained as Ca^{2+} is moved from one side of the membrane to the other by counter movement of monovalent cations. K^+ and H^+ have been suggested as the counter-ions (Moutin and Dupont 1991; Yu et al. 1994).

8.3. Functional characterization of the Ca,Mg-ATPase

For characterization, the enzyme is exposed to various inhibitors or activators, and the effect is measured as a change in either the products of ATP hydrolysis or Ca^{2+} uptake; the interpretation is assumed to be comparable. Activity of the SR Ca,Mg-ATPase from smooth muscles is very much lower than that from skeletal muscle. Thus, in smooth muscles the phosphate produced by the Ca,Mg-ATPase represents only a small portion of the total ATPase (Carsten and Miller

1977b). This Ca,Mg-ATPase, unlike the one from PM, is not stimulated by calmodulin (Wibo et al. 1981).

Vanadate is an inhibitor of Ca,Mg-ATPases. In cardiac muscle, the vanadate half-maximal inhibition (IC_{50}) at 0.6 μM of the PM enzyme is 10 times lower than the IC_{50} of the SR enzyme. The difference in smooth muscle is less (Wibo et al. 1981) and in gastric smooth muscle no difference in vanadate sensitivity was found (Raeymaekers et al. 1983). In permeabilized aortic cells in culture, vanadate inhibition occurred with an IC_{50} of 6 μM, but a small proportion of Ca^{2+} uptake was insensitive even to 2 mM vanadate and was also resistant to other inhibitors of the Ca,Mg-ATPase (Missiaen et al. 1991). The nature of the resistant Ca^{2+} uptake has not been worked out, and the compartment may be different from the SR.

Thapsigargin, a plant-derived sesquiterpene lactone that specifically binds to the Ca,Mg-ATPase, is the most potent inhibitor of intracellular Ca^{2+} accumulation (Thastrup et al. 1989). Its inhibition is antagonized by high Ca^{2+} concentrations, which was interpreted to mean that thapsigargin inhibits the formation of E-Ca_2 noted in Table 1 (Sagara et al. 1992; Wictome et al. 1992). The transport of Ca^{2+} and phosphorylation of the Ca,Mg-ATPase of SR by either ATP or P_i are inhibited at nanomolar concentrations, whereas 100 times the concentration has minimal effect on the PM Ca,Mg-ATPase or Na,K-ATPase (Lytton et al. 1991). At the very low inhibitor concentrations needed, there may have been too much enzyme present to allow for the required 1:1 stoichiometry (Lytton et al. 1991). This may have accounted for some earlier reports that the inhibitor was ineffective against the skeletal muscle enzyme (Thastrup et al. 1989). In some systems, even high concentrations do not cause full inhibition (Thomas and Hanley 1994); there is a consequent tendency to use excessive amounts. As with all inhibitors, specificity is lost at higher concentrations. At micromolar concentrations, effects on the voltage-dependent Ca^{2+} channel have been described by Buryi et al. (1995).

Several other Ca,Mg-ATPase inhibitors have been investigated. These include cyclopiazonic acid, a specific competitive reversible inhibitor at the ATP binding site of the Ca,Mg-ATPase, as well as 2,5-di(ter-butyl)hydroquinone, which seems to reduce the formation of the E-Ca_2 (Table 1). Both agents may also block other aspects of Ca^{2+} signaling including inhibition of the Ca^{2+} leak. Their specificity and mechanism of action are not as clear as those of thapsigargin (Thomas and Hanley 1994).

8.4. Structure of the SR Ca,Mg-ATPases from various muscles

The kinetic properties of the Ca,Mg-ATPase and the tryptic digestion patterns of the SR membranes from aorta smooth muscle were strikingly similar to those from skeletal and cardiac SR, but displayed no immunological cross-reactivity (Chiesi et al. 1984). Using tryptic phosphoprotein fragments of different isoforms of the Ca^{2+} transport ATPase, immunologic similarities of the cardiac, smooth, and slow-twitch skeletal muscle ATPases could be shown, but all differed from the fast-twitch skeletal muscle ATPase (Wuytack et al. 1989).

The cDNAs that encode Ca^{2+}-transporting ATPases have been identified. A nomenclature has been proposed to distinguish the various gene products: SERCA (sarco[endo]plasmic reticulum Ca-ATPase). They are numbered 1–3 and alternative splices are listed as a and b. Thus, the skeletal muscle fast-twitch isoform is SERCA-1a and the neonatal fast-twitch isoform is SERCA-1b. The cardiac/slow-twitch isoform is SERCA-2a, and the nonmuscle/smooth-muscle isoform is SERCA-2b. SERCA-3 has also been identified in smooth muscle and nonmuscle. In all SERCA isoforms, approximately 75% of the amino acids are identical (for a review see the introduction in Lytton et al. 1992). The two isoforms identified from SERCA-2 encode proteins that share the first 993 amino acids. SERCA-2b encodes an enzyme that has 1,042 or 1,043 amino acids as opposed to SERCA-2a, which encodes an enzyme of 997 amino acids. The last four amino acids of SERCA-2a are replaced by 49 or 50 amino acids in SERCA-2b. A longer chain may allow for an extra transmembrane segment, thus placing the SERCA-2b "C" terminus on the inside of the SR rather than on the cytoplasmic side. The SERCA "N" terminus is always on the cytoplasmic side of the membrane (Thomas and Hanley 1994). It has been proposed, based on hydrophobicity plots, that there are ten transmembrane segments; so far, however, biochemical data have conclusively shown only eight (Shin et al. 1994). Using site-specific mutations and chimeric enzymes, the various parts of the molecule that are important for binding ATP, Ca^{2+}, and so forth have been identified (Clarke et al. 1993). All P-class ATPases hydrolyze ATP while moving cations. These enzymes show great similarity in the fourth membrane-spanning region (M4) and in the loop to M5. The phosphorylation site and the nucleotide binding site are found on this cytoplasmic-facing loop. The amino acid phosphorylated is aspartate, and that for nucleotide binding is lysine. This region is distant from the Ca^{2+} binding sites. In the SR ATPase, several membrane-spanning sites are involved in the binding of Ca^{2+}, including M4, M5, M6, and M8. At present, the best evidence is that at least 30 Å separate the Ca^{2+} and phosphorylation sites (Inesi and Kirtley 1992; Myung and Jencks 1995).

The mRNA for the Ca,Mg-ATPase is expressed in various smooth muscles to different extents, and there are also species differences. Most smooth muscles express SERCA-2a and -2b, though in different proportions. Vascular smooth muscle expresses 30–70% and uterus 10–15% the cardiac/slow-twitch isoform (Lompre et al. 1989). Cardiac and slow-twitch muscles express approximately 95% and esophagus about 80% the cardiac/slow-twitch isoform (SERCA-2a). Aorta and urinary bladder each express 85% the smooth/nonmuscle isoform with the other isoform making up the remainder (Lytton et al. 1989). Some functional differences have been found for the different gene products. Thus SERCA-2b has a slower turnover and higher Ca^{2+} affinity than SERCA-2a, and the Ca^{2+} affinity of SERCA-3 is lower than that of the other SERCA gene products (Lytton et al. 1992; Verboomen et al. 1992).

8.5. Modulation of ATP-dependent Ca^{2+} uptake

In smooth muscle SR, the role of cAMP in ATP-dependent Ca^{2+} uptake is still unresolved. Evidence for an increase or no change in activity has been published (see Carsten and Miller 1990). There is some evidence that cAMP-dependent protein kinase phosphorylates phospholamban. Phospholamban, a protein that inhibits ATP-dependent Ca^{2+} uptake in cardiac SR, has been observed in some smooth muscles but not in others. The inhibitory effect is removed when phospholamban is phosphorylated, leading to stimulation of ATP-dependent Ca^{2+} uptake (Chiesi et al. 1991). In bovine pulmonary artery, phosphorylation of a phospholamban-like protein by cAMP/cGMP-dependent protein kinase stimulated ATP-dependent Ca^{2+} uptake, though to a lesser degree than in cardiac SR (Raeymaekers et al. 1990). The amount of phospholamban mRNA varied considerably in different smooth muscles of the pig (Eggermont et al. 1990). The variability of these findings and the significance for the *in vivo* regulation of Ca^{2+} has yet to be determined. It has been suggested that phospholamban inhibition must be removed to activate the SR Ca,Mg-ATPase. There are similarities to the removal of the autoinhibitory region of the PM Ca,Mg-ATPase (Chiesi et al. 1991) (see Section 9.3). The lipid environment may also play an important role in the activity of the Ca,Mg-ATPase, though this has not been studied in smooth muscle.

8.6. Ca^{2+}-binding proteins of the SR

Smooth muscle SR has associated with it 30–50 mmol of Ca^{2+}/kg dry weight (electron probe analysis; see Broderick and Broderick 1990). Sarcoplasmic reticulum of skinned taenia caeca muscle fibers can take up 220 μmol Ca^{2+}/l of cell water. Based on a SR volume of 1.5% in this tissue, the maximum Ca^{2+} capacity is 15 mmol/l of SR and at physiologic Ca^{2+} levels approximately half this (Iino 1989). These values make sense if the wet-to-dry ratio is 3–4:1. Binding of Ca^{2+} on the inside of the SR to inorganic or organic anionic sites lowers the free Ca^{2+} concentration, and makes it difficult to estimate the true Ca^{2+} gradient. Based on studies with the Ca^{2+} fluorescent dye Furaptra, the free Ca^{2+} in the SR was approximately 0.1 mM (Sugiyama and Goldman 1995). Skeletal muscle SR contains Ca^{2+}-binding proteins, besides the ATPase, in the vesicular lumen. Calsequestrin is a low-affinity, high-capacity Ca^{2+}-binding protein found in skeletal SR; a second form has been identified in cardiac SR. Both forms have now been identified in smooth muscles (Volpe et al. 1994). The amount of each is tissue-dependent, and not all smooth muscles have significant quantities. Calreticulin, previously called high-affinity Ca^{2+}-binding protein (dissociation constant K_d = 1 μM), was found in skeletal muscle. It has now been found nearly universally, and is the major Ca^{2+}-binding protein in the smooth muscles studied (Milner et al. 1991).

9. Decrease of cytoplasmic Ca^{2+} – plasma membrane

9.1. Comparison with SR membrane

The function of the two membrane systems in the control of Ca^{2+} is very similar. It is thus not surprising that many properties of the two membranes are superficially similar. In fact, some differences are more quantitative than qualitative.

For example, the cholesterol:phospholipid ratio is higher in the PM than in the SR (Raeymaekers et al. 1983).

Like other Ca^{2+}-pumping ATPases, the Ca,Mg-ATPase of the PM has a high affinity for Ca^{2+} (Imai et al. 1990). The contribution of the PM Ca^{2+} pump to Ca^{2+} transport appears to vary from tissue to tissue; this Ca^{2+} pump was found to play a smaller role in Ca^{2+} homeostasis in the bovine pulmonary artery than in the pig stomach (Eggermont et al. 1988). A more comprehensive comparison of the importance of the PM Ca,Mg-ATPase of various tissues is difficult because of different methods of preparation in different laboratories.

9.2. Functional characterization of the PM

Because the function of the PM Ca^{2+} pump is to remove Ca^{2+} from the cell, ideally one should show Ca^{2+} efflux from whole cells. This is experimentally difficult, so the properties of Ca,Mg-ATPase have most often been studied in microsomal preparations. The ability of the Ca,Mg-ATPase to transport Ca^{2+} to the outside of the cell is measured as ATP-dependent Ca^{2+} uptake into vesicles. Only ATP-dependent Ca^{2+} uptake by inside-out, nonleaky vesicles corresponds with Ca^{2+} transport by the PM Ca,Mg-ATPases (see the discussion in Section 8.1).

To separate the PM, microsomal fractions have been prepared from a variety of smooth muscles, such as bovine uterus (Carsten and Miller 1980), rat uterus (Matlib et al. 1979; Enyedi et al. 1988), pig and rabbit stomach (Raeymaekers et al. 1983; Grover et al. 1988), guinea pig ileum (Wibo et al. 1981), bovine pulmonary artery (Eggermont et al. 1988) and porcine aorta (Imai et al. 1990). Separation is by ultracentrifugation followed by sucrose density gradient centrifugation and calmodulin affinity chromatography (Wuytack et al. 1982) or digitonin treatment (Wibo et al. 1981; Imai et al. 1990). Purity is assessed by enzyme assays, polyacrylamide gel electrophoresis, and immunological methods.

In contrast to the SR membrane, the PM is not permeable to oxalate. Hence, stimulation of PM ATP-dependent Ca^{2+} uptake by oxalate is minimal. Furthermore, PM ATP-dependent Ca^{2+} uptake is enhanced more by phosphate than by oxalate as demonstrated in gastric and intestinal smooth muscle (Raeymaekers et al. 1983; Sharma et al. 1987).

The transport Ca,Mg-ATPase represents only a small fraction of total Ca,Mg-ATPases. A high-affinity Ca-ATPase that does not transport Ca^{2+} is present with properties quite different from the Ca^{2+} transporter (Kwan et al. 1986; Enyedi et al. 1988; Sun et al. 1990). Mg-ATPase activity nonspecifically uses nucleotide triphosphates as substrates, whereas Ca^{2+} uptake activity appears to require ATP as shown in ileum (Sharma et al. 1987) and myometrium (Enyedi et al. 1988). However, Ca^{2+} uptake in PM from gastric smooth muscle shows much less nucleotide specificity and is similar to the SR ATPase (Raeymaekers et al. 1983). It is not clear whether preferential use of ATP over other nucleotides for Ca^{2+} uptake is an intrinsic property of the enzyme or whether it is membrane-specific or species-specific.

The phosphorylated intermediate of the Ca,Mg-ATPase of the PM is similar to that of the SR (Herscher et al. 1994). In the presence of Ca^{2+} and Mg^{2+}, phosphorylation of the ATPase occurs at an aspartyl residue with an acylphos-

Table 2. *Comparison of Ca,Mg-ATPases*

	Plasma membrane [A]	Sarcoplasmic reticulum	
Distribution	Isoform-specific	Isoform-specific	[F]
Molecular mass	134 kDA	110–115 kDA	[D]
Inhibitors	Vanadate, La^{3+}	Thapsigargin, cyclopiazonic acid	[E]
Mechanism	Aspartyl-phosphate	Aspartyl-phosphate	
Charge balance	H exchange	H exchange	
Ca/ATP stoichiometry	1:1	2:1	[A]
Ca affinity	K_m = 0.5 μM (high affinity) 10–20 μM (low affinity)	0.44 μM isoform 1 0.38 μM isoform 2a 0.27 μM isoform 2b 1.10 μM isoform 3	[B]
Calmodulin	Induces high affinity	No effect	
Purification	Calmodulin affinity chromatography	Sucrose density centrifugation	
Activators (endogenous)	Calmodulin association, −charged phospholipids	Phospholamban dissociation	[C]
Activator stoichiometry	1:1 calmodulin	5:1 phospholamban	[C]
Number of genes	4	3	
Transcripts	20	5	
Effect of La	Increases phosphoprotein	Decreases phosphoprotein	

Sources: [A] Carafoli and Stauffer (1994); [B] Lytton et al. (1992); [C] Chiesi et al. (1991); [D] Inesi and Kirtley (1992); [E] Thomas and Hanley (1994); [F] see text.

phate bond (Wuytack et al. 1982; Imai et al. 1990). Mercurial reagents completely inhibit ATP-dependent Ca^{2+} uptake (Enyedi et al. 1988).

9.3. Structure of the Ca,Mg-ATPase of the plasma membrane

Although the Ca,Mg-ATPases of SR and PM are similar in function, large differences between them were found at the molecular level (Table 2). The structure of the smooth muscle PM Ca,Mg-ATPase resembles the erythrocyte enzyme, but is not identical (Verma et al. 1988). This is also evident from the observations that an antibody to the erythrocyte Ca,Mg-ATPase specifically cross-reacted with the PM fraction and inhibited both Ca,Mg-ATPase activity and Ca^{2+} uptake (Verbist et al. 1985).

The cDNAs coding for a PM Ca^{2+} pump were isolated from a human teratoma library and sequenced. This translated into 1,220 amino acids with a calculated molecular weight of 135 kDa. Considering that both the SR and PM Ca,Mg-ATPases are members of the P class of ion motive ATPases, they are surprisingly different in structure. All P-class ATPases have a similar region surrounding the aspartic acid where the phosphorylated intermediate is formed. In the PM Ca,Mg-

ATPase this is Asp-475. The only area shared uniquely with the skeletal muscle SR Ca,Mg-ATPase is the hydrophobic region (885–905 in the PM Ca,Mg-ATPase). Possibly this region confers Ca^{2+} specificity and/or is engaged in Ca^{2+} transport (Verma et al. 1988).

A nomenclature has been suggested for the PM enzymes, with PMCA standing for PM Ca,Mg-ATPase (Shull and Greeb 1988). There are at least four genes; the most widely expressed is the PMCA-1 gene. This gene may be spliced, and the resulting isoforms are labeled a, b, c, and d. The cDNAs coding for the PM Ca^{2+} pump have been isolated from a pig smooth muscle cDNA library; a protein of 1,220 amino acids was encoded. Only one gene was detected coding for the PMCA-1b isoform, although at least two enzymes have been displayed by immunological analysis (De Jaegere et al. 1990). In other species, all gene products have been found in smooth muscle except the PMCA-3. Various splicing products are found and seem to differ from tissue to tissue, and even from muscle to muscle. The uterus seems to have much more PMCA-2b than other smooth muscles (Keeton et al. 1993; Carafoli and Stauffer 1994).

Unlike the SR Ca,Mg-ATPase, the C-terminal end of the PM Ca,Mg-ATPase confers an autoinhibitory action. The inhibitory domain, the C domain, consists of 28 amino acids. Another domain downstream apparently contributes to the inhibition. The binding of calmodulin to the C domain largely removes the inhibition (Verma et al. 1988, 1994). The C domain has been shown to bind to an area of the enzyme between the phosphoenzyme-forming aspartic acid and the nucleotide binding site (Verma et al. 1988 and De Jaegere et al. 1990; for a diagram of structure see Hilfiker et al. 1993). A site that can be phosphorylated by cyclic AMP-dependent protein kinase is also present in this part of the molecule (Verma et al. 1988; De Jaegere et al. 1990), although no function for this site has been proven in smooth muscle (Vrolix et al. 1988).

9.4. Calmodulin regulation of PM Ca,Mg-ATPase

A notable property of the plasma membrane Ca^{2+} pump is its stimulation by calmodulin (Wuytack et al. 1980). Calmodulin is a highly conserved, soluble, acidic Ca^{2+}-binding protein of 148 amino acid residues, with a molecular weight of 16,680 (Watterson et al. 1980). It contains four homologous domains, each binding one Ca^{2+}. The reported affinity of the four binding sites for Ca^{2+} is controversial. However, the affinity can be expressed as a single binding constant of $K_d = 2.6 \times 10^{-7}$ M, based on the assumption that there are two high-affinity sites with a K_d of 2.0×10^{-7} M that display cooperativity, and two lower-affinity sites with a K_d of 2.0×10^{-6} M (Wang 1985). The binding of Ca^{2+} to the two-high affinity Ca^{2+} sites exposes the binding sites of calmodulin for the PM Ca,Mg-ATPase. With calmodulin bound to this enzyme, the remaining two Ca^{2+} binding sites become high-affinity sites. This explains why the activation occurs *in vivo* at Ca^{2+} concentrations present in the cell during a contraction (for a more detailed discussion, see Vogel 1988).

In the presence of Ca^{2+}, the PM Ca,Mg-ATPase interacts with calmodulin, increasing both Ca^{2+} affinity and maximum velocity of the Ca,Mg-ATPase. A dissociation constant of 17 nM was calculated for calmodulin dissociation from the purified enzyme (De Schutter et al. 1984). The calmodulin binding domain is thought to be an autoinhibitory region of the Ca,Mg-ATPase; when bound to calmodulin, the inhibition is removed.

Calmodulin inhibitors such as trifluoperazine and calmidazolium affect Ca^{2+} transport in PM vesicles from gastric smooth muscle when used for very brief periods. However, when used for longer periods, large increases in Ca^{2+} permeability occurred, which prohibited the evaluation of these effects on the PM Ca,Mg-ATPase (Lucchesi and Scheid 1988). Furthermore, these authors suggested that the two calmodulin inhibitors have different mechanisms of action. Trifluoperazine competes directly with calmodulin for its binding site on the ATPase, whereas calmidazolium appears to interact with the calmodulin-ATPase complex in the PM.

$InsP_3$ has been shown to inhibit the PM Ca,Mg-ATPase in cardiac cells (Kuo and Tsang 1988) and in erythrocytes (Davis et al. 1991); inositol 4,5-bisphosphate ($InsP_2$) was also active. $InsP_4$ and 1,4-$InsP_2$ were inactive as inhibitors. The apparent mechanism was inhibition of calmodulin binding (Davis et al. 1991). Whether $InsP_3$ inhibits the Ca,Mg-ATPase in smooth muscle has yet to be shown. (For the major actions of $InsP_3$, see Section 10.2).

Calmodulin is present in relatively large amounts in the cytoplasm. The calmodulin yield was approximately 11 mg per 100 g of uterine tissue and about half this value in other smooth muscles (Grand et al. 1979). Little is known about high-affinity cytosolic Ca^{2+}-complexing proteins that contribute to the Ca^{2+} buffering capacity of the smooth muscle cell. Whether calmodulin can be considered as such a protein is questionable. The difficulty with considering calmodulin as a Ca^{2+} buffer is that a true buffer should prevent Ca^{2+} from activating the cell, whereas Ca^{2+} binding to calmodulin results in stimulation of Ca^{2+}-dependent processes.

9.5. Other regulation of the PM Ca^{2+} pump

Attempts have been made to show that a variety of physiologic and pharmacologic agents causing relaxation work through regulation of the PM Ca^{2+} pump (for a review not limited to smooth muscle, see Monteith and Roufogalis 1995).

No effect of cAMP on the Ca^{2+} pump could be demonstrated (Furukawa et al. 1988). Nor was there a direct stimulation by protein kinase C (Imai et al. 1990). However, the function of cGMP in Ca^{2+} homeostasis is still under investigation. The PM Ca^{2+} pump appears to be regulated by cGMP in cultured vascular smooth muscle cells (Furukawa et al. 1988). Cyclic GMP is the common pathway for vasodilation by endothelium-derived relaxing factor and nitrovasodilators. It has been hypothesized that this action of cGMP is mediated through a decrease in intracellular free Ca^{2+} (Ignarro 1989). Stimulation of the Ca,Mg-ATPase by cGMP-dependent protein kinase has been reported (Furukawa and Nakamura 1987). Since phosphorylation of the Ca,Mg-ATPase could not be confirmed, it was suggested that a contaminating protein indirectly stimulated the reaction (Baltensperger et al. 1988; Vrolix et al. 1988). The phosphorylation of a 240-kDa protein correlates with increased ATPase activity (Yoshida et al. 1991).

There is a phospholipid binding site on the Ca,Mg-ATPase consisting of many positively charged amino acids in the loop between the second and third transmembrane domains (Hilfiker et al. 1993). Ca,Mg-ATPase function requires a lipid environment. The nature of this lipid may influence the activity of the PM Ca,Mg-ATPase. Phosphatidyl inositides have been shown to increase Ca,Mg-ATPase affinity for Ca^{2+} (Missiaen et al. 1989). Since the concentration of these lipids is increased during relaxation, one could speculate that they augment relaxation by activating the PM Ca^{2+} pump and lowering the Ca^{2+} concentration.

In microsomes and myometrial strips, the PM Ca,Mg-ATPase is inhibited by oxytocin (Akerman and Wikstrom 1979; Soloff and Sweet 1982), and this action is dependent on the hormonal status of the animal (Enyedi et al. 1989). Partial inhibition of the enzyme by pre-incubation of microsomes in a carbachol-containing solution has been reported; no mechanism was given (Missiaen et al. 1988).

The phosphorylation of the PM Ca,Mg-ATPase is increased by lanthanum (100 μM), whereas under the same conditions the phosphorylation of the SR Ca,Mg-ATPase is decreased (Wuytack et al. 1982). At much lower concentrations in skeletal muscle SR, lanthanum prevents phosphoenzyme decay by binding at several sites unrelated to the Ca^{2+} transport site. The amount of phosphorylation is dependent on the relative concentrations of lanthanum and divalent cations (Fujimori and Jencks 1990).

9.6. Na^+/Ca^{2+} exchange

A second mechanism responsible for Ca^{2+} extrusion from the cell is Na^+/Ca^{2+} exchange. An antiporter in the PM carries Na^+ in one direction and Ca^{2+} in the other. This mechanism was first described in cardiac muscle (Reuter and Seitz 1968), and is now accepted for smooth muscles. The Na^+/Ca^{2+} exchanger has a lower affinity for Ca^{2+} than Ca,Mg-ATPase but a much higher turnover rate and hence a greater capacity (Blaustein et al. 1992). The physiologic importance of Na^+/Ca^{2+} exchange in smooth muscles has been difficult to verify. For example, the occurrence of Na^+/Ca^{2+} exchange seems to vary with the type of smooth muscle. The mechanism was shown to be present in single taenia coli cells but not in coronary artery cells (Hirata et al. 1981). Specific blocking of the Na^+/Ca^{2+} exchange is experimentally difficult. Other mechanisms that transport Ca^{2+} and Na^+, such as membrane depolarization and voltage-dependent Ca^{2+} channels, may account for false observations of Na^+ and Ca^{2+} exchange.

Theoretically, the Na^+/Ca^{2+} exchanger can work either to raise or lower intracellular Ca^{2+}. Physiological conditions favor extrusion of Ca^{2+} by the Na^+/Ca^{2+} exchanger (Smith et al. 1989b). However, experiments to document the presence of Na^+/Ca^{2+} exchange are more easily performed under conditions that would lead to a rise in intracellular Ca^{2+}. Extremely low external Na^+, or high internal Na^+ obtained by poisoning the Na,K-ATPase with digitalis, can be used to increase the movement of Ca^{2+} into the cell.

Early indications for a role of Na^+ in Ca^{2+} fluxes and tension development came from work in vascular smooth muscle. Reduction of the external Na^+ concentration caused tonic contraction (Bohr et al. 1969). This is thought to be due to Na^+ leaving the cell and Ca^{2+} coming into the cell in exchange. Microelectrodes, isotopic Na^+ (Aickin et al. 1987), and benzofuran isophthalate, a fluorescent Na^+ indicator (Johnson et al. 1991), have all been used to document the Na^+ part of the exchange. The Ca^{2+} part of the exchange has been shown using muscular tension, tracer $^{45}Ca^{2+}$, and direct measurement of Ca^{2+} by Fura-2 (Blaustein et al. 1992). When the Na^+/K^+ pump was inhibited, Na^+/Ca^{2+} exchange was demonstrated as a major regulatory mechanism. Moreover, such putative inhibitors of Na^+/Ca^{2+} exchange as quinacrine, dichlorobenzamil, and dodecylamine blocked the contracture of ouabain-treated tissue to Na^+ withdrawal (Aickin et al. 1987). In single cells of the guinea-pig ureter, Na^+/Ca^{2+} exchange was demonstrated when the Na^+ gradient, Ca^{2+} inside, and the membrane potential were within physiological range. But major changes in Ca^{2+} inside occurred only when the Na^+ gradient was drastically changed. It was concluded that Na^+/Ca^{2+} exchange was not a major physiologic mechanism for Ca^{2+} extrusion (Aaronson and Benham 1989).

Microsomal preparations would seem ideal to document the presence of Na^+/Ca^{2+} exchange. However, a common problem is that exchange of Na^+ for Ca^{2+} at cation binding sites may occur. This is best handled by using La^{3+} to remove the Ca^{2+} bound to the external surface. The direction of the Ca^{2+} movement by Na^+/Ca^{2+} exchange must be opposite to that of Ca^{2+} carried through the opening of a channel, and most convincingly the accumulated Ca^{2+} must be releasable upon addition of a Ca^{2+} ionophore. In a control experiment, a Na^+ ionophore such as monensin is added to destroy the Na^+ gradient and thereby prevent the Na^+/Ca^{2+} exchange. Though all these controls have been performed occasionally (Morel and Godfraind 1984), more often one or more is missing (Grover et al. 1983; Grover and Kwan 1987; Matlib 1988).

Morel and Godfraind (1984) noted that changes in Na^+ permeability during membrane preparation, loss or lack of an adequate Na^+ gradient, or impure preparations may account for the many low values obtained. In addition, Na^+ permeability may vary from tissue to tissue. The measurement of permeability using isotope tracers has inherent problems of variations in the size and exchange properties of the intrinsic ion pools.

More precise experiments have successfully demonstrated the presence of Na^+/Ca^{2+} exchange. Activity of the Na^+/Ca^{2+} exchanger, greater than that of the PM Ca,Mg-ATPase, was found in microsomes from PM of several smooth muscles. $^{45}Ca^{2+}$ taken up by Na^+/Ca^{2+} exchange was rapidly released by the ionophore A23187 in the presence of EGTA, indicating that the Ca^{2+} was inside the vesicles (Slaughter et al. 1987, 1989). Inhibition of the Na^+/Ca^{2+} exchange by amiloride and quinidine supports the veracity of the demonstrated exchange (Khan et al. 1988). The Na^+/Ca^{2+} exchanger of the mesenteric artery was solubilized and incorporated into phospholipid vesicles. The rate and magnitude of Na^+/Ca^{2+} exchange were greater in the reconstituted than in the native vesicles (Matlib and Reeves 1987).

ATP is not required in cardiac muscle, and three Na^+ are exchanged for one Ca^{2+}. Thus, the exchange is electrogenic. If more than two Na^+ are exchanged for one Ca^{2+} then an electrogenic potential will build up, and this should decrease Na^+/Ca^{2+} exchange. In the presence of K^+, the K^+ ionophore valinomycin prevents the establishment of an electrogenic potential and increases exchange of

Ca^{2+} for Na^+ in Na^+-loaded vesicles. This establishes that smooth muscle Na^+/Ca^{2+} exchange is electrogenic (Slaughter et al. 1987; Matlib 1988).

The interpretation of the results of Na^+/Ca^{2+} exchange experiments is made more difficult by the findings that follow: Na^+ withdrawal generates $InsP_3$ (Smith et al. 1989a); phorbol esters activate Na^+/Ca^{2+} exchange (Vigne et al. 1988); and the absence of ATP inhibits Na^+/Ca^{2+} exchange (Smith and Smith 1990).

Recent work has focused on the localization and structure of the Na^+/Ca^{2+} exchange protein within the smooth muscle. It has great similarity to the cardiac Na^+/Ca^{2+} exchanger. The protein has several isoforms, all with approximately 900 amino acids (Nakasaki et al. 1993). Antibodies to the exchanger were found to label areas of the PM adjacent to the SR, an area with high intracellular [Ca^{2+}], which could increase the activity of the exchanger. Thus the activity of the Na^+/Ca^{2+} exchanger appears to be closely coupled to availability of Ca^{2+} from intracellular stores (Moore et al. 1993; Juhaszova et al. 1994).

10. Increase of cytoplasmic Ca^{2+} – sarcoplasmic reticulum

10.1. Ca^{2+} pump

The all-important event for contraction to occur is the temporary rise in intracellular free Ca^{2+}. Early work found it difficult to produce Ca^{2+} release from isolated SR. Thus it was hypothesized that endogenous ionophores caused Ca^{2+} release. This idea has since been abandoned (see Carsten and Miller 1990).

Other experiments attempted to induce Ca^{2+} release by reversal of the Ca^{2+} pump. A phosphorylated intermediate (see Table 1) can be formed when Ca^{2+}-loaded SR vesicles from skeletal muscle are incubated at very low Ca^{2+} and high ADP and inorganic phosphate concentrations. However, a rapid Ca^{2+} release from this intermediate with the concomitant synthesis of ATP is unlikely to take place under physiological conditions (for a review see Inesi 1985). Thus, neither of these two experiments simulated Ca^{2+} release for smooth muscle contraction in the living cell.

10.2. $InsP_3$ formation and metabolism

In most smooth muscles, the physiological agent that induces Ca^{2+} release for contraction upon agonist stimulation is $InsP_3$. $InsP_3$ was demonstrated to act as a second messenger to release Ca^{2+} from intracellular stores, now identified as SR (for a review see Berridge 1993). D-myo-inositol 1,4,5-trisphosphate ($InsP_3$) and also sn-1,2-diacylglycerol originate from a receptor-activated hydrolysis of phosphatidylinositol 4,5-bisphosphate (Berridge 1984), a minor constituent of PM phospholipids (Abdel-Latif 1986). A specific phospholipase C (PLC) catalyzes the hydrolysis of the phosphate that is bound to the third carbon of the glycerol of phosphatidylinositol 4,5-bisphosphate in the presence of physiologic concentrations of Ca^{2+}. The enzyme preferentially uses the bisphosphate as substrate rather than the unphosphorylated phosphatidylinositol. It does not hydrolyze phosphatidylcholine or phosphatidylethanolamine (Abdel-Latif 1986). There are at least three immunologically distinct enzymes, PLC-β, PLC-γ, and

PLC-δ, with nominal molecular masses of 150, 145, and 85 kDa, respectively. They have very different amino acid sequences and Ca^{2+} requirements. A previously named PLC-α has now been rejected (Rhee et al. 1991). In at least one smooth muscle, the PLC-β is membrane bound and associated with receptors and G protein. The PLC-γ and PLC-δ are cytosolic proteins (Zhou et al. 1994). Of the four subtypes of PLC-β, an antibody to the β1 isoform inhibits 85% of contraction, with an antibody to β3 blocking the remaining activity in circular intestinal smooth muscle (Murthy and Makhlouf 1995).

In the presence of physiological concentrations of Ca^{2+} and Mg^{2+}, $InsP_3$ is inactivated by a stepwise dephosphorylation to inositol that is reincorporated into the phosphoinositide pool. Phosphatidylinositol 4,5-bisphosphate is then resynthesized (Berridge 1984). A second metabolic pathway for $InsP_3$ is provided by phosphorylating the 3-hydroxyl group to inositol tetrakisphosphate (1,3,4,5-$InsP_4$) by an ATP-dependent kinase (Irvine et al. 1986), shown in ileal smooth muscle (Bielkiewicz-Vollrath et al. 1987). The enzyme is Ca^{2+}- and calmodulin-dependent (Yamaguchi et al. 1988). With the rise in Ca^{2+} concentration, the metabolic pathway is directed away from dephosphorylation and toward $InsP_4$ generation. $InsP_4$ is then dephosphorylated to the inactive 1,3,4-$InsP_3$ and sequentially to inositol.

Formation of $InsP_3$, subsequent Ca^{2+} release from intracellular stores, and smooth muscle contraction are initiated by agonist binding to specific receptors in the PM. This sequence of events has been shown with many agonists and in many smooth muscles, and on occasion the whole chain of events has been shown in one system. Concentration of agonist, increase in free Ca^{2+}, and contractile response correlate qualitatively (Bitar et al. 1986). Stored Ca^{2+} can also be released by the application of $InsP_3$ to permeabilized cells or microsomes. The rise in intracellular free [Ca^{2+}] and contraction are similar whether they are induced by $InsP_3$ or by agonist. The intracellular Ca^{2+} release can support both phasic and tonic contractions, and repetitive applications of $InsP_3$ generate repeated appropriate contractions (Somlyo et al. 1985). However, there are observations inconsistent with this singular mechanism for Ca^{2+} mobilization (see Section 10.7). Furthermore, the rise in $InsP_3$ in some studies occurs at ten times the concentration of agonist that initiates muscle contraction (Best et al. 1985; Grandordy et al. 1986; Howe et al. 1986; Goureau et al. 1992). This lack of correlation would occur if a minute rise of $InsP_3$ produced a maximal effect. Alternatively, methodologic problems could be the cause. They include low sensitivity of the inositol phosphate assay, measurement of inositol phosphates taken many minutes instead of a few seconds after the event, and assay of total inositol phosphates in the presence of lithium instead of $InsP_3$. Not all agonists that release intracellular Ca^{2+} work through $InsP_3$. Although it has long been known that prostaglandins release Ca^{2+} from SR preparations (Carsten 1974; Carsten and Miller 1977b), whether $InsP_3$ is the mechanism by which prostaglandins release intracellular Ca^{2+} remains controversial (Goreau et al. 1992; Molnar and Hertelendy, 1990). Thus, although $InsP_3$ is a major factor in Ca^{2+} release from internal stores, it is not the only available mechanism.

Ca^{2+} release by $InsP_3$ is specifically and competitively inhibited by heparin at low concentration (<1 μg/ml, ~0.2 μM). The effect is specific because another sulfated polysaccharide, chondroitin sulfate, does not block the action of $InsP_3$

(Worley et al. 1987; Kobayashi et al. 1989). Heparin appears to compete for the $InsP_3$ binding sites (Worley et al. 1987; Ghosh et al. 1988), a mechanism that has been confirmed in smooth muscles (Kobayashi et al. 1989). At comparable concentrations, heparin does not affect Ca^{2+} release triggered by $InsP_4$, caffeine, GTP (Ghosh et al. 1988) or prostaglandin $F_{2\alpha}$ (Molnar and Hertelendy 1990). Heparin does not inhibit in the intact cell, since it does not cross the PM. Heparin is not entirely selective; other reactions may obscure the results. Heparin has been reported to inhibit phospholipase A_2 (Diccianni et al. 1990), to bind to L-type Ca^{2+} channels (Knaus et al. 1992) and to uncouple the α_1 adreno receptor from its G protein (Dasso and Taylor 1991). A second antagonist is decavanadate, but it has low specificity and binds to all $InsP_3$ and $InsP_4$ recognition sites (Potter and Nahorski 1992).

Other inhibitors interfere with the phosphoinositide cycle. 2,3-diphosphoglycerate inhibits the hydrolysis of $InsP_3$ and at higher concentrations competes for the specific $InsP_3$ binding site (Downes et al. 1982; Guillemette et al. 1987). Synthetic nonhydrolyzable phosphonate analogs of phosphatidylinositol inhibit the PLC (Shashidhar et al. 1990). NCDC at 50 μM (a serine esterase inhibitor) or U73122 at 100 nM (an inhibitor of signal transduction in PLC pathways) prevent the formation of $InsP_3$ on agonist stimulation (Molnar and Hertelendy 1992). On the other hand, neomycin, once thought to be a specific inhibitor of $InsP_3$ formation and action, can bind to many polyphosphorylated compounds including ATP. Thus, the inhibition of Ca^{2+} release by neomycin is not evidence of the involvement of $InsP_3$ (Prentki et al. 1986).

When using permeabilized cells, it must be emphasized that the released Ca^{2+} could originate from any intracellular organelle or from Ca^{2+} bound to the outside of the PM. Mitochondrial Ca^{2+} release cannot be excluded even in the presence of inhibitors of mitochondrial oxidative phosphorylation. Thus, microsomal preparations of SR give more specific information. $InsP_3$ releases Ca^{2+} previously taken up by the SR (10 nmol/mg protein) in the presence of ATP. The Ca^{2+} release is a function of the $InsP_3$ concentration and, at 5 μM $InsP_3$, amounts to approximately 40% of stored Ca^{2+} (Carsten and Miller 1985).

Contraction follows agonist stimulation very rapidly; hence $InsP_3$ formation and Ca^{2+} release must also occur in this time frame, and the events must occur in proper sequence. Many earlier studies used long incubation times; therefore, the sequential order of events was not apparent. Measurements of $InsP_3$ were often made after 5–10 minutes, sometimes even after 30 minutes, although the peak rise in $InsP_3$ occurs in less than 2–5 seconds (Berridge 1983; Salmon and Bolton 1988). To prove that $InsP_3$ is the physiological messenger of Ca^{2+} release, $InsP_3$ must appear first, and the inactive $InsP_2$ and inositol monophosphate metabolites follow. This fact was often obscured by including Li^+ in the incubations and measuring total inositol phosphates. Accurate kinetics of agonist action and messenger release can be determined using *caged compounds*, inactive photosensitive precursors of effector compounds that are cleaved and activated by flash photolysis. Thus, it was possible to record the rise of force after photolysis of the caged α_1-adrenergic agent phenylephrine in an intact muscle strip. Similar recording was made of caged $InsP_3$ in a permeabilized muscle strip of guinea-pig portal vein. The onset of contraction followed the release of phenylephrine by 1.5 sec, whereas it followed the release of $InsP_3$ by 0.5 sec. One may conclude that, after agonist stimulation, $InsP_3$ formation and Ca^{2+} release occur rapidly and sequentially. In fact, using Fluo-3, the time from photolysis of caged $InsP_3$ to the rise in intracellular free Ca^{2+} was only 0.03 sec. There is another 0.2–0.3 sec interval between the intracellular free Ca^{2+} rise and the contraction of intact smooth muscle; this time is also dependent on the ATP concentration (Somlyo and Somlyo 1990).

Early observations that low temperature does not inhibit $InsP_3$-induced Ca^{2+} release are consistent with the opening of a Ca^{2+} channel. In contrast, ion carriers are expected to be inhibited by low temperatures because of increased viscosity of the lipid bilayer (Smith et al. 1985). It has subsequently been shown that $InsP_3$ opens Ca^{2+} channels in canine aortic smooth muscle SR incorporated into planar lipid bilayers (Ehrlich and Watras 1988).

The $InsP_3$ receptor ($InsP_3R$), purified from bovine aorta smooth muscle microsomes, is a homotetrameric glycoprotein. Each of the four subunits has a molecular weight of 224,000. The receptor bound 2.7 ± 0.18 nmol of $InsP_3$ per milligram of protein (approximately 0.7 mol $InsP_3$ per mole of receptor monomer) and had a K_d of 2.4 ± 0.24 nM (Chadwick et al. 1990). Similar results were obtained using receptor purified from rat vas deferens (Mourey et al. 1990). Based on cDNA data, each monomer has one $InsP_3$ binding site, two phosphorylation sites, and two nucleotide binding sites – all on the cytoplasmic side of the membrane. Peptides from the aortic smooth muscle $InsP_3R$ were compared with the known structure of brain $InsP_3R$ and were almost identical (Marks et al. 1990). The purified $InsP_3$ receptor was incorporated into lipid bilayers and formed $InsP_3$-gated Ca^{2+} channels (Mayrleitner et al. 1991). Different isoforms are encoded in several distinct genes located in different chromosomes (Mikoshiba et al. 1994). The $InsP_3$ and ryanodine receptors (see Section 10.7) consisting of four protomers, share fourfold symmetry and have homologous transmembrane regions thought to form part of the Ca^{2+} release channel. Yet both receptors are structurally distinct from the PM Ca^{2+} channel (for a review see Taylor and Richardson 1991 or Mikoshiba et al. 1994).

The rate of $InsP_3$-induced Ca^{2+} release is modulated by ATP and Ca^{2+}. Each shows a peak stimulation that drops off while still within the physiological range. Thus it was found that Ca^{2+} release was six times faster at 0.3 than at either 0 or 3 μM [Ca^{2+}] in saponin-skinned smooth muscle fibers (Iino 1990). In preparations from portal vein, $InsP_3$-induced Ca^{2+} release appears to be stimulated as the ATP concentration is raised from 0 to approximately 0.5 mM, but declines at physiological levels (about 1 mM). At still higher concentrations, ATP acts as a weak competitor of $InsP_3$ (Iino 1991). The ATP effect is due to binding to the receptor; ADP and nonhydrolyzable analogs of ATP can substitute, but AMP or GTP cannot (Ferris et al. 1990; Iino 1991).

Protein kinases regulated by cAMP (PKA), cGMP (PKG), Ca^{2+}-calmodulin (CaM kinase II) and diacylglycerol (PKC) phosphorylate the rat brain receptor. All phosphorylate a serine residue, though there are differences in phosphopeptides. More recently, the isolated receptor protein was found to autophosphorylate (Ferris et al. 1992). However, the effects of phosphorylation on the release of Ca^{2+} by $InsP_3$ remain controversial. Both increased and decreased release of Ca^{2+} by $InsP_3$ have been shown after phosphorylation of the $InsP_3R$. These divergent findings may be caused by the presence of at least three receptor types, each of which has a unique phosphorylation site (Mikoshiba et al. 1994).

Other than $InsP_3$, the only inositol phosphate presumed to be associated with Ca^{2+} homeostasis is 1,3,4,5-$InsP_4$, which is thought to open Ca^{2+} channels in the PM, thereby stimulating Ca^{2+} entry and refilling the SR (Irvine 1992). The physiological role of $InsP_4$ has not been fully documented. Only a few investigations on smooth muscle are available. No change in intracellular Ca^{2+} on application of $InsP_4$ was found in one study (Watras et al. 1989). However, another study, using a patch-clamp technique, showed that $InsP_4$ in the presence of Ca^{2+} induced an outward current carried by K^+, a possible indicator of locally increased Ca^{2+} (Molleman et al. 1991). A specific binding site for $InsP_4$ has been identified in colonic smooth muscle membrane preparations. The $InsP_4$ binding site is different from the one for $InsP_3$, with a different pH optimum. $InsP_3$ could replace only 60% of the bound $InsP_4$ and showed a much lower affinity. Both the PM and intracellular compartments have binding sites for $InsP_4$ (Zhang et al. 1993). Other putative roles for $InsP_4$ action are that it prolongs the action of $InsP_3$ by inhibiting its hydrolysis and that $InsP_4$ replenishes the $InsP_3$-sensitive Ca^{2+} store. However, Ca^{2+} can be replenished without $InsP_4$ and without opening of voltage-operated Ca^{2+} channels (Nelemans et al. 1990). Although many proposed actions of $InsP_4$ are related to potentiation of $InsP_3$ action and Ca^{2+}, the precise mechanism and site of action of $InsP_4$ on Ca^{2+} homeostasis are still unknown. More work in this area is required, especially in smooth muscle.

Inositol pentakisphosphate ($InsP_5$) and inositol hexakisphosphate ($InsP_6$) are also present in animal tissue. No increase in either occurred when cells from a smooth muscle cell line were stimulated with norepinephrine (Nelemans et al. 1990). Cyclic inositol phosphates have been found in many tissues. Under physiological conditions, their contribution to Ca^{2+} mobilization seems insignificant (Willcocks et al. 1989).

Synthetic analogs serve to identify function–activity relationships for the $InsP_3R$. Most of the information on specificity was obtained from nonsmooth muscle systems (DeLisle et al. 1994). The phosphates in position 4 and 5 are important for Ca^{2+} release. Potency is reduced by a factor of 60 by removing the 1 phosphate, and by a factor of five by shifting it to the 2 position (Irvine et al. 1984). Additional groups on the 2 position cause little change in binding to the receptor or Ca^{2+} release (Hirata et al. 1989, 1990a). Inositol trisphosphorothioate ($InsP_3$ $[S]_3$) shows full $InsP_3$ agonist activity but is minimally affected by either the kinase or phosphatase, and so exhibits a prolonged effect and empties Ca^{2+} stores (Taylor et al. 1989). The 1 isomer of $InsP_3$ does not have significant activity.

10.3. Cyclic nucleotides in Ca^{2+} regulation

Modulation of many regulators of Ca^{2+} has been attributed to cyclic nucleotides. These are discussed in detail in the sections dealing with the specific Ca^{2+} control mechanisms. In summary, we find support for the effects of cyclic nucleotides on PM Ca,Mg-ATPase and phospholamban (already discussed in Section 8.5) and on the $InsP_3$ generating system.

In smooth muscles there is much suggestive evidence for an interaction of cyclic nucleotides and the $InsP_3$ system. It is still not known whether the inhibition by cyclic nucleotides occurs at the agonist receptor, the G protein, the $InsP_3R$, or the PLC. The literature is replete with contradictory findings, some of which may relate to organ, agonist, and even isoform specificity. In tracheal, uterine, and vascular smooth muscles stimulated by histamine, oxytocin, or endothelin, agents such as cAMP (or analogs), isoproterenol, forskolin or relaxin inhibited $InsP_3$ generation (Madison and Brown 1988; Anwer et al. 1989; Xuan et al. 1991). In rat uterine muscle, oxytocin-stimulated $InsP_3$ generation and intracellular free Ca^{2+} rise were inhibited by a cAMP analog. This effect could be reversed by an inhibitor of cAMP-dependent protein kinase but not by an inhibitor of protein kinase C (Anwer et al. 1990). However, carbachol-induced inositol phosphate generation was not inhibited in the tracheal muscle (Madison and Brown 1988; Hall et al. 1990). In membrane fractions from the iris sphincter, pretreatment with isoproterenol reduced carbachol-stimulated inositol phosphate production, suggesting the G protein as the target of protein kinase A (Zhou et al. 1994).

Many experiments, however, use pharmacologic agents that lack specificity. Thus, inhibition of $InsP_3$ generation in the presence of theophylline may represent a phosphodiesterase inhibitory effect, but the stimulation of purinergic receptors may play a greater role. Similarly, when agents that generate cAMP are used, it is not clear whether cAMP kinase phosphorylates a protein in the $InsP_3$ generating system or whether activation of the cAMP generating system (or even membrane potential changes) interfere with $InsP_3$ generation. At present, investigation of the interaction of the cyclic nucleotide and $InsP_3$ systems is still in its infancy (Chilvers et al. 1994).

The relation of cGMP to phosphoinositide formation and turnover is even less clear. A carbachol-stimulated Ca^{2+} rise in the tracheal muscle was counteracted by cGMP (Felbel et al. 1988). In blood vessels, cGMP inhibited formation of inositol phosphates. It was suggested that phosphorylation by ATP inhibited G-protein activation and G-protein coupling to PLC (Hirata et al. 1990b).

10.4. Role of G proteins in agonist-stimulated Ca^{2+} release

GTP is required for agonists to activate PLC at physiological cation concentrations (Sasaguri et al. 1985; Fulle et al. 1987). Nonhydrolyzable analogs of GTP can also release $InsP_3$ by themselves, as shown in vascular smooth muscle (Sasaguri et al. 1985; LaBelle and Murray 1990) and in circular, but not longitudinal, intestinal muscle cells (Murthy et al. 1992). In skinned vascular muscle, neither GTP nor $InsP_3$ alone releases Ca^{2+} but together they do, and Gpp(NH)p can substitute for GTP (Saida and van Breemen 1987). The requirement for GTP suggests that a G protein links agonist receptor occupancy and PLC activation. The G-protein system was best described for the generation of cAMP. G proteins consist of a heterotrimeric protein with $\alpha\beta\gamma$ subunits. The α subunit, when separated from the other subunits, binds GTP and serves a regulatory role on effector proteins, such as activating the adenylate cyclase. Since both $InsP_3$ and cyclic nucleotide generation are G protein–mediated events, one may want to investigate whether there is interaction at the G protein level. Cross-talk could be generated if the responses of the two messenger systems were coordinated by the activation/inhibition of similar G proteins. Difficulties in interpreting these findings relate to the fact that increased levels of cAMP interfere with the production and effect of $InsP_3$, as discussed previously.

A classic activator of G protein–mediated processes is fluoride in the presence of aluminum. Increased release of $InsP_3$ upon addition of aluminum and fluoride further substantiates the involvement of G proteins in this process (Marc et al. 1988; LaBelle and Murray 1990; Socorro et al. 1990; Zhou et al. 1994).

The G protein that activates adenyl cyclase is called Gs. Another family of G proteins, Gi, is inhibitory. The presence of Gs and Gi were explained after the actions of pertussis toxin and cholera toxin were understood. Pertussis toxin prevents the receptor-mediated activation of Gi, and cholera toxin activates Gs. Finding that these toxins did not consistently work in the $InsP_3$ generating system led to the present increase in the number of recognized G proteins. Thus Gq is a family of G proteins (including α_q, α_{11}, α_{14}, α_{15}, and α_{16}) thought to increase the synthesis of $InsP_3$. Furthermore, the effects of the PLC may be activated by the $G\beta\gamma$ subunit and not just by the $G\alpha$ subunit. At present a large variety of α, β, and γ subunits have been identified. The α subunit is the one used in naming the G protein; some selectivity of the other two subunits for the α subunit is observed. However, the selectivity is not well understood. Conflicting results have been obtained as to the nature of the G proteins involved in the PLC system of smooth muscles. Some observed differences are tissue-specific and may be related to the fact that only the PLC-β is G protein–sensitive (for a review see Birnbaumer and Birnbaumer 1995). Multiple actions of stimulators such as fluoride can lead to inconsistent interpretation of results, since fluoride may also stimulate Ca^{2+} entry (Murthy and Makhlouf 1994; Prestwich and Bolton 1995).

Although cholera toxin causes activation of the stimulatory Gs protein in cAMP formation, it is without effect on the $InsP_3$ generation induced by oxytocin in the guinea-pig uterus (Marc et al. 1988). On the other hand, cholera toxin inhibited oxytocin-induced $InsP_3$ generation in the rat myometrium (Ruzycky and Crankshaw 1988). In vascular muscle cells, it inhibited angiotensin II–induced $InsP_3$ generation. A suggested mechanism for cholera toxin inhibition is that it decreases the number of agonist receptors without decreasing receptor affinity (Socorro et al. 1990). Whether this is specific to cholera toxin or a cAMP effect is debated (cf. the preceding discussion of the cAMP effect on $InsP_3$ formation).

Pertussis toxin inhibits the inhibitory G protein (G_i) in cAMP formation, thus increasing cAMP. However, it was without effect on $InsP_3$ generation induced by oxytocin or carbachol in the guinea-pig uterus (Marc et al. 1988) and by norepinephrine or angiotensin II or endothelin in vascular smooth muscle (LaBelle and Murray 1990; Socorro et al. 1990; Takuwa et al. 1990). On the other hand, it inhibited $InsP_3$ generation induced by oxytocin in the rat myometrium (Ruzycky and Crankshaw 1988; Anwer et al. 1989) and by angiotensin II in vascular muscle cells (Gaul et al. 1988). The reason for these disparities is not known at present. However, from studies in other tissues it has become apparent that several different G proteins may couple receptors to PLC in one cell (Martin 1991).

10.5. Ca^{2+} release by GTP

A different Ca^{2+} release mechanism is evoked by GTP in the presence of ATP. This mechanism differs from that of $InsP_3$ inasmuch as it is slow and temperature-dependent (Gill et al. 1992). Both mechanisms release Ca^{2+} from the SR, and in some systems GTP releases as much or more Ca^{2+} than does $InsP_3$. It was suggested that GTP promotes a conveyance of Ca^{2+} between organelles with different oxalate permeabilities and $InsP_3$ sensitivities (Bian et al. 1991; for a review see Gill et al. 1992). Some evidence suggests that a multicompartment Ca^{2+} store is seen only in cultured permeabilized suspended cells and not in attached cells even if permeabilized (Short et al. 1993; Hajnoczky et al. 1994).

10.6. Ca^{2+} pools

To explore the functioning of the intracellular Ca^{2+} pool, investigators have used multiple inhibitors and releasing agents. This has led to the definition of multiple Ca^{2+} pools. It has been suggested that these represent pools with different functions and locations within the cell (Leijten et al. 1985). The evidence supporting these suggestions has been slow to appear. Presumably these pools are neither nuclear, mitochondrial, nor associated with structural proteins. ATP is uniformly required to sequester Ca^{2+} into the pools, and all are membrane-limited as shown by the release of Ca^{2+} in the presence of a Ca^{2+} ionophore. Pools are differentiated by:

(1) Ca^{2+}-transport ATPase sensitivity to vanadate (Missiaen et al. 1991) or thapsigargin (Bian et al. 1991; Missiaen et al. 1991);
(2) Ca^{2+} affinity of the ATPase (Missiaen et al. 1991);
(3) responsiveness of Ca^{2+} release to GTP and $InsP_3$ (Bian et al. 1991; Missiaen et al. 1991), caffeine (Leijten et al. 1985; Iino, 1989), and cyclic adenosine diphosphate-ribose (Lee 1994); and
(4) presence of Ca^{2+}-induced Ca^{2+} release (Leijten et al. 1985; Iino 1989).

It is not clear whether these represent one pool with a spectrum of sensitivities or multiple independent structures. The number of hypothesized pools has progressively increased.

10.7. Ca^{2+}-induced Ca^{2+} release and the ryanodine receptor

The concept of Ca^{2+}-induced Ca^{2+} release was introduced to account for events in cardiac muscle (Fabiato 1983). The minute amounts of Ca^{2+} that enter the cell on stimulation do not directly activate contraction, but rather trigger release of larger amounts of Ca^{2+} from the SR. At this time it is not certain how important the physiologic role of Ca^{2+}-induced Ca^{2+} release is in smooth muscle. Evidence from different muscles shows great differences in importance. For example, longitudinal intestinal muscle shows Ca^{2+}-induced Ca^{2+} release whereas circular muscle does not (Kuemmerle et al. 1994).

In skeletal and cardiac muscle, a Ca^{2+}-gated Ca^{2+} release channel from the SR has been identified. This channel is the receptor for the plant alkaloid ryanodine. The complementary DNA of the skeletal muscle ryanodine receptor was cloned and sequenced (Takeshima et al. 1989). Ca^{2+} and ATP fully activate this channel, whereas Mg^{2+} is inhibitory (Smith et al. 1986). A protein isolated from aorta and imbedded in a lipid bilayer shows characteristics of the skeletal muscle ryanodine receptor. These include the formation of a Ca^{2+}-conducting channel that is activated by both Ca^{2+} and caffeine and inhibited by ruthenium red. Ryanodine opens the channel at submicromolar concentrations but closes the channel at higher concentrations (Herrmann-Frank et al. 1991). Several isoforms of the ryanodine receptor have been identified in mammalian cells (Ledbetter et al. 1994). In smooth muscles RyR3 is expressed, as opposed to RyR1 in skeletal and RyR2 in cardiac muscles (Giannini et al. 1995).

Activation of the ryanodine calcium release channel in the SR adjacent to the sarcolemma has been shown to cause short-lived intracellular Ca^{2+} increases in small cerebral arteries. These localized increases, called Ca^{2+} *sparks*, may open Ca^{2+}-activated K^+ channels and hyperpolarize the cell. Thus, an increase in Ca^{2+} will result in relaxation rather than contraction (Nelson et al. 1995). These localized effects may be quite different from whole-cell effects, and arise because Ca^{2+} is poorly diffusible within the cell (Allbritton et al. 1992).

10.8. Caffeine effects on Ca^{2+} homeostasis

Rather controversial is the effect of caffeine on smooth muscle. Caffeine has a dual action, contracting and relaxing.

10.8.1. Contracting

Caffeine can release Ca^{2+} for contraction from internal stores in taenia caeca and aorta (Ahn et al. 1988) and in skinned mesenteric artery fibers (Kanmura et al. 1989). A part of the Ca^{2+} store can be released with either caffeine or $InsP_3$, whereas another part can only be released with $InsP_3$ (Endo et al. 1990). It was further suggested that caffeine acted through Ca^{2+}-induced Ca^{2+} release, and this Ca^{2+} could also be released by ryanodine (Endo et al. 1990; Iino 1990). Inhibition of Ca^{2+} release by ruthenium red was interpreted to mean that Ca^{2+}-induced Ca^{2+} release channels were present (Kanmura et al. 1989). However, the physiological role of Ca^{2+}-induced Ca^{2+} release in smooth muscle has been doubted in view of the low Ca^{2+} sensitivity observed (Iino 1989).

10.8.2. Relaxing

Caffeine relaxed the noradrenaline-induced contraction in the aorta and the carbachol-induced contraction in the taenia coli, and also caused a concentration-dependent increase in cAMP content in the aorta (Ahn et al. 1988). Furthermore, caffeine was found to have a relaxing effect in skinned uterine fibers (Savineau 1988). The relaxing action of caffeine in smooth muscles can partially be explained by its action on inhibiting cAMP breakdown. Caffeine may also inhibit the Ca^{2+} influx stimulated by high K^+, apparently by binding to a Ca^{2+} channel (Ahn et al. 1988; Martin et al. 1989).

Caffeine is a methylxanthine like theophylline. Both are thought of as phosphodiesterase inhibitors. However, they are more potent as antagonists at the two receptor subtypes, A_1 and A_2, of the adenosine receptor P_1 (Burnstock 1990). Since activation of A_1 and A_2 produces opposite effects on the adenylate cyclase, either an increase or decrease in cAMP may occur. This may explain the discrepancies in action of caffeine in different smooth muscle systems.

10.9. Ca^{2+} release by cyclic ADP-Ribose (cADPR)

Ca^{2+} can be released by cADPR, a metabolic product of β-NAD, possibly through the Ca^{2+}-induced Ca^{2+} release mechanism. Neither heparin (an inhibitor of $InsP_3$-induced Ca^{2+} release) nor depletion of the $InsP_3$-releasable Ca^{2+} pool interferes with Ca^{2+} release by this agent (Lee 1994). There appears to be no similar action on permeabilized vas deferens (Nixon et al. 1994), but in intestinal longitudinal muscle cADPR increased Ca^{2+}-induced Ca^{2+} release (Kuemmerle and Makhlouf 1995).

10.10. Ca^{2+} oscillations

Agonist-evoked rise in intracellular free Ca^{2+} characteristically displays oscillations. Periodicity varies, but is often between 5 and 60 seconds. The frequency of the oscillations varies with the type of cell, type of agonist, agonist concentration, and method of detection. Ca^{2+} oscillations are observed with activation of the phosphoinositide messenger system by photo-released $InsP_3$ or with activation of a G protein by injecting $GTP_\gamma S$. Several mechanisms have been suggested as being responsible for the oscillations (for a review see Stucki and Somogyi 1994). Since the frequency of oscillations is directly related to the agonist concentration, the concept evolved that the tissue response is a function of the frequency of oscillations rather than of the absolute Ca^{2+} concentration (Berridge et al. 1988). Although this is an attractive hypothesis, direct evidence for the frequency-related response has been slow to appear.

10.11. Protein phosphorylation by tyrosine kinase

The involvement of tyrosine kinase in smooth muscle contraction is under intense investigation. Though there is suggestive evidence that protein phosphorylation at a tyrosine residue causes increased contraction and increased free intracellular $[Ca^{2+}]$, the relative importance of changes in Ca^{2+} sensitivity versus changes in intracellular free $[Ca^{2+}]$ is controversial (Gould et al. 1995; Semenchuk and Di Salvo 1995).

11. Increase of cytoplasmic Ca^{2+} – plasma membrane

Ion channels were first described for sodium and potassium in nerve and characterized by their electrical properties (Hodgkin and Huxley 1952). Studies have recorded changes in membrane conductance following stimulation. Because extracellular Ca^{2+} is mandatory for *continued* contraction of smooth muscles, Ca^{2+} channels became a focus for investigation. The use of the voltage-clamp method and the development of the patch-clamp technique have greatly improved our understanding of these channels. (For a more detailed description of these electrical properties, see Chapter 3 in this volume.) This section will detail what is known about the structure and function of Ca^{2+} channels in smooth muscles, and concentrate on how these channels affect Ca^{2+} homeostasis.

Ion channels are ion-selective pores formed by transmembrane proteins. All ion channels have two features in common, a gate and a selectivity filter; in order to function, the gate must be opened. The selectivity filter determines which ion can pass. Open and closed forms of the channel do not account for all of the properties of a channel and so a third, ''inactivated'' form was proposed. In the inactivated form, the channel does not allow passage of ions but still has properties different from the closed form. The probability of a channel being in any given state, and transition from one state to any other, is dependent on such factors as the type of channel, membrane potential, and so forth. There are several types of Ca^{2+} channels, the best documented being voltage-operated. Others include receptor– and second-messenger–operated, capacitative, and possibly G-protein–gated channels. It is not clear at present whether several of these describe the same phenomenon. Channels may be activated and opened by more than one mechanism.

11.1. Voltage-operated Ca^{2+} channels

Voltage-operated Ca^{2+} channels have been found in all smooth muscles examined. They are the main site of Ca^{2+} entry into the smooth muscle cell and the primary source of the action potential. The current can also be carried by Ba^{2+} or Sr^{2+} (but not by Mg^{2+}), and can be suppressed by other inorganic ions such as La^{3+}, Co^{2+}, Mn^{2+}, and Ni^{2+}. Two extensive reviews of Ca^{2+} channels have been published (Godfraind 1994; McDonald et al. 1994). In smooth muscle, two types of Ca^{2+} channels have been described; an L-type and a T-type Ca^{2+} channel. The L-type Ca^{2+} channel inactivates slowly, is permeant more to Ba^{2+} than to Ca^{2+}, and is highly sensitive to dihydropyridine Ca^{2+} channel blockers or agonists. The T-type Ca^{2+} channel inactivates rapidly, is equally permeant to Ca^{2+} and Ba^{2+}, and is less sensitive to dihydropyridine Ca^{2+} channel blockers. Ca^{2+} channels with properties that differ from the L and T types have been described in other tissues but seem to have minimal importance for smooth muscle.

11.2. Structure of the L-Type Ca^{2+} channel

L-type Ca^{2+} channels in skeletal muscle consist of five subunits: α_1, α_2, β, γ, and δ. The α_1 subunit is the channel's main functional subunit; it forms the ion conducting pore and contains the receptor for the dihydropyridine Ca^{2+} channel blockers. cDNA clones of the α_1 subunit from aorta and lung are similar but not identical to the cardiac muscle α_1 subunit (Biel et al. 1990; Koch et al. 1990). The skeletal muscle α_1 subunit is derived from a separate gene but with a similar transmembrane topology. There are four repeating motifs, each with six transmembrane-spanning regions. Each motif has a positively charged membrane-spanning unit, thought to be a voltage sensor. Incorporation of the synthetic RNA into oocytes induces the formation of voltage-sensitive Ca^{2+} channels. The Ca^{2+} channel in smooth muscle has a mass of 242 kDa and shows a homology of 65% to that of the skeletal muscle (Biel et al. 1990). There are several isoforms of the α_1 subunit that, along with other subunits of the Ca^{2+} channel, are thought to be responsible for tissue specificity of the channel function. In smooth muscles there are at least two genes that produce proteins, homologous to the skeletal muscle β subunit (approximate 55 kDa). When incorporated into oocytes they potentiate the α_1 channel activity but confer different drug sensitivity and kinetics (Hullin et al. 1992). The α_2 subunit (140–155 kDa) is very similar (if not identical) to the skeletal muscle subunit, and is thought to be encoded by the $\alpha_2\delta$ gene. The δ subunit is a <28-kDa peptide linked by a disulfide bond to the α_2 subunit. Both are heavily glycosylated, and the molecular mass is variable. The γ subunit cDNA encodes for 222 amino acids and a molecular mass of 25 kDa when unglycosylated (for a review of molecular properties see Godfraind 1994 or McDonald et al. 1994).

Binding of agents to the dihydropyridine site may either impede or enhance Ca^{2+} passage through the channel and thereby affect Ca^{2+} homeostasis. Ca^{2+} channel activators and antagonists interact competitively at an identical high-affinity binding site, located close to the extracellular surface of the PM. In fact, Ca^{2+} channel blockers that do not enter the cell have access to this site (Kass et al. 1991). Changes in the receptor conformation change receptor affinities for the Ca^{2+} channel blocker. It appears that specific binding of dihydropyridine to its receptor occurs preferentially to the inactivated form instead of the closed or open forms. The K_d is significantly lower (binding affinity is greater) in the depolarized state (Ca^{2+} channels inactivated) than at resting membrane potential. When binding was measured in depolarized intact cells, there was good correlation of binding affinity with pharmacologic inhibition as described by the IC_{50} (Godfraind 1994).

In addition, smooth muscles have receptors for agents like phenylalkylamines (verapamil, bepridil) and benzothiazepines. While also inhibiting Ca^{2+} entry, these agents bind to distinct binding sites on the channel protein (for a review see Godfraind 1994). The sites interact allosterically with each other (McDonald et al. 1994). Comparative studies of dihydropyridines with diphenylpiperazines (cinnarizine and flunarizine) showed that blocking of Ca^{2+} influx differs between various blood vessels. Inhibition of Ca^{2+} entry correlated quantitatively with the blockade of smooth muscle contraction (Godfraind 1994).

Modifying Ca^{2+} homeostasis through control of Ca^{2+} channel activity has been used clinically to control blood pressure, as well as to prevent cerebral ischemia after intracranial hemorrhage, prevent premature labor, and prevent coronary ischemia.

11.3. Regulation of voltage-operated Ca^{2+} channels

What modulates the voltage-operated Ca^{2+} channel? Experiments to define such factors are not very satisfying because Ca^{2+} movement is frequently assessed indirectly. To prove that a specific mechanism modulates the voltage-operated effect, all other variables must be held constant. This may prove difficult, since many of these are interrelated and affect the membrane potential. The action of a modulating agent on the Ca^{2+} channels is frequently opposite its known physiologic effect on Ca^{2+} influx, as demonstrated in whole cells. Suggested modulators include biochemical reactions involving phosphorylation by protein kinases, dephosphorylation by phosphatases, direct G protein interaction with channel proteins, and direct hormonal effects.

cAMP has been shown to regulate L-type Ca^{2+} channels in heart and skeletal muscles. In skeletal muscle, increased levels of intracellular cAMP caused increased Ca^{2+} entry through voltage-dependent Ca^{2+} channels; the α_1 and the β subunits were substrates for cAMP-dependent protein kinase, and the rate of Ca^{2+} influx was proportional to the degree of phosphorylation up to a stoichiometry of 1 (Nunoki et al. 1989; for a review see McDonald et al. 1994). Most studies have failed to show a cAMP effect on smooth muscle L-type Ca^{2+} channels. The lack of effect of cAMP on these channels as compared to skeletal muscle channels may be related to differences in the α and/or β subunits. The α_1 subunit of the smooth muscle channel, though very similar to the one in skeletal muscle, is missing one of the six phosphorylatable sites (for a review see Godfraind 1994). When smooth muscle α_1 was expressed in oocytes there was no change on addition of cAMP. However, when the skeletal muscle β subunit was also expressed, cAMP increased the activity of the channel. It was concluded that differences in the β subunit account for the lack of responsiveness of vascular smooth muscle L channels to cAMP (Klöckner et al. 1992). However, cAMP seems to decrease Ca^{2+} currents in some vascular tissues but not in uterine smooth muscle cells (see Xiong and Sperelakis 1995). We conclude that the action of cAMP on L-type calcium channels is tissue-and species-specific.

The late-onset contractions produced by phorbol esters activating protein kinase C are partially dependent on extracellular Ca^{2+}. This Ca^{2+} may enter the cell through voltage-operated Ca^{2+} channels (for a review see McDonald et al. 1994 or Xiong and Sperelakis 1995). Whether these reactions occur *in vivo* and whether they play a physiological role in modifying channel function is unknown at present. Furthermore, a membrane receptor may directly modulate Ca^{2+} channel activity through a G protein without involvement of second messengers or protein kinases (for a review see Birnbaumer et al. 1990 or McDonald et al. 1994).

11.4. Receptor-operated Ca^{2+} channels

Receptor-operated Ca^{2+} channels are characterized by Ca^{2+} entry through the PM without previous membrane depolarization. It is often difficult to distinguish modulation of voltage-operated Ca^{2+} channels, release of internally stored Ca^{2+}, second-messenger–operated channels, and true receptor-operated channels. One of the assumed attributes that distinguish between receptor- and voltage-operated channels is that the former are not inhibited by organic or inorganic Ca^{2+} channel blockers. Inhibitor specificity, however, is dependent on concentration (Rico et al. 1990).

Little is known about receptor-operated Ca^{2+} channels in smooth muscles.

Indeed, it has been questioned whether there are true receptor-operated channels rather than receptor modification of voltage-operated channels. Only the purinergic$_2$ receptor-operated Ca^{2+} channel has been convincingly shown in smooth muscle cells. In the experiment, ATP activated a Ca^{2+} current in arterial smooth muscle cells at an extracellular site. The current was not inhibited by nifedipine, and a rise in intracellular Ca^{2+} was demonstrated (Benham and Tsien 1987).

Many agonists appear to have a dual action: causing a rise in intracellular Ca^{2+} released from internal stores, and opening of receptor-operated Ca^{2+} channels (Suzuki et al. 1990). Whether these represent two aspects of the same event or activation of different types of receptors by the same agonist is not yet clear. The dual action may also be a consequence of the generation of inositol phosphates, but this possibility has not been investigated. Second-messenger–operated channels must be opened from the inside of the PM. Although there is considerable speculation that this occurs, there is a paucity of evidence. It is still uncertain whether $InsP_4$ opens Ca^{2+} channels in the PM of smooth muscles.

The foregoing may be explained by "capacitative Ca^{2+} entry." This term, coined by Putney (1990), describes the inverse relation between the rate of Ca^{2+} entry through the PM and the amount of Ca^{2+} present in the intracellular store. Both, an as-yet-unidentified diffusible factor and a direct protein–protein interaction between the SR and the PM have been hypothesized to account for the observation. Though $InsP_3$ influences the magnitude of stored Ca^{2+}, it is not otherwise involved with this system (Putney 1990). Trp (transient receptor potential), a protein originally identified in *Drosophila*, has now been found in mammals and may be the channel through which calcium enters in response to emptying of the intracellular Ca^{2+} stores (Berridge 1995).

The still-unanswered question is whether the receptor-operated and the voltage-operated Ca^{2+} channels are two different structures or one and the same structure. The distinguishing characteristics between receptor modulation of voltage-operated channels and true receptor-operated channels are not at all clear. They could be one structure operated by different mechanisms under various conditions. Further experiments to distinguish capacitative entry, second-messenger opened channels, and receptor-operated channels are very difficult to design. Clearly, much more work is needed on this subject.

12. Perspective

From the foregoing it is obvious that Ca^{2+} homeostasis is maintained by a variety of mechanisms situated at several cellular locations. Many of these mechanisms are interacting and overlapping. Membranes that define compartments have mechanisms for the active transport of Ca^{2+} in one direction and a separate mechanism for the movement of Ca^{2+} in the other direction. Frequently there is more than one pathway in each direction, and several modulators may be present. One agonist has multiple effects – not all consistent – that achieve Ca^{2+} homeostasis, and the effects vary between different smooth muscles.

Not long ago, Ca^{2+} influx and efflux across the PM – and ATP-dependent Ca^{2+} uptake into the SR – seemed to require only the phosphoinositide cycle and G proteins to complete our picture of Ca^{2+} homeostasis. But the lines so carefully drawn are now blurring. We need to learn much more about G proteins, receptor-operated Ca^{2+} channels, and second-messenger–operated Ca^{2+} channels. We cannot focus on just one aspect of the problem, since such experiments do not rule out other alternatives. Our success in this endeavor will depend to a large extent on the refinement of present technology and the development of new technology. As to the basic physiological mechanisms that maintain Ca^{2+} homeostasis, nature has yet to reveal all secrets and it is still up to us to solve the remaining riddles.

Acknowledgment

The authors wish to express their gratitude to Mr. William M. Bowers for his expert editorial assistance.

References

Aaronson, P. I., and Benham, C. D. (1989). Alterations in $[Ca^{2+}]_i$ mediated by sodium–calcium exchange in smooth muscle cells isolated from the guinea-pig ureter. *J. Physiol.* 416: 1–18.

Abdel-Latif, A. A. (1986). Calcium-mobilizing receptors, polyphosphoinositides, and the generation of second messengers. *Pharmacol. Rev.* 38: 227–72.

Ahn, H. Y., Karaki, H., and Urakawa, N. (1988). Inhibitory effects of caffeine on contractions and calcium movement in vascular and intestinal smooth muscle. *Br. J. Pharmacol.* 93: 267–74.

Aickin, C. C., Brading, A. F., and Walmsley, D. (1987). An investigation of sodium-calcium exchange in the smooth muscle of guinea-pig ureter. *J. Physiol.* 391: 325–46.

Akerman, K. E. O., and Wikstrom, M. K. F. (1979). (Ca^{2+} + Mg^{2+})-stimulated ATPase activity of rabbit myometrium plasma membrane is blocked by oxytocin. *FEBS Lett.* 97: 283–5.

Allbritton, N. L., Meyer, T., and Stryer, L. (1992). Range of messenger action of calcium ion and inositol 1,4,5-trisphosphate. *Science* 258: 1812–5.

Anwer, K., Hovington, J. A., and Sanborn, B. M. (1989). Antagonism of contractants and relaxants at the level of intracellular calcium and phosphoinositide turnover in rat uterus. *Endocrinology* 124: 2995–3002.

Anwer, K., Hovington, J. A., and Sanborn, B. M. (1990). Involvement of protein kinase A in the regulation of intracellular free calcium and phosphoinositide turnover in rat myometrium. *Biol. Reprod.* 43: 851–9.

Baltensperger, K., Carafoli E., and Chiesi, M. (1988). The Ca^{2+}-pumping ATPase and the major substrates of the cGMP-dependent protein kinase in smooth muscle sarcolemma are distinct entities. *Eur. J. Biochem.* 172: 7–16.

Baudet, S., Hove-Madsen, L., and Bers, D. M. (1994). How to make and use calcium-specific mini-and micro-electrodes. *Methods Cell Biol.* 40: 93–113.

Benham, C. D., and Tsien, R. W. (1987). A novel receptor-operated Ca^{2+}-permeable channel activated by ATP in smooth muscle. *Nature* 328: 275–8.

Berridge, M. J. (1983). Rapid accumulation of inositol trisphosphate reveals that agonists hydrolyse polyphosphoinositides instead of phosphatidylinositol. *Biochem. J.* 212: 849–58.

Berridge, M. J. (1984). Inositol trisphosphate and diacylglycerol as second messengers. *Biochem. J.* 220: 345–60.

Berridge, M. J. (1993). Inositol trisphosphate and calcium signalling. *Nature* 361: 315–25.

Berridge, M. J. (1995). Capacitative calcium entry. *Biochem. J.* 312: 1–11.

Berridge, M. J., Cobbold, P. H., and Cuthbertson, K. S. R. (1988). Spatial and temporal aspects of cell signalling. *Phil. Trans. Roy. Soc. Lond. B* 320: 325–43.

Bers, D. M., Patton, C. W., and Nuccitelli, R. (1994). A practical guide to the preparation of Ca^{2+} buffers. *Methods Cell Biol.* 40: 3–29.

Best, L., Brooks, K. J., and Bolton, T. B. (1985). Relationship between stimulated inositol lipid hydrolysis and contractility in guinea-pig visceral longitudinal smooth muscle. *Biochem. Pharmacol.* 34: 2297–301.

Bian, J., Ghosh, T. K., Wang, J., and Gill, D. L. (1991). Identification of intracellular calcium pools. *J. Biol. Chem.* 266: 8801–6.

Biel, M., Ruth, P., Bosse, E., Hullin, R., Stuhmer, W., Flockerzi, V., and Hofmann, F. (1990). Primary structure and functional expression of a high voltage activated calcium channel from rabbit lung. *FEBS Letts.* 269: 409–12.

Bielkiewicz-Vollrath, B., Carpenter, J. R., Schulz, R., and Cook, D. A. (1987). Early production of 1,4,5-inositol trisphosphate and 1,3,4,5-inositol tetrakisphosphate by histamine and carbachol in ileal smooth muscle. *Mol. Pharmacol.* 31: 513–22.

Birnbaumer, L., Abramowitz, J., and Brown, A. M. (1990). Receptor-effector coupling by G proteins. *Biochim. Biophys. Acta* 1031: 163–224.

Birnbaumer, L., and Birnbaumer, M. (1995). Signal transduction by G proteins. *J. Recept. Signal Transduction Res.* 15: 213–52.

Bitar, K. N., Bradford, P. G., Putney, J. W., and Makhlouf, G. M. (1986). Stoichiometry of contraction and Ca^{2+} mobilization by inositol 1,4,5-trisphosphate in isolated gastric smooth muscle cells. *J. Biol. Chem.* 261: 16591–6.

Blatter, L. A., and Wier, W. G. (1990). Intracellular diffusion, binding, and compartmentalization of the fluorescent calcium indicators indo-1 and fura-2. *Biophys. J.* 58: 1491–9.

Blaustein, M. P., Ambesi, A., Bloch, R. J., Goldman, W. F., Juhaszova, M., and Lindenmayer, G. E. (1992). Regulation of vascular smooth muscle contractility: Roles of the sarcoplasmic reticulum (SR) and the sodium/calcium exchanger. *Jpn. J. Pharmacol.* 58: 107P–114P.

Bohr, D. F., Seidel, C., and Sobieski, J. (1969). Possible role of sodium-calcium pumps in tension development of vascular smooth muscle. *Microvasc. Res.* 1: 335–43.

Bond, M., Kitazawa, T., Somlyo, A. P., and Somlyo, A. V. (1984a). Release and recycling of calcium by the sarcoplasmic reticulum in guinea-pig portal vein smooth muscle. *J. Physiol.* 355: 677–95

Bond, M., Shuman, H., Somlyo, A. P., and Somlyo, A. V. (1984b). Total cytoplasmic calcium in relaxed and maximally contracted rabbit portal vein smooth muscle. *J. Physiol.* 357: 185–201.

Bouscarel, B., Augert, G., Taylor, S. J., and Exton, J. H. (1990). Alterations in vasopressin and angiotensin II receptors and responses during culture of rat liver cells. *Biochim. Biophys. Acta* 1055: 265–,272.

Broderick, R., and Broderick, K. A. (1990). Ultrastructure and calcium stores in the myometrium. In: *Uterine Function – Molecular and Cellular Aspects* (M. E. Carsten and J. D. Miller, eds.). New York: Plenum Press, pp. 1–33.

Burnstock, G. (1990). Purinergic mechanisms. *Ann. N. Y. Acad. Sci.* 603: 1–18.

Buryi, V., Morel, N., Salomone, S., Kerger, S., and Godfraind, T. (1995). Evidence for

a direct interaction of thapsigargin with voltage-dependent Ca^{2+} channel. *Naunyn-Schmiedeberg's Arch. Pharmacol.* 35: 40–5.
Carafoli, E., and Stauffer, T. (1994). The plasma membrane calcium pump: Functional domains, regulation of the activity, and tissue specificity of isoform expression. *J. Neurobiol.* 25: 312–24.
Carsten, M. E. (1969). Role of calcium binding by sarcoplasmic reticulum in the contraction and relaxation of uterine smooth muscle. *J. Gen Physiol.* 53: 414–26.
Carsten, M. E. (1974). Prostaglandins and oxytocin: their effects on uterine smooth muscle. *Prostaglandins* 5: 33–40.
Carsten, M. E., and Miller, J. D. (1977a). Purification and characterization of microsomal fractions from smooth muscle. In: *Excitation-Contraction Coupling in Smooth Muscle*, (R. Casteels, T. Godfraind, and J. C. Ruegg, eds.). Amsterdam: Elsevier/North Holland, pp. 155–63.
Carsten, M. E., and Miller, J. D. (1977b). Effects of prostaglandins and oxytocin on calcium release from a uterine microsomal fraction. *J. Biol. Chem.* 252: 1576–81.
Carsten, M. E., and Miller, J. D. (1980). Characterization of cell membrane and sarcoplasmic reticulum from bovine uterine smooth muscle. *Arch. Biochem. Biophys.* 204: 404–12.
Carsten, M. E., and Miller, J. D. (1984). Properties of a phosphorylated intermediate of the Ca,Mg-activated ATPase of microsomal vesicles from uterine smooth muscle. *Arch. Biochem. Biophys.* 232: 616–23.
Carsten, M. E., and Miller, J. D. (1985). Ca^{2+} release by inositol trisphosphate from Ca^{2+}-transporting microsomes derived from uterine sarcoplasmic reticulum. *Biochem. Biophys. Res. Commun.* 130: 1027–31.
Carsten, M. E., and Miller, J. D. (1990). Calcium control mechanisms in the myometrial cell and the role of the phosphoinositide cycle. In: *Uterine Function – Molecular and Cellular Aspects* (M. E. Carsten and J. D. Miller, eds.). New York: Plenum Press.
Cassidy, P., Hoar, P. E., and Kerrick, W. G. L. (1979). Irreversible thiophosphorylation and activation of tension in functionally skinned rabbit ileum strips by [^{35}S]ATPgammaS. *J. Biol. Chem.* 254: 11148–53.
Chadwick, C. C., Saito, A., and Fleischer, S. (1990). Isolation and characterization of the inositol trisphosphate receptor from smooth muscle. *Proc. Nat. Acad. Sci. USA* 87: 2132–6.
Chiesi, M., Gasser, J., and Carafoli, E. (1984). Properties of the Ca-pumping ATPase of sarcoplasmic reticulum from vascular smooth muscle. *Biochem. Biophys. Res. Commun.* 124: 797–806.
Chiesi, M., Vorherr, T., Falchetto, R., Waelchli, C., and Carafoli, E. (1991). Phospholamban is related to the autoinhibitory domain of the plasma membrane Ca^{2+}-pumping ATPase. *Biochemistry* 30: 7978–83.
Chilvers, E. R., Lynch, B. J., and Chaliss, R. A. J. (1994). Phosphoinositide metabolism in airway smooth muscle. *Pharmac. Ther.* 62: 221–45.
Clarke, D. M., Loo, T. W., Rice, W. J., Andersen, J. P., Vilsen, B., and MacLennan, D. H. (1993). Functional consequences of alterations to hydrophobic amino acids located in the M_4 transmembrane sector of the Ca^{2+}-ATPase of sarcoplasmic reticulum. *J. Biol. Chem.* 268: 18359–64.
Cobbold, P. H., and Rink, T. J. (1987). Fluorescence and bioluminescence measurement of cytoplasmic free calcium. *Biochem. J.* 248: 313–28.
Dasso, L. L. T., and Taylor, C. W. (1991). Heparin and other polyanions uncouple α_1-adrenoceptors from G-proteins. *Biochem. J.* 280: 791–5.
Davis, F. B., Davis, P. J., Lawrence, W. D., and Blas, S. D. (1991). Specific inositol phosphates inhibit basal and calmodulin-stimulated Ca^{2+}-ATPase activity in hu-

man erythrocyte membranes in vitro and inhibit binding of calmodulin to membranes. *FASEB J.* 5: 2992–5.

Defeo, T. T., and Morgan, K. G. (1986). A comparison of two different indicators: quin 2 and aequorin in isolated single cells and intact strips of ferret portal vein. *Pflügers Arch.* 406: 427–9.

De Jaegere, S., Wuytack, F., Eggermont, J. A., Verboomen, H., and Casteels, R. (1990). Molecular cloning and sequencing of the plasma-membrane Ca^{2+} pump of pig smooth muscle. *Biochem. J.* 271: 655–60.

DeLisle, S., Radenberg, T., Wintermantel, M. R., Tietz, C., Parys, J. B., Pittet, D., Welsh, M. J., and Mayr, G. W. (1994). Second messenger specificity of the inositol trisphosphate receptor: reappraisal based on novel inositol phosphates. *Am. J. Physiol.* 266: C429–C436.

De Schutter, G., Wuytack, F., Verbist, J., and Casteels, R. (1984). Tissue levels and purification by affinity chromatography of the calmodulin-stimulated Ca^{2+}-transport ATPase in pig antrum smooth muscle. *Biochim. Biophys. Acta* 773: 1–10.

Devine, C. E., Somlyo, A. V., and Somlyo, A. P. (1972). Sarcoplasmic reticulum and excitation-contraction coupling in mammalian smooth muscles. *J. Cell Biol.* 52: 690–718.

Diccianni, M. B., Mistry, M. J., Hug, K., and Harmony, J. A. K. (1990). Inhibition of phospholipase A_2 by heparin. *Biochim. Biophys. Acta* 1046: 242–8.

Downes, C. P., Mussat, M. C., and Michell, R. H. (1982). The inositol trisphosphate phosphomonoesterase of the human erythrocyte membrane. *Biochem. J.* 203: 169–77.

Ebashi, S., and Lipmann, F. (1962). Adenosine triphosphate linked concentration of calcium ions in a particulate fraction of rabbit muscle. *J. Cell Biol.* 14: 389–400.

Edman, K. A. P., and Schild, H. O. (1963). Calcium and the stimulant and inhibitory effects of adrenaline in depolarized smooth muscle. *J. Physiol.* 169: 404–11.

Eggermont, J. A., Vrolix, M., Raeymaekers, L., Wuytack, F., and Casteels, R. (1988). Ca^{2+} transport ATPases of vascular smooth muscle. *Circ. Res.* 62: 266–78.

Eggermont, J. A., Wuytack, F., Verbist, J., and Casteels, R. (1990). Expression of endoplasmic-reticulum Ca^{2+}-pump isoforms and of phospholamban in pig smooth-muscle tissues. *Biochem. J.* 271: 649–53.

Ehrlich, B. E., and Watras, J. (1988). Inositol 1,4,5-trisphosphate activates a channel from smooth muscle sarcoplasmic reticulum. *Nature* 336: 583–6.

Endo, M., Iino, M., Kobayashi, T., and Yamamoto, T. (1990). Control of calcium release in smooth muscle cells. In: *Frontiers in Smooth Muscle Research* (Progress in Clinical and Biological research, vol. 327), (N. Sperelakis and J. D. Wood, eds.). New York: Wiley-Liss.

Enyedi, A., Brandt, J., Minami, J., and Penniston, J. T. (1989). Oxytocin regulates the plasma membrane Ca^{2+} transport in rat myometrium. *Biochem. J.* 261: 23–8.

Enyedi, A., Minami, J., Caride, A. J., and Penniston, J. T. (1988). Characteristics of the Ca^{2+} pump and Ca^{2+}-ATPase in the plasma membrane of rat myometrium. *Biochem. J.* 252: 215–20.

Fabiato, A. (1983). Calcium-induced release of calcium from the cardiac sarcoplasmic reticulum. *Am. J. Physiol.* 245: C1–C14.

Felbel, J., Trockur, B., Ecker, T., Landgraf, W., and Hofmann, F. (1988). Regulation of cytosolic calcium by cAMP and cGMP in freshly isolated smooth muscle cells from bovine trachea. *J. Biol. Chem.* 263: 16764–71.

Ferris, C. D., Cameron, A. M., Bredt, D. S., Huganir, R. L., and Snyder, S. H. (1992). Autophosphorylation of inositol 1,4,5-trisphosphate receptors. *J. Biol. Chem.* 267: 7036–41.

Ferris, C. D., Huganir, R. L., and Snyder, S. H. (1990). Calcium flux mediated by purified inositol 1,4,5-trisphosphate receptor in reconstituted lipid vesicles is allosterically regulated by adenine nucleotides. *Proc. Natl. Acad. Sci. USA* 87: 2147–51.

Filo, R. S., Bohr, D. F., and Ruegg, J. C. (1965). Glycerinated skeletal and smooth muscle: Calcium and magnesium dependence. *Science* 147: 1581–3.

Fujimori, T., and Jencks, W. P. (1990). Lanthanum inhibits steady-state turnover of the sarcoplasmic reticulum calcium ATPase by replacing magnesium as the catalytic ion. *J. Biol. Chem.* 265: 16262–70.

Fulle, H.-J., Hoer, D., Lache, W., Rosenthal, W., Schultz, G., and Oberdisse, E. (1987). In vitro synthesis of ^{32}P-labelled phosphatidylinositol 4,5-bisphosphate and its hydrolysis by smooth muscle membrane-bound phospholipase C. *Biochem. Biophys. Res. Commun.* 145: 673–9.

Furukawa, K.-I., and Nakamura, H. (1987). Cyclic GMP Regulation of the plasma membrane (Ca^{2+}-Mg^{2+})ATPase in vascular smooth muscle. *J. Biochem.* 101: 287–90.

Furukawa, K.-I., Tawada, Y., and Shigekawa, M. (1988). Regulation of the plasma membrane Ca^{2+} pump by cyclic nucleotides in cultured vascular smooth muscle cells. *J. Biol. Chem.* 263: 8058–65.

Gabella, G. (1971). Caveolae intracellulares and sarcoplasmic reticulum in smooth muscle. *J. Cell Sci.* 8: 602–9.

Ganz, M. B., Rasmussen, J., Bollag, W. B., and Rasmussen H. (1990). Effect of buffer systems and pH_i on the measurement of $[Ca^{2+}]_i$ with fura 2. *FASEB J.* 4: 1638–44.

Gaui, G., Gierschik, P., and Marme, D. (1988). Pertussis toxin inhibits angiotensin II-mediated phosphatidylinositol breakdown and ADP-ribosylates a 40 kDa protein in cultured smooth muscle cells. *Biochem. Biophys. Res. Commun.* 150: 841–7.

Ghosh, T. K., Eis, P. S., Mullaney, J. M., Ebert, C. L., and Gill, D. L. (1988). Competitive, reversible, and potent antagonism of inositol 1,4,5-trisphosphate-activated calcium release by heparin. *J. Biol. Chem.* 263: 11075–9.

Giannini, G., Conti, A., Mammarella, S., Scrobogna, M., and Sorrentino, V. (1995). The ryanodine receptor/calcium channel genes are widely and differentially expressed in murine brain and peripheral tissues. *J. Cell Biol.* 128: 893–904.

Gilchrist, J. S. C., Czubryt, M. P., and Pierce, G. N. (1994). Calcium and calcium-binding proteins in the nucleus. *Mol. Cell. Biochem.* 135: 79–88.

Gill, D. L., Ghosh, T. K., Bian, J., Short, A. D., Waldron, R. T., and Rybak, S. L. (1992). Function and organization of the inositol 1,4,5-trisphosphate-sensitive calcium pool. *Adv. Second Messenger Phosphoprotein Res.* 26: 265–308.

Godfraind, T. (1994). Calcium antagonists and vasodilatation. *Pharmac. Ther.* 64: 37–75.

Godfraind, T., Sturbois, X., and Verbeke, N. (1976). Calcium incorporation by smooth muscle microsomes. *Biochim. Biophys. Acta* 455: 254–68.

Gould, E. M., Rembold, C. M., and Murphy, R. A. (1995). Genistein, a tyrosine kinase inhibitor, reduces Ca^{2+} mobilization in swine carotid media. *Am. J. Physiol.* 268: C1425–C1429.

Goureau, O., Tanfin, Z., Marc, S., and Harbon, S. (1992). Diverse prostaglandin receptors activate distinct signal transduction pathways in rat myometrium. *Am. J. Physiol.* 263: C257–C265.

Grand, R. J. A., Perry, S. V., and Weeks, R. A. (1979). Troponin C-like proteins (calmodulins) from mammalian smooth muscle and other tissues. *Biochem. J.* 177: 521–9.

Grandordy, B. M., Cuss, F. M., Sampson, A. S., Palmer, J. B., and Barnes, P. J. (1986). Phosphatidylinositol response to cholinergic agonists in airway smooth muscle: Relationship to contraction and muscarinic receptor occupancy. *J. Pharmacol. Exp. Ther.* 238: 273–9.

Grover, A. K., Boonstra, I., Garfield, R. E., and Campbell, K. P. (1988). Ca pumps in rabbit stomach smooth muscle plasma membrane and endoplasmic reticulum. *Biochem. Arch.* 4: 169–79.

Grover, A. K., and Kwan, C. Y. (1987). Sodium-calcium exchange in rat myometrium. *Biochem. Arch.* 3: 353–62.

Grover, A. K., Kwan, C. Y., Rangachari, P. K., and Daniel, E. E. (1983). Na-Ca exchange in a smooth muscle plasma membrane-enriched fraction. *Am. J. Physiol.* 244: C158–C165.

Grynkiewicz, G., Poenie, M., and Tsien, R. Y. (1985). A new generation of Ca^{2+} indicators with greatly improved fluorescence properties. *J. Biochem. Chem.* 260: 3440–50.

Guillemette, G., Balla, T., Baukal, A. J., Spat, A., and Catt, K. J. (1987). Intracellular receptors for inositol 1,4,5-trisphosphate in angiotensin II target tissues. *J. Biol. Chem.* 262: 1010–15.

Hajnoczky, G., Lin, C., and Thomas, A. P. (1994). Luminal communication between intracellular calcium stores modulated by GTP and the cytoskeleton. *J. Biol. Chem.* 269: 10280–7.

Hall, I. P., Donaldson, J., and Hill, S. J. (1990). Modulation of carbachol-induced inositol phosphate formation in bovine tracheal smooth muscle by cyclic AMP phosphodiesterase inhibitors. *Biochem. Pharmacol.* 39: 1357–63.

Hasselbach, W., and Makinose, M. (1961). Die Calciumpumpe der ''Erschlaffungs-grana'' des Muskels und ihre Abhangigkeit von der ATP-Spaltung. *Biochem. Z.* 333: 518–28.

Hayashi, H., and Miyata, H. (1994). Fluorescence imaging of intracellular Ca^{2+}. *J. Pharmacol. Toxicol. Methods* 31: 1–10.

Herrmann-Frank, A., Darling, E., and Meissner, G. (1991). Functional characterization of the Ca^{2+}-gated Ca^{2+} release channel of vascular smooth muscle sarcoplasmic reticulum. *Pflügers Arch.* 418: 353–9.

Herscher, C. J., Rega, A. F., and Garrhan, P. J. (1994). The dephosphorylation reaction of the Ca^{2+}-ATPase from plasma membranes. *J. Biol. Chem.* 269: 10400–6.

Hilfiker, H., Strehler-Page, M. A., Stauffer, T. P., Carafoli, E., and Strehler, E. E. (1993). Structure of the gene encoding the human plasma membrane calcium pump isoform 1. *J. Biol. Chem.* 268: 19717–25.

Hill, A. V. (1948). On the times required for diffusion and its relation to processes in muscle. *Proc. Roy. Soc. London B* 135: 446–53.

Himpens, B., De Smedt, H., and Casteels, R. (1994). Subcellular Ca^{2+}-gradients in A735 vascular smooth muscle. *Cell Calcium* 15: 55–65.

Hirata, M., Itoh, T., and Kuriyama, H. (1981). Effects of external cations on calcium efflux from single cells of the guinea-pig taenia coli and porcine coronary artery. *J. Physiol.* 310: 321–36.

Hirata, M., Kohse, K. P., Chang, C.-H., Ikebe, T., and Murad, F. (1990b). Mechanism of cyclic GMP inhibition of inositol phosphate formation in rat aorta segments and cultured bovine aortic smooth muscle cells. *J. Biol. Chem.* 265: 1268–73.

Hirata, M., Watanabe, Y., Ishimatsu, T., Kidebe, T., Kimura, Y., Yamaguchi, K., Ozaki, S., and Koga, T. (1989). Synthetic inositol trisphosphate analogs and their effects on phosphatase, kinase, and the release of Ca^{2+}. *J. Biol. Chem.* 264: 20303–8.

Hirata, M., Yanaga, F., Koga, T., Ogasawara, T., Watanabe, Y., and Ozaki, S. (1990a). Stereospecific recognition of inositol 1,4,5-trisphosphate analogs by the phosphatase, kinase, and binding proteins. *J. Biol. Chem.* 265: 8404–7.

Hodgkin, A. L., and Huxley, A. F. (1952). A quantitative description of membrane current and its application to conduction and excitation in nerve. *J. Physiol.* 117: 500–44.

Howe, P. H., Akhtar, R. A., Naderi, S., and Abdel-Latif, A. A. (1986). Correlative studies on the effect of carbachol on myo-inositol trisphosphate accumulation, myosin light chain phosphorylation and contraction in sphincter smooth muscle of rabbit iris. *J. Pharmacol. Exp. Ther.* 239: 574–83.

Hullin, R., Singer-Lahat, D., Freichel, M., Biel, M., Dascal, N., Hofmann, F., and Flockerzi, V. (1992). Calcium channel β subunit heterogeneity: functional expression of cloned cDNA from heart and aorta and brain. *EMBO* 11: 885–90.

Ignarro, L. J. (1989). Endothelium-derived nitric oxide: Actions and properties. *FASEB J.* 3: 31–6.

Iino, M. (1989). Calcium-induced calcium release mechanism in guinea pig taenia caeci. *J. Gen. Physiol.* 94: 363–83.

Iino, M. (1990). Biphasic Ca^{2+} dependence of inositol 1,4,5-trisphosphate-induced Ca release in smooth muscle cells of the guinea pig taenia caeci. *J. Gen. Physiol.* 95: 1103–22.

Iino, M. (1991). Effects of adenine nucleotides in inositol 1,4,5-trisphosphate-induced calcium release in vascular smooth muscle cells. *J. Gen. Physiol.* 98: 681–98.

Imai, S., Yoshida, Y., and Sun, H.-T. (1990). Sarcolemmal (Ca^{2+} + Mg^{2+})-ATPase of vascular smooth muscle and the effects of protein kinases thereupon. *J. Biochem.* 107: 755–61.

Inesi, G. (1985). Mechanism of calcium transport. *Annu. Rev. Physiol.* 47: 573–601.

Inesi, G., and Kirtley, M. R. (1992). Structural features of cation transport ATPases. *J. Bioenerg. Biomemb.* 24: 271–83.

Inesi, G., Sumbilla C., and Kirtley M. E. (1990): Relationships of molecular structure and function in Ca^{2+}-transport ATPase. *Physiol. Rev.* 70: 749–60.

Irvine, R. F. (1992). Inositol phosphates and Ca^{2+} entry: Toward a proliferation or a simplification? *FASEB J.* 6: 3085–91.

Irvine, R. F., Brown, K. D., and Berridge, M. J. (1984). Specificity of inositol trisphosphate-induced calcium release from permeabilized swiss-mouse 3T3 cells. *Biochem. J.* 221: 269–72.

Irvine, R. F., Letcher, A. J., Heslop, J. P., and Berridge, M. J. (1986). The inositol tris/tetrakisphosphate pathway – Demonstration of Ins(1,4,5)P_3 3-kinase activity in animal tissues. *Nature* 320: 531–4.

Johnson, E. M., Theler, J. M., Capponi, A. M., and Vallotton, M. B. (1991). Characterization of oscillations in cytosolic free Ca^{2+} concentration and measurement of cytosolic Na^+ concentration changes evoked by angiotensin-II and vasopressin in individual rat aortic smooth muscle cells. *J. Biol. Chem.* 266: 12618–26.

Juhaszova, M., Ambesi, A., Lindenmayer, G. E., Bloch, R. J., and Blaustein, M. P. (1994). Na^+-Ca^{2+} exchanger in arteries: identification by immunoblotting and immunofluorescence microscopy. *Am. J. Physiol.* 266: C234–C242.

Kanmura, Y., Missiaen, L., Raeymaekers, L., and Casteels, R. (1988). Ryanodine reduces the amount of calcium in intracellular stores of smooth-muscle cells of the rabbit ear artery. *Pflügers Arch.* 413: 153–9.

Kanmura, Y., Raeymaekers, L., and Casteels, R. (1989). Effects of doxorubicin and ruthenium red on intracellular Ca^{2+} stores in skinned rabbit mesenteric smooth-muscle fibers. *Cell Calcium* 10: 433–9.

Kargacin, G. J., and Fay, F. S. (1987). Physiological and structural properties of saponin-skinned single smooth muscle cells. *J. Gen. Physiol.* 90: 49–73.

Kargacin, M. E., Scheid, C. R., and Honeyman, T. W. (1988). Continuous monitoring of Ca^{2+} uptake in membrane vesicles with fura-2. *Am. J. Physiol.* 245: C694–C698.

Kass, R. S., Arena, J. P., and Chin, S. (1991). Block of L-type calcium channels by charged dihydropyridines. Sensitivity to side of application and calcium. *J. Gen. Physiol.* 98: 63–75.

Keeton, T. P., Bork, S. E., and Shull, G. E. (1993). Alternative splicing of exons encoding the calmodulin-binding domains and C termini of plasma membrane Ca^{2+}-ATPase isoforms 1, 2, 3, and 4. *J. Biol. Chem.* 268: 2740–8.

Khan, A. M., Allen, J. C., and Shelat, H. (1988). Na^{+}-Ca^{2+} exchange in sarcolemmal vesicles from bovine superior mesenteric artery. *Am. J. Physiol.* 254: C441–C449.

Kitazawa, T., Kobayashi, S., Horiuti, K., Somlyo, A. V., and Somlyo, A. P. (1989). Receptor-coupled, permeabilized smooth muscle. *J. Biol. Chem.* 264: 5339–42.

Klöckner, U., Itagaki, K., Bodi, I., and Schwartz, A. (1992). β-Subunit expression is required for cAMP-dependent increase of cloned cardiac and vascular calcium channel currents. *Pflügers Arch.* 420: 413–5.

Knaus, H. G., Moshammer, T., Friedrich, K., Kang, H. C., Haugland, R. P., and Glossman, H. (1992). *In vivo* labeling of L-type Ca^{2+} channels by fluorescent dihydropyridines: Evidence for a functional, extracellular heparin-binding site. *Proc. Nat. Acad. Sci. USA* 89: 3586–90.

Kobayashi, S., Kitazawa, T., Somlyo, A. V., and Somlyo, A. P. (1989). Cytosolic heparin inhibits muscarinic and α-adrenergic Ca^{2+} release in smooth muscle. *J. Biol. Chem.* 264: 17997–18004.

Koch, W. J., Ellinor, P., and Schwartz, A. (1990). cDNA cloning of a dihydropyridine-sensitive calcium channel from rat aorta. *J. Biol. Chem.* 265: 17786–91.

Kowarski, D., Shuman, H., Somlyo, A. P., and Somylo, A. V. (1985). Calcium release by noradrenaline from central sarcoplasmic reticulum in rabbit main pulmonary artery smooth muscle. *J. Physiol.* 366: 153–75.

Kuemmerle, J. F., and Makhlouf, G. M. (1995). Agonist-stimulated cyclic ADP ribose. *J. Biol. Chem.* 270: 25488–94.

Kuemmerle, J. F., Murthy, K. S., and Mahklouf, G. M. (1994). Agonist-activated, ryanodine-sensitive, IP_3-insensitive Ca^{2+} release channels in longitudinal muscle of intestine. *Am. J. Physiol.* 266: C1421–C1431.

Kuo, T. H., and Tsang, W. (1988). Guanine nucleotide-and inositol trisphosphate-induced inhibition of the Ca^{2+} pump in rat heart sarcolemmal vesicles. *Biochem. Biophys. Res. Commun.* 152: 1111–16.

Kwan, C. Y., Kostka, P., Grover, A. K., Law, J. S., and Daniel, E. E. (1986). Calmodulin stimulation of plasmalemmal Ca^{2+}-pump of canine aortic smooth muscle. *Blood Vessels* 23: 22–33.

Kwan, C. Y., and Lee, R. M. K. W. (1984). Interaction of saponin with microsomal membrane fraction isolated from the smooth muscle of rat vas deferens. *Mol. Physiol.* 5: 105–14.

LaBelle, E. F., and Murray, B. M. (1990). G protein control of inositol lipids in intact vascular smooth muscle. *FEBS Letts.* 268: 91–4.

Ledbetter, M. W., Preiner, J. K., Louis, C. F., and Mickelson, J. R. (1994). Tissue distribution of ryanodine receptor isoforms and alleles determined by reverse transcription polymerase chain reaction. *J. Biol. Chem.* 269: 31544–51.

Lee, H. C. (1994). Cyclic ADP-ribose: A new member of a super family of signalling cyclic nucleotides. *Cell. Signal.* 6: 591–600.

Leijten, P., Cauvin, C., Lodge, N., Saida, K., and Van Breemen, C. (1985). Ca^{2+} sources mobilized by α_1-receptor activation in vascular smooth muscle. *Clin. Sci.* 68 (suppl. 10): 47s–50s.

Lompre, A.-M., de la Bastie, D., Boheler, K. R., and Schwartz, K. (1989). Characterization and expression of the rat heart sarcoplasmic reticulum Ca^{2+}-ATPase mRNA. *FEBS Letts.* 249: 35–41.

Lucchesi, P. A., and Scheid, C. R. (1988). Effects of the anti-calmodulin drugs calmidazolium and trifluoperazine on ^{45}Ca transport in plasmalemmal vesicles from gastric smooth muscle. *Cell Calcium* 9: 87–94.

Lytton, J., Westlin, M., Burk, S. E., Shull, G. E., and MacLennan, D. H. (1992). Functional comparisons between isoforms of the sarcoplasmic or endoplasmic reticulum family of calcium pumps. *J. Biol. Chem.* 267: 14483–9.

Lytton, J., Westlin, M., and Hanley, M. R. (1991). Thapsigarcin inhibits the sarcoplasmic or endoplasmic reticulum Ca-ATPase family of calcium pumps. *J. Biol. Chem.* 266: 17067–71.

Lytton, J., Zarain-Herzberg, A., Periasamy, M., and MacLennan, D. H. (1989). Molecular cloning of the mammalian smooth muscle sarco(endo)plasmic reticulum Ca^{2+}-ATPase. *J. Biol. Chem.* 264: 7059–65.

Madison, J. M., and Brown, J. K. (1988). Differential inhibitory effects of forskolin, isoproterenol, and dibutyryl cyclic adenosine monophosphate on phosphoinositide hydrolysis in canine tracheal smooth muscle. *J. Clin. Invest.* 82: 1462–5.

Marc, S., Leiber, D., and Harbon, S. (1988). Fluoroaluminates mimic muscarinic- and oxytocin-receptor-mediated generation of inositol phosphates and contraction in the guinea pig myometrium. *Biochem. J.* 255: 705–13.

Marks, A. R., Tempst, P., Chadwick, C. C., Riviere, L., Fleischer, S., and Nadal-Ginard, B. (1990). Smooth muscle and brain inositol 1,4,5-trisphosphate receptors are structurally and functionally similar. *J. Biol. Chem.* 265: 20719–22.

Martin, C., Dacquet, C., Mironneau, C., and Mironneau, J. (1989). Caffeine-induced inhibition of calcium channel current in cultured smooth muscle cells from pregnant rat myometrium. *Br. J. Pharmacol.* 98: 493–8.

Martin, T. F. J. (1991). Receptor regulation of phosphoinositidase C. *Pharmac. Ther.* 49: 329–45.

Matlib, M. A. (1988). Na^{+}-Ca^{2+} exchange in sarcolemmal membrane vesicles of dog mesenteric artery. *Am. J. Physiol.* 255: C323–C330.

Matlib, M. A., Crankshaw, J., Garfield, R. E., Crankshaw, D. J., Kwan, C.-Y., Branda, L. A., and Daniel, E. E. (1979). Characterization of membrane fractions and isolation of purified plasma membranes from rat myometrium. *J. Biol. Chem.* 254: 1834–40.

Matlib, M. A., and Reeves, J. P. (1987). Solubilization and reconstitution of the sarcolemmal Na^{+}-Ca^{2+} exchange system of vascular smooth muscle. *Biochim. Biophys. Acta* 904: 145–8.

Mayrleitner, M., Chadwick, C. C., Timerman, A. P., Fleischer, S., and Schindler, H. (1991). Purified IP_3 receptor from smooth muscle forms an IP_3 gated and heparin sensitive Ca^{2+} channel in planar bilayers. *Cell Calcium* 12: 506–14.

McDonald, T. F., Pelzer, S., Trautwein, W., and Pelzer, D. J. (1994). Regulation and modulation of calcium channels in cardiac, skeletal, and smooth muscle cells. *Physiol. Rev.*, 74: 365–507.

Mikoshiba, K., Furuichi, T., and Miyawaki, A. (1994). Structure and function of IP_3 receptors. *Sem. Cell Biol.* 5: 273–81.

Miller, J. D., and Carsten, M. E. (1984). Magnesium binding to uterine microsomes and the effect of calcium uptake. *INSERM* 124: 265–72.

Miller, J. D., and Carsten, M. E. (1987). The binding of Ca^{2+} to Ca^{2+}-transporting microsomes derived from bovine uterine sarcoplasmic reticulum. *Biochem. Biophys. Res. Commun.* 147: 13–17.

Milner, R. E., Baksh, S., Shemanko, C., Carpenter, M. R., Smillie, L., Vance, J. E., Opas, M., and Michalak, M. (1991). Calreticulin, and not calsequestrin, is the major calcium binding protein of smooth muscle sarcoplasmic reticulum and liver endoplasmic reticulum. *J. Biol. Chem.* 266: 7155–65.

Minta, A., Kao, J. P. Y., and Tsien, R. Y. (1989). Fluorescent indicators for cytosolic calcium based on rhodamine and fluorescein chromophores. *J. Biol. Chem.* 264: 8171–8.

Missiaen, L., De Smedt, H., Droogman, G., Declerck I., Plessers, L., and Casteels, R. (1991). Uptake characteristics of the $InsP_3$-sensitive and-insensitive Ca^{2+} pools in porcine aortic smooth-muscle cells: Different Ca^{2+} sensitivity of the Ca^{2+}-uptake mechanism. *Biochem. Biophys. Res. Commun.* 174: 1183–8.

Missiaen, L., Kanmura, Y., Wuytack, F., and Casteels, R. (1988). Carbachol partially inhibits the plasma-membrane Ca^{2+}-pump in microsomes from pig stomach smooth muscle. *Biochem. Biophys. Res. Commun.* 150: 681–6.

Missiaen L., Raeymaekers, L., Wuytack, F., Vrolix, M., De Smedt, H., and Casteels, R. (1989). Phospholipid-protein interactions of the plasma-membrane Ca^{2+}-transporting ATPase. *Biochem. J.* 263: 687– 94.

Mitsui, M., Abe, A., Tajimi, M., and Karaki, H. (1993). Leakage of the fluorescent Ca^{2+} indicator Fura-2 in smooth muscle. *Jpn. J. Pharmacol.* 61: 165–70.

Molleman, A., Hoiting, B., Duin, M., van den Akker, J., Nelemans, A., and Den Hertog, A. (1991). Potassium channels regulated by inositol 1,3,4,5-tetrakisphosphate and internal calcium in DDT_1 MF-2 smooth muscle cells. *J. Biol. Chem.* 266: 5658–63.

Molnar, M., and Hertelendy, F. (1990). Regulation of intracellular free calcium in human myometrial cells by prostaglandin $F_{2\alpha}$: Comparison with oxytocin. *J. Clin. Endocrinol. Metab.* 71: 1243–50.

Molnar, M., and Hertelendy, F. (1992). Platelet-activating factor activates the phosphoinositide cycle and promotes Ca^{2+} mobilization in human myometrial cells: Comparison with $PGF_{2\alpha}$. *J. Mat.-Fet. Med.* 1: 1–6.

Monteith, G. R., and Roufogalis, B. D. (1995). The plasma membrane calcium pump – a physiological perspective on its regulation. *Cell Calcium* 18: 459–70.

Moore, E. D. W., Becker, P. L., Fogarty, K. E., Williams, D. A., and Fay, F. S. (1990). Ca^{2+} imaging in single living cells: Theoretical and practical issues. *Cell Calcium* 11: 157–79.

Moore, E. D. W., Etter, E. F., Philipson, K. D., Carrington, W. A., Fogarty, K. E., Lifshitz, L. M., and Fay, F. S. (1993). Coupling of the Na^+/Ca^{2+} exchanger, Na^+/K^+ pump and sarcoplasmic reticulum in smooth muscle. *Nature* 365: 657–60.

Morel, N., and Godfraind, T. (1984). Sodium/calcium exchange in smooth muscle microsomal fractions. *Biochem. J.* 218: 421–7.

Morel, N., Wibo, M., and Godfraind, T. (1981). A calmodulin-stimulated Ca^{2+} pump in rat aorta plasma membranes. *Biochim. Biophys. Acta* 644: 82–8.

Mourey, R. J., Verma, A., Supattapone, S., and Snyder, S. H. (1990). Purification and characterization of the inositol 1,4,5-trisphosphate receptor protein from rat vas deferens. *Biochem. J.* 272: 383–9.

Moutin, M.-J., and Dupont, Y. (1991). Interaction of potassium and magnesium with the high affinity calcium-binding sites of the sarcoplasmic reticulum calcium-ATPase. *J. Biol. Chem.* 266: 5580–6.

Murthy, K. S., Grider, J. R., and Makhlouf, G. M. (1992). Receptor-coupled G proteins mediate contraction and Ca^{++} mobilization in isolated intestinal muscle cells. *J. Pharmacol. Exp. Ther.* 260: 90–7.

Murthy, K. S., and Makhlouf, G. M. (1994). Fluoride activates G protein-dependent and -independent pathways in dispersed intestinal smooth muscle cells. *Biochem. Biophys. Res. Commun.* 202: 1681–7.

Murthy, K. S., and Makhlouf, G. M. (1995). Functional characterization of phosphoinositide-specific phospholipase C-β1 and -β3 in intestinal smooth muscle. *Am. J. Physiol.* 269: C969–C978.

Myung, J., and Jencks, W. P. (1994). Lumenal and cytoplasmic sites for calcium on the calcium ATPase of sarcoplasmic reticulum are different and independent. *Biochemistry* 33: 8775–85.

Myung, J., and Jencks, W. P. (1995). There is only one phosphoenzyme intermediate with bound calcium on the reaction pathway of the sarcoplasmic reticulum calcium ATPase. *Biochemistry* 34: 3077–83.

Nakasaki, Y., Iwamoto, T., Hanada, H., Imagawa, T., and Shigekawa, M. (1993). Cloning of the rat aortic smooth muscle Na^+/Ca^{2+} exchanger and tissue-specific expression of isoforms. *J. Biochem.* 114: 528–34.

Nelemans, A., Hoiting, B., Molleman, A., Duin, M., and Den Hertog, A. (1990). α-Adrenoceptor regulation of inositol phosphates, internal calcium and membrane current in DDT_1MF-2 smooth muscle cells. *Eur. J. Pharmacol.* 189: 41–9.

Nelson, M. T., Cheng, H., Rubart, M., Santana, L. F., Bonev, A. D., Khnot, H. J., and Lederer, W. J. (1995). Relaxation of arterial smooth muscle by calcium sparks. *Science* 270: 633–7.

Nixon, G. F., Mignery, G. A., and Somlyo, A. V. (1994). Immunogold localization of inositol 1,4,5-trisphosphate receptors in phasic and tonic smooth muscle. *J. Muscle Res. Cell Motil.* 15: 682–700.

Nunoki, K., Florio, V., and Catterall, W. A. (1989). Activation of purified calcium channels by stoichiometric protein phosphorylation. *Proc. Nat. Acad. Sci. USA* 86: 6816–20.

Potter, B. V. L., and Nahorski, S. R. (1992). Synthesis and biology of inositol polyphosphate analogues. *Biochem. Soc. Trans.* 20: 434–42.

Prentki, M., Deeney, J. T., Matschinsky, F. M., and Joseph, S. K. (1986). Neomycin: A specific drug to study the inositol-phospholipid signalling system? *FEBS Lett.* 197: 285–8.

Prestwich, S. A., and Bolton, T. B. (1995). G-protein involvement in muscarinic receptor stimulation of inositol phosphates in longitudinal smooth muscle from the small intestine of the guinea-pig. *Br. J. Pharmacol.* 114: 119–26.

Putney, J. W., Jr. (1990). Capacitative calcium entry revisited. *Cell Calcium* 11: 611–24.

Raeymaekers, L., Agostini, B., and Hasselbach, W. (1981). The formation of intravesicular calcium phosphate deposits in microsomes of smooth muscle. *Histochemistry* 70: 139–50.

Raeymaekers, L., Eggermont, J. A., Wuytack, F., and Casteels, R. (1990). Effects of cyclic nucleotide dependent protein kinases on the endoplasmic reticulum Ca^{2+} pump of bovine pulmonary artery. *Cell Calcium* 11: 261–8.

Raeymaekers, L., and Hasselbach, W. (1981). Ca^{2+} uptake, Ca^{2+}-ATPase activity, phosphoprotein formation and phosphate turnover in a microsomal fraction of smooth muscle. *Eur. J. Biochem.* 116: 373–8.

Raeymaekers, L., Wuytack, F., Eggermont, J., De Schutter, G., and Casteels, R. (1983). Isolation of a plasma-membrane fraction from gastric smooth muscle. Comparison of the calcium uptake with that in endoplasmic reticulum. *Biochem. J.* 210: 315–22.

Reuter, H., and Seitz, N. (1968). The dependence of calcium efflux from cardiac muscle on temperature and external ion composition. *J. Physiol.* 195: 451–70.

Reynolds, E. E., and Dubyak, G. R. (1986). Agonist-induced calcium transients in cultured smooth muscle cells: Measurements with fura-2 loaded monolayers. *Biochem. Biophys. Res. Commun.* 136: 927–34.

Rhee, S. G., Kim, H., Suh, P.-G., and Choi, W. C. (1991). Multiple forms of phosphoinositide-specific phospholipase C and different modes of activation. *Biochem. Soc. Trans.* 19: 337–41.

Rico, I., Alonso, M. J., Salaices, M., and Marin, J. (1990). Pharmacological dissection of Ca^{2+} channels in the rat aorta by Ca^{2+} entry modulators. *Pharmacology* 40: 330–42.

Ringer, S. (1882). A further contribution regarding the influence of the different constituents of the blood on the contraction of the heart. *J. Physiol.* 4: 29–42.

Rizzuto, R., Bastianutto, C., Brini, M., Murgia, M., and Pozzan, T. (1994). Mitochondrial Ca^{2+} homeostasis in intact cells. *J. Cell Biol.* 126: 1183–94.

Ruzycky, A. L., and Crankshaw, D. J. (1988). Role of inositol phospholipid hydrolysis in the initiation of agonist-induced contractions of rat uterus: Effects of domination by 17β-estradiol and progesterone. *Can. J. Physiol. Pharmacol.* 66: 10–17.

Sagara, Y., Fernandez-Belda, F., Demeis, L., and Inesi, G. (1992). Characterization of the inhibition of intracellular Ca^{2+} transport ATPases by thapsigargin. *J. Biol. Chem.* 267: 12606–13.

Saida, K., and van Breemen, C. (1987). GTP requirement for inositol 1,4,5 trisphosphate-induced Ca^{2+} release from sarcoplasmic reticulum in smooth muscle. *Biochem. Biophys. Res. Commun.* 144: 1313–16.

Salmon, D. M. W., and Bolton, T. B. (1988). Early events in inositol phosphate metabolism in longitudinal smooth muscle from guinea-pig intestine stimulated with carbachol. *Biochem. J.* 254: 553–7.

Sasaguri, T., Hirata, M., and Kuriyama, H. (1985). Dependence on Ca^{2+} of the activities of phosphatidylinositol 4,5-bisphosphate phosphodiesterase and inositol 1,4,5-trisphosphate phosphatase in smooth muscles of the porcine coronary artery. *Biochem. J.* 231: 497–503.

Savineau, J. P. (1988). Caffeine does not contract skinned uterine fibers with a functional Ca store. *Eur. J. Pharmacol.* 149: 187–90.

Savineau, J. P., Mironneau, J., and Mironneau, C. (1988). Contractile properties of chemically skinned fibers from pregnant rat myometrium: Existence of an internal Ca store. *Pflügers Arch.* 411: 296–303.

Semenchuk, L. A., and Di Salvo, J. (1995). Receptor-activated increases in intracellular calcium and protein tyrosine phosphorylation in vascular smooth muscle cells. *FEBS Letts.* 370: 127–30.

Sharma, R. V., Butters, C. A., McEldoon, J. P., and Bhalla, R. C. (1987). Characterization of Ca^{2+} uptake in plasma membrane vesicles isolated from guinea pig ileum smooth muscle. *Cell Calcium* 8: 65–77.

Shashidhar, M. S., Volwerk, J. J., Keana, J. F. W., and Griffith, O. H. (1990). Inhibition of phosphatidylinositol-specific phospholipase C by phosphonate substrate analogues. *Biochim. Biophys. Acta* 1042: 410–12.

Shin, J. M., Kajimura, M., Arguello, J. M., Kaplan, J. H., and Sachs, G. (1994). Biochemical identification of transmembrane segments of the Ca^{2+}-ATPase of sarcoplasmic reticulum. *J. Biol. Chem.* 269: 22533–7.

Short, A. D., Klein, M. G., Schneider, M. F., and Gill, D. L. (1993). Inositol 1,4,5-trisphosphate-mediated quantal Ca^{2+} release measured by high resolution imaging of Ca^{2+} within organelles. *J. Biol. Chem.* 268: 25887–93.

Shull, G. E., and Greeb, J. (1988). Molecular cloning of two isoforms of the plasma membrane Ca^{2+}-transporting ATPase from rat brain. *J. Biol. Chem.* 263: 8646–57.

Shuttleworth, T. J., and Thompson, J. L. (1991). Effect of temperature on receptor-activated changes in $[Ca^{2+}]_i$ and their determination using fluorescent probes. *J. Biol. Chem.* 266: 1410–14.

Slaughter, R. S., Shevell, J. L., Felix, J. P., Garcia, M. L., and Kaczorowski, G. J. (1989). High levels of sodium-calcium exchange in vascular smooth muscle sarcolemmal membrane vesicles. *Biochemistry* 28: 3995–4002.

Slaughter, R. S., Welton, A. F., and Morgan, D. W. (1987). Sodium-calcium exchange in sarcolemmal vesicles from tracheal smooth muscle. *Biochim. Biophys. Acta* 904: 92–104.

Smith, J. B., Coronado, R., and Meissner, G. (1986). Single channel measurements of the calcium release channel from skeletal muscle sarcoplasmic reticulum. *J. Gen. Physiol.* 88: 573–88.

Smith, J. B., Dwyer, S. D., and Smith, L. (1989a). Decreasing extracellular Na^+ concentration triggers inositol polyphosphate production and Ca^{2+} mobilization. *J. Biol. Chem.* 264: 831–7.

Smith, J. B., and Smith, L. (1990). Energy dependence of sodium-calcium exchange in vascular smooth muscle cells. *Am. J. Physiol.* 259: C302–C309.

Smith, J. B., Smith, L., and Higgins, B. L. (1985). Temperature and nucleotide dependence of calcium release by myo-inositol 1,4,5-trisphosphate in cultured vascular smooth muscle cells. *J. Biol. Chem.* 260: 14413–16.

Smith, J. B., Zheng, T., and Smith, L. (1989b). Relationship between cytosolic free Ca and Na^+-Ca^{2+} exchange in aortic muscle cells. *Am. J. Physiol.* 256: C147–C154.

Socorro, L., Alexander, R. W., and Griendling, K. K. (1990). Cholera toxin modulation of angiotensin II-stimulated inositol phosphate production in cultured vascular smooth muscle cells. *Biochem. J.* 265: 799–807.

Soloff, S., and Sweet, P. (1982). Oxytocin inhibition of (Ca^{2+}+Mg^{2+})-ATPase activity in rat myometrial plasma membranes. *J. Biol. Chem.* 257: 10687–93.

Somlyo, A. P., and Somlyo, A. V. (1990). Flash photolysis studies of excitation-contraction coupling, regulation, and contraction in smooth muscle. *Annu. Rev. Physiol.* 52: 857–74.

Somlyo, A. V., Bond, M., Somlyo, A. P., and Scarpa, S. M. (1985). Inositol trisphosphate-induced calcium release and contraction in vascular smooth muscle. *Proc. Nat. Acad. Sci. USA* 82: 5231–5.

Somlyo, A. V., and Somlyo, A. P. (1968). Electromechanical and pharmacomechanical coupling in vascular smooth muscle. *J. Pharmacol. Exp. Ther.* 159: 129–45.

Stout, M. A., and Diecke, F. P. J. (1983). Ca distribution and transport in saponin skinned vascular smooth muscle. *J. Pharmacol. Exp. Ther.* 225: 102–11.

Stucki, J. W., and Somogyi, R. (1994). A dialogue on Ca^{2+} oscillations. *Biochim. Biophys. Acta* 1183: 453–72.

Sugiyama, T., and Goldman, W. F. (1995). Conversion between permeability states of IP_3 receptors in cultured smooth muscle cells. *Am. J. Physiol.* 269: C813–C818.

Sun, H.-T., Yoshida, Y., and Imai, S. (1990). A Ca^{2+}-activated, Mg^{2+}-dependent ATPase with high affinities for both Ca^{2+} and Mg^{2+} in vascular smooth muscle microsomes: Comparison with plasma membrane Ca^{2+}-pump ATPase. *J. Biochem.* 108: 730–6.

Suzuki, E., Tsujimoto, G., Tamura, K., and Hashimoto, K. (1990). Two pharmacologically distinct α_1-adrenoceptor subtypes in the contraction of rabbit aorta. *Mol. Pharmacol.* 38: 725–36.

Takeshima, H., Nishimura, S., Matsumoto, T., Ishida, H., Kangawa, K., Minamino, N., Matsuo, H., Ueda, M., Hanaoka, M., Hirose, T., and Numa, S. (1989). Primary structure and expression from complementary DNA of skeletal muscle ryanodine receptor. *Nature* 339: 439–45.

Takuwa, Y., Kasuya, Y., Takuwa, N., Kudo, M., Yanagisawa, M., Goto, K., Masaki,

T., and Yamashita, K. (1990). Endothelin receptor is coupled to phospholipase C via a pertussis toxin-insensitive guanine nucleotide-binding regulatory protein in vascular smooth muscle cells. *J. Clin. Invest.* 85: 653–8.

Taylor, C. W., Berridge, M. J., Cooke, A. M., and Potter, B. V. L. (1989). Inositol 1,4,5-trisphosphorothioate a stable analogue of inositol trisphosphate which mobilized intracellular calcium. *Biochem. J.* 259: 645–50.

Taylor, C. W., and Richardson, A. (1991). Structure and function of inositol trisphosphate receptors. *Pharmac. Ther.* 51: 97–137.

Thastrup, O., Dawson, A. P., Scharff, O., Foder, B., Cullen, P. J., Drobak, B. K., Bjerrum, P. J., Christensen, S. B., and Hanley, M. R. (1989). Thapsigargin, a novel molecular probe for studying intracellular calcium release and storage. *Agents and Actions* 27: 17–23.

Thomas, D., and Hanley, M. R. (1994). Pharmacological tool for perturbing intracellular calcium storage. *Methods Cell Biol.* 40: 65–89.

Verbist, J., Wuytack, F., Raeymaekers, L., and Casteels, R. (1985). Inhibitory antibodies to plasmalemmal Ca^{2+}-transporting ATPases. *Biochem. J.* 231: 737–42.

Verboomen, H., Wuytack, F., De Smedt, H., Himpens, B., and Casteels, R. (1992). Functional difference between SERCA2a and SERCA2b Ca^{2+} pumps and their modulation by phospholamban. *Biochem. J.* 286: 591–6.

Verma, A. K., Enyedi, A., Filoteo, A. G., and Penniston, J. T. (1994). Regulatory region of the plasma membrane Ca^{2+} pump. *J. Biol. Chem.* 269: 1687–91.

Verma, A. K., Filoteo, A. G., Stanford, D. R., Wieben, E. D., Penniston, J. T., Strehler, E. E., Fischer, R., Heim, R., Vogel, G., Mathews, S., Strehler-Page, M. A., James, P., Vorherr, T., Krebs, J., and Carafoli, E. (1988). Complete primary structure of a human plasma membrane Ca^{2+} pump. *J. Biol. Chem.* 263: 14152–9.

Vigne, P., Breitmayer, J.-P., Duval, D., Frelin, C., and Lazdunski, M. (1988). The Na^{+} /Ca^{2+} antiporter in aortic smooth muscle cells – Characterization and demonstration of an activation by phorbol esters. *J. Biol. Chem.* 263: 8078–83.

Vogel, H. J. (1988). Ligand-binding sites on calmodulin. In: *Calcium in Drug Action (Handbook of Experimental Pharmacology)*, (P. F. Baker, ed.). New York: Springer-Verlag, pp. 57–87.

Volpe, P., Martini, A., Furlan, S., and Meldolesi, J. (1994). Calsequestrin is a component of smooth muscles: The skeletal- and cardiac-muscle isoforms are both present, although in highly variable amounts and ratios. *Biochem. J.* 301: 465–9.

Vrolix, M., Raeymaekers, L., Wuytack, F., Hofmann, F., and Casteels, R. (1988). Cyclic GMP-dependent protein kinase stimulates the plasmalemmal Ca^{2+} pump of smooth muscle via phosphorylation of phosphatidylinositol. *Biochem. J.* 255: 855–63.

Wang, C.-L. A. (1985). A note on Ca^{2+} binding to calmodulin. *Biochem. Biophys. Res. Commun.* 130: 426–30.

Watras, J., Benevolensky, D., and Childs, C. (1989). Calcium release from aortic sarcoplasmic reticulum. *J. Mol. Cell. Cardiol.* 21 (suppl. 1): 125–30.

Watterson, D. M., Sharief, F., and Vanaman, T. C. (1980). The complete amino acid sequence of the Ca^{2+}-dependent modulator protein (calmodulin) of bovine brain. *J. Biol. Chem.* 255: 962–71.

Wibo, M., Morel, N., and Godfraind, T. (1981). Differentiation of Ca^{2+} pumps linked to plasma membrane and endoplasmic reticulum in the microsomal fraction from intestinal smooth muscle. *Biochim. Biophys. Acta* 649: 651–60.

Wictome, M., Henderson, I., Lee, A. G., and East, J. M. (1992). Mechanism of inhibition of the calcium pump of the sarcoplasmic reticulum by thapsigargin. *Biochem. J.* 283: 525–9.

Willcocks, A. L., Strupish, J., Irvine, R. F., and Nahorski, S. R. (1989). Inositol 1:2-

cyclic,4,5-trisphosphate is only a weak agonist at inositol 1,4,5-trisphosphate receptors. *Biochem. J.* 257: 297–300.

Worley, P. F., Baraban, J. M., Supattapone, S., Wilson, V. S., and Snyder, S. H. (1987). Characterization of inositol trisphosphate receptor binding in brain – regulation by pH and calcium. *J. Biol. Chem.* 262: 12132–6.

Wuytack, F., De Schutter, G., and Casteels, R. (1980). The effect of calmodulin on the active calcium-ion transport and (Ca^{2+} + Mg^{2+})-dependent ATPase in microsomal fractions of smooth muscle compared with that in erythrocytes and cardiac muscle. *Biochem. J.* 190: 827–31.

Wuytack, F., Kanmura, Y., Eggermont, J. A., Raeymaekers, L., Verbist, J., Hartweg, D., Gietzen, K., and Casteels, R. (1989). Smooth muscle expresses a cardiac/slow muscle isoform of the Ca^{2+}-transport ATPase in its endoplasmic reticulum. *Biochem. J.* 257: 117–23.

Wuytack, F., Raeymaekers, L., De Schutter, G., and Casteels, R. (1982). Demonstration of the phosphorylated intermediates of the Ca^{2+} transport ATPase in a microsomal fraction and in a (Ca^{2+} + Mg^{2+}) ATPase purified from smooth muscle by means of calmodulin affinity chromatography. *Biochim. Biophys. Acta* 693: 45–52.

Xiong, Z., and Sperelakis, N. (1995). Regulation of L-type calcium channels of vascular smooth muscle cells. *J. Mol. Cell. Cardiol.* 27: 75–91.

Xuan, Y.-T., Watkins, W. D., and Whorton, A. R. (1991). Regulation of endothelin-mediated calcium mobilization in vascular smooth muscle cells by isoproterenol. *Am. J. Physiol.* 260: C492–C502.

Yamaguchi, K., Hirata, M., and Kuriyama, H. (1988). Purification and characterization of inositol 1,4,5-trisphosphate 3-kinase from pig aortic smooth muscle. *Biochem. J.* 251: 129–34.

Yamamoto, Y., Hu, S. L., and Kao, C. Y. (1989). Inward current in single smooth muscle cells of the guinea pig taenia coli. *J. Gen. Physiol.* 93: 521–550.

Yoshida,Y., Sun, H.-T., Cai, J.-Q., and Imai, S. (1991). Cyclic GMP-dependent protein kinase stimulates the plasma membrane Ca^{2+} pump ATPase of vascular smooth muscle via phosphorylation of a 240-kDa protein. *J. Biol. Chem.* 266: 19819–25.

Yu, X., Hao, L., and Inesi, G. (1994). A pK change of acidic residues contributes to cation countertransport in the Ca-ATPase of sarcoplasmic reticulum. *J. Biol. Chem.* 269: 16656–61.

Zhang, L., Bradley, M. E., Khoyi, M., Westfall, D. P., and Buxton, I. L. O. (1993). Inositol 1,4,5-trisphosphate and inositol 1,3,4,5-tetrakisphosphate binding sites in smooth muscle. *Br. J. Pharmacol.* 109: 905–12.

Zhou, C. J., Akhtar, R. A., and Abdel-Latif, A. A. (1994). Identification of phosphoinositide-specific phospholipase C-β1 and GTP-binding protein, $G_{q\alpha}$, in bovine iris sphincter membranes. *Exp. Eye Res.* 59: 377–84.

3

Ionic Channel Functions in Some Visceral Smooth Myocytes

C. Y. KAO

1. Introduction

Smooth muscles subserve important physiological functions via the processes of membrane excitation, excitation–contraction coupling, and contraction. Until recently, their excitability properties have usually been understood only in broad generalities, because their small individual cells have been difficult to study with advanced electrophysiological methods. Much valuable information has been obtained through the use of impalement microelectrodes on cells enmeshed in their natural environment. However, the underlying ionic currents are poorly known, because applications of voltage-clamp methods to multicellular preparations are fraught with insuperable intrinsic obstacles. Successful isolation of viable individual smooth myocytes from amphibian tissues (Bagby et al. 1971; Fay and Delise 1973) and mammalian tissues (Momose and Gomi, 1980) opened the way to studying their cellular properties, first by use of impalement microelectrodes (Walsh and Singer 1980), and now more extensively by use of the tight-seal patch-clamp method (Hamill et al. 1981).

With rapid blossoming of investigations on cellular ionic channel functions of smooth myocytes, it is opportune to assess extant information in order to guide our perspectives. Because many new discoveries remain to be made, it may be well to recall that smooth myocytes do not function as isolated cells in their natural environment, and that a full view of the functions of the parent tissue or organ still requires considerations of the interplay of multitudes of single myocytes with one another, with their innervations, and with other types of cells among them. Moreover, because smooth myocytes are highly diverse in their properties, a true appreciation of the functions of a smooth muscle tissue requires knowledge of tissue-specific properties beyond general common features. Transplanted information from another smooth muscle or from other tissues is as likely to be incompatible as probable. This chapter will deal with the voltage-gated ionic channels of several mammalian visceral smooth muscles for which some details are known.

1.1. Methodology

Under normal physiological conditions, electrical excitability of smooth muscle tissues are manifested as voltage functions: resting potentials and action potentials, which are due to movements of electrical charges (ions) across resistances (Ohm's law). These are the properties observed in impalement microelectrode studies, in which fine-tipped (<0.5 μm) but high resistance (~100 MΩ) electrodes are thrust into myocytes with minimal injury. In current understanding, flow of ionic currents depend on two factors: electrochemical driving forces bearing on any specific ion, and properties of the pathway (channel) through which that ion must move. The latter can be properly studied only if the former can be fully controlled – that is, by the voltage-clamp method (Hodgkin, Huxley, and Katz 1952; Cole 1968). The pathways are now known to be specific proteins, traversing the plasma membrane. When activated by appropriate voltage changes or by interactions with specific ligands, the ionic channel proteins change conformation from a state in which an intrinsic aqueous path is closed (when in the resting state) to one in which it is open and conducting, allowing movement of ions and water. An ionic channel can also be in an inactivated state in which it is not openable. Major ionic channels have not only been isolated, but their amino acid sequences have been deduced from cDNA data (cloned). From appropriate mRNAs, some of them have also been functionally expressed in heterologous systems, as have their point-mutated products. General discourses can be found in Hille (1992), and examples of specific studies on the sodium channel may be found in various articles compiled in Kao and Levinson (1986).

Voltage clamps of excitable cells with responses in the tens of milliseconds or less were difficult to accomplish through electrodes of 100 MΩ resistance, because insufficient current could be pushed through to charge the membrane capacity and eliminate the complicating capacitive current. Moreover, no smooth muscle cell is large enough to sustain the impalement of two microelectrodes, one for recording and the other for passing current, without substantial injury. For smooth muscle tissues, the voltage-clamp method was first applied to small bundles of isolated uterine muscle from estrogen-primed nonpregnant rats by use of a double sucrose-gap technique (Anderson 1969). The method depended on separating the multicellular bundle into electrically connected segments by small cuffs of high-resistance sucrose in the extracellular fluid. Current was then forced to flow primarily by intracellular pathways, providing some degree of voltage control. From the outset, limitations of the method were recognized to arise chiefly from electrolytes in the narrow intercellular clefts, which presented as a high resistance in series with that of the membrane. As the clamping current of the voltage command must flow through both, the series resistance posed difficult problems in attaining proper voltage control.

These problems are now mostly obviated in single dissociated smooth myocytes by use of the tight-seal patch-clamp method (Hamill et al. 1981) in which electrodes (also referred to as pipettes) of <5 MΩ resistance form seals of high electrical resistance ($>10^9$ Ω) with appropriately clean cell surfaces. An important advantage of the high seal resistance is that it permits recording of very small signals, such as those from single ionic channels in the patch of cell membrane confined within the tip lumen. The membrane can be left on the cell in a *cell-attached* mode; it can also be detached from the cell in either an *inside-out* configuration, with the normal inside of the membrane facing the bath, or in an *outside-out* configuration. The membrane patch is ruptured to gain access to the cell interior in the *whole-cell* configuration of patch clamping. The advantages are that the access has a low electrical resistance, and the filling solution in the electrode can diffuse into the cell to exert some influence over its composition. A disadvantage is that vital cellular ingredients can also diffuse into the much larger pipette volume and lead to malfunctions (known as ''run-down'' effects).

In this chapter, ionic currents of alimentary, uterine, and urinary tract smooth muscles will be discussed. Because of differences in experimental procedures used (e.g., use of Ba^{2+} instead of Ca^{2+} as charge carrier to study Ca^{2+} channel properties, differences in holding potentials, availability of information on cell sizes), some variance in results is to be expected. The main properties to be compared are the charge carriers, the steady-state activation/inactivation properties, and some kinetic properties. Actions of pharmacological agents are dealt with only lightly, and usually only for specific blocking agents. The size of the cell influences the total current, so current densities based on unit capacitance or estimated cell surface area are used when comparing different kinds of myocytes. Kinetic processes often follow exponential time courses, and can be compared on the basis of time constants τ. Steady-state activation and inactivation properties are compared on the basis of their Boltzmann distribution functions, which state the voltage dependencies of the probabilities that a channel is in a particular state (see Hille 1992). The properties most useful for such comparisons are the voltage $V_{0.5}$ at which one-half of a process (activation or inactivation) is attained, and the slope K of the relation, which indicated the voltage sensitivity of the process.

In single-channel studies, the probability p_o of a channel being in an open state is the usual benchmark, varying from 0 when always closed to 1 when always open. In the Boltzmann functions, $V_{0.5}$ of the p_o is the voltage at which a channel is as likely to be open as closed. The unitary conductance in picosiemens (pS) is useful, but two qualifying conditions are specified. For channels that show rectification (a nonlinear current–voltage relation), the voltage at which the conductance is derived is stated. For K^+ channels, the concentration gradients are specified, whether asymmetrical, mimicking physiological distributions (e.g., 140 mM inside/6 mM outside), or symmetrical (e.g., 140 mM inside and outside). For large-conductance Ca^{2+}-activated K^+ channels, the unitary conductance is larger under symmetrical than asymmetrical K^+ distributions. Even under symmetrical distributions, the unit conductance increases as the square root of the concentration, and unusually large conductance values reported for 250 mM K^+ may turn out to be nothing extraordinary under the more physiological conditions of 140 mM.

1.2. Suitability of dissociated myocytes as models

Where insuperable obstacles in multicellular preparations preclude clear answers, dissociated individual myocytes provide the platform for detailed studies of cellular ionic channel functions. As important as many new observations are, a rarely addressed issue is the appropriateness of the isolated myocyte as a model of the myocyte in its physiological environment. Although the issue has been extensively examined for cardiac myocytes (Lieberman et al. 1987; Piper and Isenberg 1989), for smooth myocytes there is much less soul-searching. The isolation relies on actions of various enzymes and some mechanical disruptions to free individual myocytes from enmeshing connective tissue and other extracellular matrix. In addition to collagenase, a whole array of enzymes with different proteolytic actions have been used, usually on an entirely empirical basis. Indeed, there are concessions that some damage does occur during isolation (Bolton et al. 1985), and that some postdigestion treatment is as crucial to securing functional ionic channels in dissociated myocytes as the digestion process is in securing dissociated myocytes (Sanders 1989). Hence, in some studies, freshly dissociated myocytes were placed into primary cultures and then used as substitutes for freshly dissociated myocytes. At the center of the question of how normally ionic channels function in dissociated smooth myocytes is the problem that, for the most part, there are no standards for comparison, except for some arbitrary expectations (see the discussion in Sanders 1989). Our present understanding of ionic channel functions in mammalian smooth myocytes is shaped, to an astonishing degree, by such expectations, a position that requires recognition if not realignment.

For dissociated myocytes from two viscera, however, there are some studies on multicellular preparations to serve as bases for comparison. The myocytes in those preparations were not subjected to enzyme digestion, and their interior was not exposed to artificial Ca^{2+} buffers. As shall become clear in the sections that follow, unbeknownst to advocates (Bolton, Tomita, and Vassort 1981) of generic criticisms of multicellular studies, not only are many basic aspects of ionic currents in dissociated myocytes similar to myocytes enmeshed in their usual environment, but so are some unique cell-specific properties. For instance, in small multicellular strips of intestinal muscles, the inward current was carried by Ca^{2+} without any evidence of a Na^+ component, while the outward K^+ current was influenced by $[Ca^{2+}]_o$ (Inomata and Kao 1976; Weigel, Conner, and Prosser 1979). The Ca^{2+} channel was blocked by Co^{2+} and Mn^{2+}, but passed Sr^{2+} and Ba^{2+} with greater ease than Ca^{2+} (Inomata and Kao 1976, 1979, 1984). Both Sr^{2+} and Ba^{2+} slowed the inactivation of the inward current and influenced repolarization of the action potential (see also Weigel et al. 1979). The steady-state availability of the inward-current channel was the same in Ba^{2+} and Ca^{2+}. Ba^{2+} also blocked the predominant K^+ outward current in an unusual way by displacing the activation of the K^+ conductance to more positive voltages (Inomata and Kao 1984). Some of these properties are now confirmed not only in myocytes freshly dissociated from taenia coli preparations (Yamamoto, Hu, and Kao 1989a,b), but also in several other types of intestinal myocytes (Bolton et al. 1985; Benham and Bolton 1986a; Langton, Burke, and Sanders 1989; Bielefeld, Hume, and Krier 1990; Vogalis et al. 1993).

In multicellular preparations of pregnant rat myometrium, the inward current contained a Na^+ component in addition to a Ca^{2+} component (Anderson, Ramon, and Snyder 1971; Kao and McCullough 1975; Kao 1978), and the ratio of the Na^+/Ca^{2+} components rose as pregnancy progressed toward term (Nakai and Kao 1983). In contrast to the situation in the taenia coli, the predominant K^+ current in the late-pregnant myometrium was not influenced by $[Ca^{2+}]_o$ (Wakui and Kao 1981). These observations have also been confirmed in freshly dissociated uterine myocytes (Kao et al. 1989; Ohya and Sperelakis 1989; Yoshino, Wang, and Kao 1990; Inoue and Sperelakis 1991; Miyoshi, Urabe, and Fujiwara 1991; Wang and Kao 1993; Wang et al. 1996). Such experiences suggest that, when properly prepared, dissociated smooth myocytes do possess many cell-specific excitability properties, and can serve as models for studying cellular ionic channels.

The question of whether tissue-cultured cells can serve as models is different, and can only be addressed by comparing ionic channel functions in tissue-cultured cells with those of freshly dissociated myocytes. Fortunately, most mammalian visceral smooth myocytes have been studied in a freshly dissociated state. The lone exception is the uterine myocyte, which apparently has been difficult to isolate for patch-clamp studies (Mollard et al. 1986; Piedras-Renteria, Toro, and Stefani 1991), prompting some investigators to rely on cultured material as models (Mollard et al. 1986; Amedee, Mironneau, and Mironneau 1987; Toro, Stefani, and Erulkar 1990b; Rendt et al. 1992). Because these studies were based entirely on tissue-cultured material without any comparisons with freshly dissociated myocytes, they must be recognized as noncontrolled studies of artificial systems. When ionic channels of the cultured material are compared with those of freshly dissociated uterine myocytes (see Section 3), so many discrepancies become apparent that claims of similarities ring hollow on factual contradictions.

Because smooth myocytes are generally thin and long fusiform spindles, there is concern whether uniform voltages can be imposed over the entire cell. The question has usually been addressed by showing that: (a) the space constant is many times longer than the half-myocyte length, since the voltage command is applied in the middle of the cell; (b) the input resistance of the myocyte is usually greater than 1 GΩ; and (c) the ionic currents are usually small. In some cases, such as myocytes from the small intestine (see Conner et al. 1977), these conditions are only marginally satisfied, if at all. On urinary bladder myocytes (Klöckner and Isenberg 1985b) and taenia coli myocytes (Yamamoto et al. 1989a), voltage uniformity over the entire cell has been demonstrated experimentally.

2. Alimentary tract myocytes

Diversity among smooth myocytes is well illustrated by myocytes from different parts of the alimentary tract. Although the charge carriers are similar, the channels through which they pass show many variations – not only in different parts of the tract, but also in different muscle layers of the same region.

2.1. Esophageal myocytes

The musculature of the esophagus is complex, consisting of variable amounts of skeletal and smooth muscles in different species and in different locations. Myocytes from the circular muscles of the lower esophagus of the cat (Sims et al. 1990) and the opossum (Akbarali and Goyal 1994), and from the muscularis mucosa of the rabbit (Akbarali and Giles 1993) have been studied in the freshly dissociated form. Under current-clamp conditions, the opossum myocytes produced repetitive action potentials with rapid repolarization phases (Akbarali and Goyal 1994).

2.1.1. Inward currents

L-type Ca^{2+} channels were present in all these myocytes; additionally, some T-type Ca^{2+} channels have been seen in feline myocytes (Sims et al. 1990). Key properties of the major Ca^{2+} current were similar to those seen in other smooth myocytes, such as first occurrence at around −30 mV, maximum current at 0–10 mV peaking at 25–30 ms at room temperature, and susceptibility to blockade by Co^{2+}, Cd^{2+}, Ni^{2+}, and dihydropyridines. In the opossum and rabbit myocytes, I_{Ca} had an average density of ~6 μA/cm². Half-maximal inactivation of I_{Ca} occurred at −25 mV in opposum myocytes and at −30 mV in rabbit myocytes. At the usual resting potential of −50 mV, over 90% of the Ca^{2+} channels were available. In the rabbit myocyte, half-maximal activation occurred at −17 mV. Between −40 mV and 10 mV, activation and inactivation curves overlapped, suggesting the existence of some non-inactivating I_{Ca} (''window current''), which reached 25–30% of the maximum I_{Ca} (Abkarali and Giles 1993).

2.1.2. Outward currents

In the feline and opossum myocytes, the outward current consisted primarily of K^+ currents, whereas in the rabbit submucosal myocytes, a Cl^- current was also involved. The I_K in the opossum myocyte had three components: a transient outward current (I_{TO}), a Ca^{2+}-activated I_K, and a Ca^{2+}-insensitive delayed rectifier current (Akbarali et al. 1995). The I_{TO} had a rather negative inactivation voltage range, with half-maximal inactivation at −57 mV. Under physiological conditions, it may act to counter the inward I_{Ca}. The Cl^- current in the rabbit myocyte was Ca^{2+}-activated, and might be involved in neurohumoral responses (Akbarali and Giles 1993).

2.2. Gastric myocytes

Electrophysiological properties of the stomach have been studied descriptively for decades, and much is known about them (see Sanders and Publicover 1989). Slow waves, consisting of 20–30-mV depolarizations lasting several seconds, are prominent aspects of the bioelectric phenomena throughout most of the stomach, but fast, spikelike action potentials are seen only in the antrum. Dissociated myocytes from the stomach of the toad *Bufo marinus* were the first single smooth myocytes successfully studied electrophysiologically (Walsh and Singer 1980). Among mammalian species, direct recordings of ionic currents have been made on freshly dissociated myocytes of the guinea-pig (Mitra and Morad 1985; Katzka and Morad 1989), rabbit (Mitra and Morad 1985), and the corpus of the dog stomach (Carl and Sanders 1989; Carl et al. 1990; Sims 1992; Vogalis, Publicover, and Sanders 1992). In the guinea-pig myocytes (portion of stomach unspecified), rapidly repolarizing action potentials were seen either in spontaneous discharges or when elicited by short outward-current pulses (Mitra and Morad 1985). However, in cells from the circular muscle of canine gastric corpus, no action potentials could be elicited, even in myocytes with resting potentials of approximately −60 mV and input resistance of 3–4 GΩ (Sims 1992).

2.2.1. Inward currents

The inward current was carried entirely by Ca^{2+}, without any demonstrable Na^+ component (Katzka and Morad 1989), through L-type Ca^{2+} channels (see also Sims 1992 and Vogalis et al. 1992). I_{Ca} activated around −30 mV, and reached a maximum between 0mV and 10 mV. The channel was also permeant to some other divalent cations, with a selectivity of $Ba^{2+} > Ca^{2+} \geq Sr^{2+} > Mg^{2+}$ (Katzka and Morad 1989). Nifedipine blocked the channel, as did methyl verapamil (D600) and diltiazem (Katzka and Morad 1989).

The I_{Ca} decayed in two phases, with time constants τ_f for the rapid phase and τ_s for the slower phase, influenced differently by voltage and by $[Ca^{2+}]_i$. At 25°C (Sims 1992) or 35°C (Vogalis et al. 1992), the I_{Ca} of dog gastric corpus myocytes decayed with a τ_f of ~10 ms and a τ_s of ~70 ms at 0–10 mV. The faster phase was dependent on $[Ca^{2+}]_i$, whereas the slower phase was not (Vogalis et al. 1992). Thus, when the charge carrier was Ba^{2+} or Na^+ (when $[Ca^{2+}]_o$ was submicromolar), the inward current decayed as a single exponential with kinetics equivalent to τ_s. In the guinea-pig gastric myocyte, I_{Ca} also decayed in two phases, but the τ_f was 52 ms and the τ_s 129 ms (Katzka and Morad 1989). The difference from the dog myocyte may be caused, in part, by a different Ca^{2+}-buffering system in the intracellular phase: the guinea-pig myocytes were studied with 10 mM EGTA in the pipette solution, whereas the dog myocytes were studied with a buffering of 0.1 mM Indo-1 (Vogalis et al. 1992) or 1 mM EGTA (Sims 1992). When 2 mM EGTA was added to the 0.1 mM Indo-1, the τ_f and τ_s of dog myocytes were unaffected (Vogalis et al. 1992), but when 30 mM or 90 mM citrate was added to the 10 mM EGTA, the τ_f of guinea-pig myocytes was increased by a factor of two to three, and the τ_s by 20–50% (Katzka and Morad 1989). The slowing indicates that a part of the inactivation was dependent on free $[Ca^{2+}]_i$.

The steady-state inactivation of I_{Ca} was similar in the dog and guinea-pig myocytes, being half-maximal at about −40 mV in the dog myocytes (−37 mV, Sims 1992; −43 mV, Vogalis et al. 1992), and −34 mV in the guinea-pig myocytes (Katzka and Morad 1989). The Ca^{2+} channel reactivated with a τ of 1.4 s (Vogalis et al. 1992). A small portion (<10%) of the I_{Ca} did not inactivate even at 20 mV at the end of 10 s (Vogalis et al. 1992). This proportion could be a source of a sustained elevated $[Ca^{2+}]_i$, which would be necessary for maintaining a long plateau in the slow wave, and for preventing further activation of

I_{Ca}. Without such non-inactivating Ca^{2+} channels, an inactivation τ_s of ~70 ms could not account for the seconds-long plateau of the slow waves.

2.2.2. *Outward currents*

The outward current in gastric myocytes is due primarily to K^+ efflux; the shortfall of the reversal potentials from E_K in tail-current analyses (Sims 1992) might be due to uncorrected residual I_{Ca}, although a Cl^- flux had not been excluded. From a holding potential of −80 mV following the inward I_{Ca}, the outward current surged to a peak and then declined, over 2–3 s, to a more stable level of rather noisy current (Sims 1992). The current–voltage relation of the peak I_K exhibited distinct outward rectification. The surge was ascribed to a Ca^{2+}-activated K^+ current, because it was significantly reduced when I_{Ca} was blocked with Cd^{2+}. Tetraethylammonium ion (TEA, 1 or 5 mM) reduced appreciably both the peak and the stable components of the outward current. The noisiness and the outward rectification were consistent with a large-conductance Ca^{2+}-activated K^+ current. In single-channel studies of dog antral myocytes, the predominant K^+ channel had a unitary conductance of 265 pS in symmetrical K^+ media (140 mM inside/140 mM outside) and 118 pS in asymmetrical K^+ media (140/6) (Carl et al. 1990). However, in the latter medium, the channel was not highly selective for K^+, because the observed reversal potential fell short of the expected value by ~20 mV.

There may be other types of K^+ channels with significant functional roles in the gastric myocytes, but they have not been explicitly demonstrated. Thus, it is unclear whether any I_{TO} was involved in the initial surge of the outward current. Also, in the ensemble current from single-channel activities, the reconstructed macroscopic outward current showed no temporal decay over several seconds (Carl et al. 1990), an observation that differed from the rather marked decay of recorded whole-cell outward current.

2.3. *Myocytes of the small intestine*

Like the musculature of the large intestine, the predominant electrical activity of the circular and longitudinal muscles of the small intestine have long been recognized as being distinct. Although the myocytes from these layers will be discussed separately, systematic studies of these myocytes are more limited than those of the large intestine, the myocytes having been used more frequently as material for attention on one or another specific channel. Consequently, our present understanding of myocytes of the small intestine is spotty and is sometimes based on assumed similarities with myocytes of the large intestine.

From the circular muscle layer, dissociated myocytes from the jejunum of the dog and human have been studied (Farrugia et al. 1993a; Farrugia, Rae, and Szurszewski 1993b; Farrugia et al. 1995). From the longitudinal muscle layer, dissociated myocytes from the jejunum of the rabbit (Bolton et al. 1985; Benham and Bolton, 1986a,b; Benham et al. 1985, 1986), and from the ileum of the rabbit (Ohya et al. 1986), guinea pig (Droogmans and Callewart 1986), and rat (Smirnov, Zholos, and Shuba 1992) have been studied. The rabbit ileal preparation often consisted of cell fragments, which surprisingly retained properties identical to the whole myocytes (Ohya et al. 1986).

2.3.1. *Circular layer myocytes*

Although net I_{Ca} had been observed in multicellular preparations of the circular muscle of cat small intestine (Weigel et al. 1979), no I_{Ca} was seen in myocytes isolated from the circular muscles of the canine and human jejunum in physiological saline ($[Ca^{2+}]_o$ = 2.5 mM). To study properties of the channel, the inward current was enhanced by using 80 mM Ba^{2+} as the charge carrier (Farrugia et al. 1995). Because this high concentration of Ba^{2+} distorts activation and inactivation properties (Inomata and Kao 1984), the observations have little physiological informational value.

Both canine and human circular layer myocytes have large, $[Ca^{2+}]$-insensitive delayed–rectifier type outward K^+ currents, attributed to channels with deduced unitary conductance of 220 pS (Farrugia et al. 1993a). Although highly selective to K^+ with a reversal potential at the E_K, the channel accounted for only half of the conductance, the rest being attributed to a leakage component with a reversal potential at 0 mV (Farrugia et al. 1993a,b). From analyses of macroscopic tail currents, the half-maximal open-probability of the underlying single K^+ channels of canine myocytes was deduced to be at about −24 mV (Farrugia et al. 1993b), and the human channel at about −15 mV (Farrugia et al. 1993a). Some potentially significant inwardly rectifying K^+ current seen in multicellular preparations (Conner et al. 1977) were not observed in such isolated myocytes.

2.3.2. *Longitudinal layer myocytes*

In the longitudinal layer myocytes, rapidly repolarizing action potentials occurred spontaneously (Droogmans and Callewart 1986) or could be elicited by current clamp (Bolton et al. 1985; Ohya et al. 1986). In myocytes from adults of all species, the inward current was carried by Ca^{2+} through L-type channels. The channels passed Ba^{2+} with greater ease than Ca^{2+} (see Inomata and Kao 1984), accounting for the larger inward currents and higher spike amplitudes. Although Ba^{2+} is often used as the charge carrier for its larger current, it shifts activation to more negative voltages and slows activation and inactivation kinetics (Inomata and Kao 1984), making any comparison with physiological situations difficult. Hence the present summary of properties of the Ca^{2+} channel will be confined to studies in which physiological concentrations of Ca^{2+} (1–3 mM) were used as the charge carrier. The maximum current was attained at +15 mV, with an average density of 6 $\mu A/cm^2$. I_{Ca} decayed by a two-component kinetics, with τ_f of 13 ms and τ_s of 107 ms. It was half-maximally inactivated at -29 mV, and part of the inactivation was linked to $[Ca^{2+}]_i$ (Ohya et al. 1986). In the rat ileal myocyte, in addition to L-type Ca^{2+} channels, T-type Ca^{2+} channels were also seen, more prominently in the newborn than in the adult myocytes (Smirnov et al. 1992). In the latter work, the L-type Ca^{2+} channels were half-maximally inactivated at −46 mV, and the T-type channels at −61 mV. Other unusual features of the rat myocyte included an insensitivity of its L-type Ca^{2+} channels to dihydropyridines, with some of them remaining unaffected at 30 μM, as well as the presence of a tetrodotoxin (TTX)-sensitive (ED_{50} of 4.5 nM) Na^+ channel capable of passing currents as large as the I_{Ca} through coexisting L-type Ca^{2+} channels.

Several K^+ currents contributed to the outward current. In rabbit ileal myocytes, whole-cell outward current was reduced substantially when the inward I_{Ca} was blocked by Mn^{2+}, indicating the presence of a Ca^{2+}-activated K^+ current (Ohya et al. 1986). Of the remaining outward current, only a part of it was blocked by high concentrations of TEA. Thus, there are at least three different K^+ currents in the longitudinal layer myocyte of the small intestine. Two channels with unitary conductance of 50 pS and 100 pS (asymmetric K^+ distribution, 126 mM inside/6 mM outside) were seen in detached membrane patches of rabbit jejunal myocytes, with the large-conductance channel predominating (Benham and Bolton 1986a). It had the properties of large-conductance Ca^{2+}-activated K^+ channels. Because channel openings were not seen at negative voltages, these channels were thought to be uninvolved in regulating the resting potential but important for repolarizing the action potential. The properties of the smaller-conductance channel were not studied. A special type of Ca^{2+}-activated K^+ current, termed *spontaneous transient outward currents* (STOCs), has been given considerable display. They were described as resulting from episodic releases of "quantal" packets of intracellular Ca^{2+} (Benham and Bolton 1986b). However, such STOCs were rarely seen in healthy myocytes, or in myocytes with resting potentials of about −50 mV (Benham and Bolton 1986a), an observation that has been confirmed on other smooth myocytes (see e.g. Cole and Sanders 1989; Vogalis et al. 1993).

2.4. Myocytes of the large intestine

The circular and longitudinal muscles of the large intestine have different electrophysiological properties. In the circular muscle, the predominant activity is the slow wave, consisting of a non-overshooting depolarizing phase, a rapid repolarization that leads to a plateau of many seconds before a final repolarization to the resting potential. In the longitudinal muscle, the predominant activity is the spikelike single action potential with an overshoot. Freshly dissociated myocytes from the circular muscle of the guinea pig (Vogalis et al. 1993), rabbit, cat (Bielefeld et al. 1990), and dog (Carl and Sanders 1989; Langton et al. 1989; Thornbury, Ward, and Sanders 1992) have been studied. Some of the myocytes have average input resistances of less than 1 GΩ and average half-cell lengths near or exceeding the length constant of active cells (Langton et al. 1989; Bielefeld et al. 1990), raising concerns about voltage uniformity during the flow of ionic currents.

2.4.1. Circular layer myocytes

The inward current in myocytes of the circular muscle passed through L-type Ca^{2+} channels with no evidence of any sensitivity to high concentrations (10 μM) of TTX (Bielefeld et al. 1990). In the canine myocyte (Langton et al. 1989), the kinetics of I_{Ca} decay (at 22–25°C) followed two exponential terms with average τ_f of 21 ms and τ_s of 123 ms. It was half-maximally inactivated at −43 mV and half-maximally activated at −21 mV. These properties cannot be readily compared with those of the guinea-pig circular myocyte, because the Ca^{2+}-channel current in the latter was studied by using 7.5 mM Ba^{2+} as the charge carrier (Vogalis et al. 1993). Although Ba^{2+} would affect activation and some kinetics, it would not affect the steady-state inactivation (Inomata and Kao 1984); on dissociated guinea-pig circular myocyte the half-maximal inactivation occurred at −31 mV (Vogalis et al. 1993), which was appreciably positive to that in the canine myocyte. Left unresolved are at least two issues: (a) the presence or absence of any very slowly decaying or non-inactivating I_{Ca} that might contribute to the long plateau of the slow waves, because a current that decays with a τ_s of ~120 ms (Langton et al. 1989) is not likely to fulfil that function; (b) the presence or absence of Ca^{2+}-mediated Ca^{2+} inactivation, which is mechanistically associated with the rate of I_{Ca} inactivation.

Several types of K^+ channels were involved in the outward current of dissociated colonic circular layer myocytes. A rapidly activating and inactivating I_{TO} was seen in the guinea-pig myocyte (Vogalis et al. 1993), but not in the canine myocyte (Thornbury et al. 1992). The channel was half-maximally activated at 4 mV. The decay of this I_{TO} had two voltage-sensitive components: τ_f of 17 ms and τ_s of 69 ms. It was fully inactivated by depolarization, with half-maximal inactivation at −58 mV. Therefore, at the usual resting potential of approximately −50 mV, 30–40% of I_{TO} could be available to influence excitability. Two other types of K^+ current were also present in the guinea-pig myocyte: a large-conductance Ca^{2+}-activated K^+ current and a Ca^{2+}-insensitive delayed–rectifier type K^+ current. Recognition of the type of current rested heavily on use of blocking agents, such as combined Cd^{2+}-TEA and 4-aminopyridine (4-AP). The former included the so-called STOCs, which were rarely present at potentials negative to that for activating the inward I_{Ca} (−40 mV, Vogalis et al. 1993; see also Cole and Sanders 1989). The delayed rectifier current was separated from other types of K^+ currents through the combined use of Cd^{2+}-TEA-4AP. It was half-maximally activated at 2 mV. About half of this current did not inactivate, and the remainder inactivated half-maximally at −52 mV. Thus, except for the sizable non-inactivating component, its properties coincided closely with those of the I_{TO}. Whether 4-AP can be used to select out delayed rectifier channels is a moot point because in other studies it has a marked blocking effect on such channels (Hart et al. 1993; Overturf et al. 1994).

In the canine circular colonic myocyte, a Ca^{2+}-activated K^+ current was the most prominent among several K^+ currents (Cole and Sanders 1989). Another current with properties of a delayed rectifier was seen when the prominent Ca^{2+}-activated K^+ current was suppressed (Thornbury et al. 1992). It activated rapidly enough at 37°C to overlap extensively with the inward I_{Ca}, and was assigned a role in producing the non-overshooting slow waves. Although ~25% of this current did not inactivate at 0 mV, the remainder showed a half-maximal inactivation at −36 mV, significantly more positive than that described for the guinea pig circular myocyte. Two channels with properties of delayed rectifier channels have been cloned and expressed in *Xenopus* oocytes, $K_v1.2$ (Hart et al. 1993) and $K_v1.5$ (Overturf et al. 1994). They both share considerable sequence homology with other delayed rectifier channels. $K_v1.2$ is found uniquely in gastrointestinal muscle, whereas $K_v1.5$ is also found in smooth muscles of blood vessels. They have single-channel conductances of 14 pS and 10 pS respectively, and share other electrophysiological properties of delayed rectifier channels, such as rapid activation and slow decay over several hundred milliseconds. Although ~30% of each channel did not inactivate, the half-maximal inactivation of the remainder occurred at −15 mV for $K_v1.2$, and at −21 mV for K_v 1.5. Both of these values are substantially more positive than that in the intact myocyte (Thornbury et al. 1992), but the discrepancy might be due to differences in the environment in which the channels are expressed. Both cloned channels were susceptible to 4-AP blockade; the ED_{50} was 75 μM for $K_v1.2$ and 211 μM for $K_v1.5$.

Single Ca^{2+}-activated K^+ channels have been studied in excised membrane patches of the canine myocyte (Carl and Sanders 1989). This channel had a conductance of 206 pS in symmetrical K^+ distribution (140 mM inside/140 mM outside), and 96 pS in asymmetrical K^+ distribution (140/5.9). It was thought to be the most influential determinant of K^+ currents. Although its openings were highly sensitive to $[Ca^{2+}]_i$, its half-maximal activation was at 107 mV in 10^{-7} M Ca^{2+} (Carl and Sanders 1989). Such a relation would suggest that, under physiological conditions, the p_o of this channel might be quite low, possibly contributing to the plateau of the slow wave. These properties can be contrasted with those of the large-conductance Ca^{2+}-activated K^+ channels of the taenia coli myocyte of the longitudinal layer, which has a rapid repolarization phase and no plateau to its action potential (see Section 2.4.2).

2.4.2. Longitudinal layer myocytes

The longitudinal muscle of the large intestine has been extensively studied, especially the guinea-pig taenia coli. Significantly, this preparation has been studied across a broad spectrum of organizational levels, extending from the non-enzyme–treated multicellular strips (Inomata and Kao 1976, 1979, 1984) through the enzyme-dissociated single myocytes (Ganitkevitch, Shuba, and Smirnov 1986, 1987; Yamamoto et al. 1989a,b; Yoshino et al. 1989) to its single Ca^{2+} channels (Yoshino et al. 1989) and single K^+ channels (Hu, Yamamoto, and Kao 1989a,b). The similarities of all major fundamental properties of multicellular preparations and single cells furnish strong support for using the isolated myocyte as a model for studying ionic channels of smooth muscles.

The inward current was carried solely by Ca^{2+}, with no evidence of any Na^+ component (Inomata and Kao 1976), through L-type Ca^{2+} channel (Ganitkevitch et al. 1986; Yamamoto et al. 1989a; Yoshino et al. 1989). Among smooth myocytes, the taenia I_{Ca} was large, with average current density of 19.5 μA/cm² (Yamamoto et al. 1989a). The underlying channel could be of two types, a 7-pS channel that activated at about −30 mV and a 30-pS channel that activated at about 0 mV (Yoshino et al. 1989). Sr^{2+} (Inomata and Kao 1979; Yamamoto et al. 1989a) and Ba^{2+} (Inomata and Kao 1984; Yamamoto et al. 1989a) both could pass through the channel, with the affinity of a binding site for Ba^{2+} some six to ten times less than that for Ca^{2+} (Inomata and Kao 1984).

In 3-mM $[Ca^{2+}]_o$, the Ca^{2+} channel of single taenia myocytes was half-maximally activated at −1 mV (Yamamoto et al. 1989a), a value remarkably close to that seen in multicellular taenia coli preparations (−4 mV, Inomata and

Kao 1984). In 30-mM $[Ca^{2+}]_o$, the half-maximal activation of taenia myocytes was shifted ~20 mV to the positive (Yamamoto et al. 1989a). When 2.5 mM Ba^{2+} replaced 2.5 mM Ca^{2+}, the activation of multicellular taenia preparations was shifted 7 mV to the negative (Inomata and Kao 1984), an effect that could also be seen in the *I–V* relations obtained on single taenia myocytes (Yoshino et al. 1989).

At 32°C, I_{Ca} decayed nearly completely in 2 s, the decay following three exponential terms with τ_f of ~7 ms, τ_m of ~67 ms, and τ_s of ~327 ms (Yamamoto et al. 1989a), however, at about 22°C, the fastest component was not seen (Ganitkevitch et al. 1986). The various components exhibited different and complex voltage dependence. The fast component also showed distinct dependence on $[Ca^{2+}]_i$, that is, Ca^{2+}-mediated Ca^{2+} inactivation (Ganitkevitch et al. 1987; Yamamoto et al. 1989a; Yoshino et al. 1989). In the steady state, the Ca^{2+} channel of taenia myocytes was half-maximally inactivated at -30 mV (Yamamoto et al. 1989), which was also the value determined on multicellular preparations (−31 mV, Inomata and Kao 1979). Neither Sr^{2+} nor Ba^{2+} affected this property (Inomata and Kao 1979, 1984). By using a simulated action potential as the voltage command in myocytes in which K^+ currents had been blocked, I_{Ca} was integrated and the Ca^{2+} influx found to be sufficient to raise $[Ca^{2+}]_i$ to 8 μM (Yamamoto et al. 1989a).

Of the whole-cell outward K^+ currents, the most prominent was a noisy current that was appreciably reduced when the inward I_{Ca} was blocked, a Ca^{2+}-activated K^+ current (Yamamoto et al. 1989b). No fast I_{TO} was seen, but a Ca^{2+}-insensitive delayed rectifier current comprised a small part. Using a simulated action potential as the voltage command, the outward current coincided with the repolarizing phase (Yamamoto et al. 1989b). Nevertheless, because repolarization results from an interplay of inward and outward currents, changes in either can affect the direction of the net current. Thus, in multicellular taenia preparations, Sr^{2+} slowed the inactivation of the inward I_{Ca} but increased the outward K^+ current. Depending on the concentration of Sr^{2+} and the relative changes of the inward and outward currents, the repolarization of the action potential was either slowed or hastened (Inomata and Kao 1979).

Three types of K^+ channels were seen in excised patches of membrane from taenia myocytes, with unit conductances at 0 mV of ~65 pS, ~100 pS, and ~150 pS in asymmetrical K^+ distribution (135 mM inside/5 mM outside, Hu et al. 1989a). The 150-pS channel constituted over 90% of K^+-channel activities, and had all the properties of a large-conductance Ca^{2+}-activated K^+ channel. It first became active at about −30 mV at pCa 8 and 7, and at about −70 mV at pCa 6. An *e*-fold increase of p_o occurred with an average positive move of 11.3 mV. Between −30 mV and +30 mV, a voltage range of physiological interest, the increased p_o was due to a shortening of the mean closed time of the channel, especially when p_o was low; but as p_o increased, some lengthening of the open time also began to affect the activity. $[Ca^{2+}]_i$ influenced the activities of this channel. At 0 mV, p_o was 0.02 in pCa 8, 0.08 in pCa 7, and 0.52 in pCa 6. Thus, at 0 mV, the $[Ca^{2+}]_i$ needed for $p_o = 0.5$ would be 10^{-6} M, a concentration quite attainable through the influx of Ca^{2+} during an action potential (Yamamoto et al. 1989a).

The role of the 150-pS channel in repolarization is readily apparent, because its p_o, which was at a minimum at approximately −30 mV, increased steeply with positive shifts of voltage (Hu et al. 1989a). However, at voltages negative to −30 mV, p_o again rose to reach a small peak at about −50 mV. This anomalous activity, with p_o of 0.08 at −50 mV, could be a sign of involvement in regulating the resting potential (Hu et al. 1989a). Further indication of its involvement in this important function can be seen in the effects of β-adrenergic agents (e.g., isoproterenol) in hyperpolarizing intestinal muscles and causing the "inhibitory" action of classical pharmacology. The hyperpolarization resulted from increased activities of the 150-pS channel via a second-messenger pathway consisting of G_s protein-adenylate cyclase-protein kinase A (Fan, Wang, and Kao 1993).

The 150-pS channel in the taenia myocyte had a selectivity ratio of K^+:Rb^+:NH_4 of 1:0.7:0.5, and was impermeant to Cs^+, Na^+, Li^+, and TEA^+ (Hu et al. 1989b). An interesting property is the way in which Ba^{2+} blocks this channel. On multicellular preparations, a Ba^{2+} blockade of the taenia coli K^+ current (for which the 150-pS channel is chiefly responsible) consisted of a parallel positive displacement of the voltage–conductance relation such that a half-maximal activation at 15 mV in 2.5 mM Ca^{2+} shifted to 39 mV in 2.5 mM Ba^{2+}, and even more positive voltages in higher concentrations of Ba^{2+} (Inomata and Kao 1984).

3. Uterine myocytes

When voltage-clamp methods were applied to smooth muscle tissues years after they had been used to study ionic channels in nerves and in skeletal and cardiac muscles, the longitudinal muscle from the rat uterus was the first to be used (Anderson 1969; Mironneau 1974; Kao and McCullough 1975). Yet, as patch-clamp studies on various dissociated smooth myocytes are proliferating, work on dissociated uterine myocytes is sparse. The reason is that uterine myocytes are apparently difficult to isolate in a state suitable for patch clamping (Mollard et al. 1986; Piedras-Renteria et al. 1991). As a result, some investigators have turned to using tissue-cultured or other artificial systems as models for studying ionic channels of uterine myocytes. At present, there are more publications on such model systems than on true uterine myocytes. However, the suitability of such model systems and the value of the information derived from them are untested issues, not only in myometrial but also in general smooth muscle physiology. The significance of these issues should not be obfuscated by prolific publications. Nevertheless, there are enough studies on freshly dissociated uterine myocytes from different laboratories to provide consistent information showing that extant cultured material or model systems do not possess adequate tissue-specific ionic channels to define them as uterine myocytes. Therefore, the following discussion will be focussed on freshly dissociated uterine myocytes, with enough details on cultured and other model systems to demonstrate where they differ fundamentally from the former.

Without doubt, uterine myocytes are among the most complex excitable cells from a variety of functional points of view, owing to their ready and profound responses to regulatory influences of ovarian hormones and pregnancy. At the same time, because of these responses, the uterine myocyte provides a platform for studying some aspects of general physiology of ionic channels rarely available in other types of cells.

Whole-cell currents and single-channel studies have been conducted on freshly dissociated myocytes from the nonpregnant (Wang and Kao 1993) and pregnant uteri of the rat (Kao et al. 1989; Ohya and Sperelakis 1989; Yoshino, Wang, and Kao 1989, 1990; Inoue and Sperelakis 1991; Miyoshi et al. 1991; Wang and Kao 1993; Wang et al. 1996), nonpregnant rabbit (Yang, Wang, and Kao 1994) and pregnant women (Inoue et al. 1990; Young, Herndon-Smith, and Anderson 1991; Kahn et al. 1993). Although circular and longitudinal muscles of the uterus have different electrophysiological properties (Osa and Katase 1975), almost all the patch-clamp studies to date have been done on longitudinal layer myocytes. An important feature unique to uterine myocytes is that the status of ovarian hormones in the organism affects ionic channels. The hormonal status can be ascertained by monitoring vaginal cytology, or artificially controlled by administrating hormones. The latter approach is better suited for species without a natural cycle (rabbit), but is less reliable in a species with a rapid maturation and a short cycle (rat), unless oophorectomized.

By enhancing nuclear transcription, estrogen induces synthesis of proteins in the uterine myocyte. Thus, along with cellular hypertrophy, some ionic channels are synthesized and inserted into the plasma membrane. Another possible estrogen effect is some nongenomic regulatory role on the expression of some ionic channels. Under the normal sequence of events, progesterone is secreted by the corpus luteum after ovulation, and exerts its regulatory influences generally on cells that are already primed by estrogen. Whether progesterone induces synthesis of new channels is not known, but it affects the expression of some channels (Yang et al. 1994).

3.1. Inward currents

As in most smooth myocytes, Ca^{2+} currents have been observed in freshly dissociated uterine myocytes of rats (Kao et al. 1989; Ohya and Sperelakis 1989; Miyoshi et al. 1991), rabbits (S. Y. Wang and C. Y. Kao, unpublished observations), and women (Inoue, et al. 1990; Young et al. 1991; S. Y. Wang, D. Nanda, and C. Y. Kao, unpublished observations). In all the species, by its activation at approximately −30 mV, its slow inactivation, and its susceptibility to blockade by submicromolar concentrations of dihydropyridines, the channel involved was identified as L-type. In addition, some T-type channels have been described for the human uterine myocyte (Inoue et al. 1990).

Unlike most smooth myocytes in which the inward current is normally carried exclusively by Ca^{2+}, in the uterine myocyte a component of the inward current is due to Na^+ influx (see Figure 1). Participation of Na^+ in the inward current (Anderson et al. 1971; Kao and McCullough 1975) in addition to Ca^{2+} (Anderson et al. 1971; Kao 1978) was known in multicellular strips of rat myometrium.

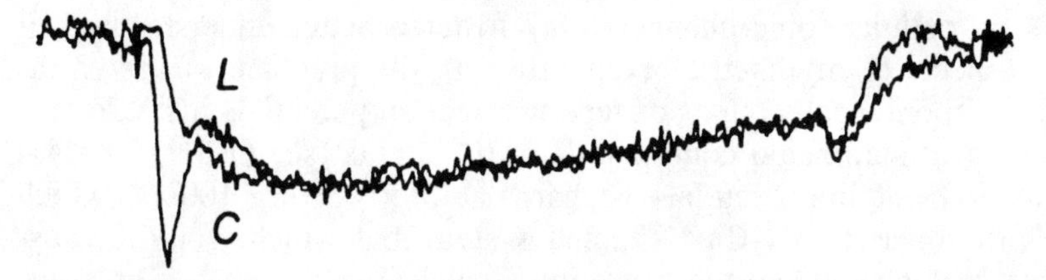

Figure 1. Inward currents from a smooth myocyte freshly dissociated from a 20-day pregnant rat uterus. Cell capacitance 125 pF; 22°C. Bath contained 140 mM Na^+ and 3 mM ca^{2+}. Pipette solution contained 135 mM Cs^+ to block all outward currents. Holding potential −80 mV; depolarized to 0 mV. Inward currents consist of two parts: a rapidly activating and inactivating component attributable to Na^+ influx, followed by a more slowly activating and inactivating component attributable to Ca^{2+} influx. Trace C, control conditions; trace L, 2 min after 0.1 mM lidocaine was introduced into the bath. I_{Na} was reduced to 43%, whereas I_{Ca} was minimally affected (visible chiefly as a smaller tail current).

Although the methods at the time precluded a direct recording of a Na^+ current (because of residual capacitive current), effects of ion substitutions left little doubt of the roles of both Na^+ and Ca^{2+}. This coexistence has been confirmed in freshly dissociated uterine myocytes.

The two components of the inward current, I_{Na} and I_{Ca}, could be distinguished from each other by ion substitutions and by pharmacological responses. I_{Na} was selectively blocked by TTX, with a K_d of 27 nM (Ohya and Sperelakis 1989). I_{Ca} was selectively blocked by dihydropyridines at 10^{-7} M. Additionally, the I_{Na} and I_{Ca} differed markedly in various activation and inactivation properties.

Na^+ current of uterine myocyte appeared first at approximately −40 mV, and reached a maximum at ~0 mV. At ~22°C, it peaked in about 1 ms, and then decayed with a voltage-dependent τ that stabilized at ~0.4 ms. It was half-maximally activated at −21 mV and half-maximally inactivated at about −60 mV (−64 mV, Ohya and Sperelakis 1989; −59 mV, Wang and Kao 1993; Yoshino et al. 1997). It reactivated as a single exponential function with a τ of 20 ms (Wang and Kao 1993; Yoshino et al. 1997).

The properties of I_{Ca} are quite different. It appeared first at about −30 mV and reached a maximum at about +10 mV. At ~22°C, it reached a maximum in ~10 ms, and then decayed as a two-term exponential function, with τ_f of 32 ms and τ_s of 132 ms. The inactivation was both voltage-dependent and sensitive to $[Ca^{2+}]_i$. Half-maximal activation occurred at -8 mV; half-maximal inactivation at about −35 mV (−39 mV, Ohya and Sperelakis 1989; −34 mV, Yoshino et al. 1997). Substantially different inactivation relations have also been reported (−61 mV for late-pregnant rat myocyte, Miyoshi et al. 1991; −54 mV for late-pregnant human myocyte, Young et al. 1991). From the data of Yoshino et al. (1997), more than 95% of I_{Ca} and less than 25% of I_{Na} were available at the usual resting potential of approximately −50 mV. I_{Ca} reactivated along a two-term exponential time course, a small part with a τ_f of 27 ms, and the greater part with a τ_s of 374 ms (nearly 20 times slower than I_{Na}).

Such details on cellular ionic channels may foster a better understanding of more complex aspects of myometrial excitability. In the pregnant rat, as parturition approaches, spontaneous bursts of repetitive action potentials increase progressively, leading to summated contractions of the uterus (see Osa and Katase 1975; Kao 1989). The action potentials discharge at intervals of ~100 ms, which would be too fast for a strictly Ca^{2+}-channel system, but which would be sustainable if some Na^+-channel system were involved.

3.1.1. *Pregnancy influence on I_{Na} and I_{Ca}*

Myometrial Na^+ channels are strongly influenced by estrogen. In myocytes from the nonpregnant rat uterus, I_{Na} was found only if the rat was in the estrus–diestrus phases, but not in the proestrus (Wang and Kao 1993; Yoshino et al. 1997). In myocytes from pregnant rat uteri, I_{Na} was recorded from the second day of gestation through to term on the twenty-second day. Throughout pregnancy, while the surface area of the myocyte increased fourfold, the density of the I_{Na} was always substantially more than a quarter of that in the nonpregnant myocyte, indicating that new Na^+ channels had been synthesized during cellular hypertrophy (genomic control). Within a few hours postpartum, I_{Na} vanished from most myocytes (Wang and Kao 1993; Yoshino et al. 1997), as if its expression had been suddenly switched off (nongenomic influence).

In small multicellular preparations of late-pregnant rat myometrium, Nakai and Kao (1983) described an increasing I_{Na}/I_{Ca} ratio as pregnancy moved toward term. On freshly dissociated myocytes, a similar but more complex phenomenon has been observed (Wang and Kao 1993; Yoshino et al. 1997). The densities of both I_{Na} and I_{Ca} of individual myocytes increased in the initial days of pregnancy, declined in mid-term, and then recovered at different rates. The net result was that the average ratio of I_{Na}/I_{Ca} of single myocytes changed from 0.5 at the beginning of pregnancy to 1.6 at term. Inoue and Sperelakis (1991) described a similar but slightly different phenomenon. They found no I_{Na} in any myocyte before the fifth day of gestation, and that the density of I_{Na} in individual myocytes remained constant from the ninth day until term. However, the fraction of myocytes expressing I_{Na} increased as pregnancy progressed. By taking their observing I_{Na} as an index of the probability of I_{Na} occurring, and by averaging the I_{Na} of all myocytes, they concluded that I_{Na} increased as pregnancy progressed. The differences between the two sets of observations and interpretations are not trivial, because they involve basic mechanisms by which estrogen and/or pregnancy factors regulate ionic channels in uterine myocytes. The data of Inoue and Sperelakis indicated that such regulations were hyperplasia-based demographic processes, with individual myocytes expressing constant densities of the respective currents, and that a subpopulation of I_{Na}-expressing myocytes proliferated with gestation. The data of Yoshino and associates showed that the regulations involved individual uterine myocyte from the earliest stages of pregnancy, through both genomic control and local regulatory influences. Further work will be needed in these respects.

An interesting feature of the myometrial Na^+ channels, with some practical implications, is that they are rather sensitive to blockade by lidocaine (Kao and Wang 1994; see also Figure 1). Because premature birth is the major perinatal public health problem in developed societies, and because no current therapy is truly effective in delaying parturition, use of lidocaine and related cardiac antiarrhythmic agents may be worth further exploration for managing threatened preterm labor.

Another development worth watching is the cloning of an ''atypical sodium channel'' from the human (George, Knittle, and Tamkun 1992) and mouse (Felipe et al. 1994) uteri, which shares about 50% homology with sodium channels in brain and in skeletal and cardiac muscle. Although the gene transcript appears to increase with pregnancy, the channel remains unexpressed. Thus, the term ''channel,'' as used, is based on sequence homology and not on function.

3.2. Outward currents

The outward current in uterine myocytes was carried by K^+ (Kao and McCullough 1975; Wang and Kao 1993; Wang et al. 1996) without participation of any Cl^- fluxes (cf. Parkington and Coleman 1990). In multicellular preparations from the estrogen-primed nonpregnant guinea-pig uterus (Vassort 1975) and from the late-pregnant rat uterus (Mironneau and Savineau 1980), two types of K^+ currents were described: an early surge following the inward current attributed to a Ca^{2+}-activated component, and a later steady component attributed to some delayed rectifier current. Although these claims are cited frequently, the evidence was suspect because rather large masses of tissue were used in the double sucrose-gap method, resulting in poor voltage control and much overlap of the inward and outward currents. This view applies especially to the pregnant rat myometrium, because small preparations of the same tissue (in which current overlaps were less) showed no such early Ca^{2+}-sensitive surges (Kao 1977), and because Ca^{2+}-activated K^+ currents are not expressed in this tissue (Kao et al. 1989; Wang et al. 1996).

Whole-cell outward currents have now been studied in freshly dissociated myocytes from nonpregnant estrus rat uterus (Piedras-Rentaria et al. 1991; Wang et al. 1996), the late-pregnant rat uterus (Kao et al. 1989; Miyoshi et al. 1991; Inoue, Shimamura, and Sperelakis 1993; Wang et al. 1996), and the pregnant human uterus (Inoue et al. 1990). Activities of single K^+ channels of the rat (Wang et al. 1996) and human myocytes (Kahn et al. 1993) have also been studied.

Even by appearance, the outward currents of nonpregnant and late-pregnant rat uterine myocytes are quite different (Figure 2). In nonpregnant myocytes, the currents were distinctly noisy, with a tendency to decay appreciably within a few hundred milliseconds. In late-pregnant myocytes, they had little noise and tended to be well-sustained within a few hundred milliseconds. Undoubtedly, different K^+ currents are concurrently active in both types of myocytes, but the tasks of sorting out their relative contributions are not yet fully accomplished. On the basis of available information, several types of K^+ channels can be discerned.

A transient outward K^+ current (I_{TO}) was seen in about half of the nonpregnant myocytes but not in late-pregnant myocytes (Wang et al. 1996). The I_{TO} activated and inactivated rapidly and had a current peak at 3–5 ms. It was followed by a

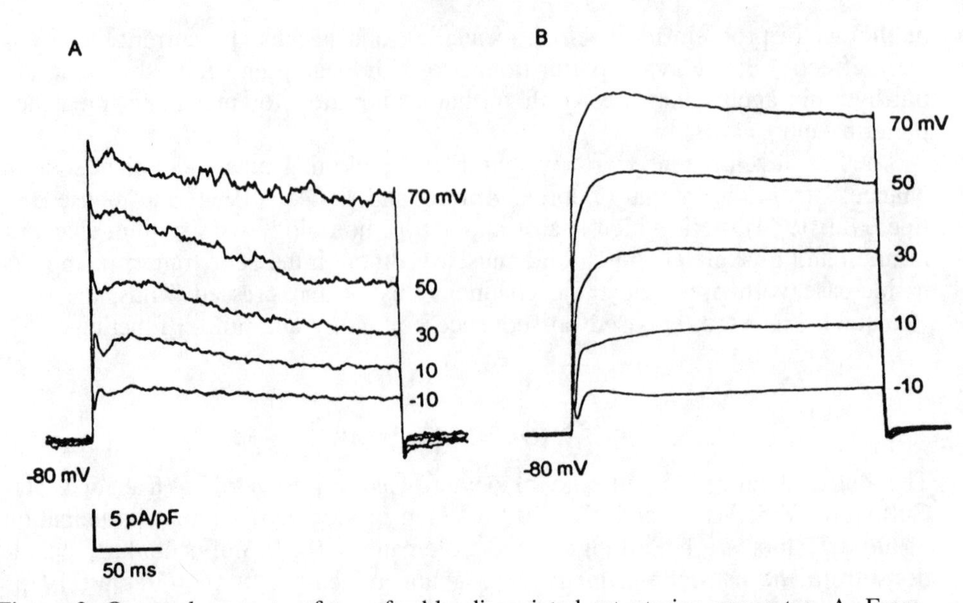

Figure 2. Outward currents of two freshly dissociated rat uterine myocytes. A. From nonpregnant estrus uterus; cell capacitance, 26 pF. B. From 17-day pregnant uterus; 144 pF and 22°C. Bath contained 1 mM Ca^{2+}. Pipette solution contained 135 mM K^+ and 1 mM EGTA. Holding potentials −80 mV; voltages at which currents are obtained are shown to right of traces. For clarity, only alternate traces are shown. Note that current densities are about the same in the two myocytes, although total currents would differ because of cell sizes. A. In nonpregnant myocyte, transient outward current (I_{TO} is followed by a noisy and decaying current that contains at least 2 types of K^+ currents. One of those is responsive to changes in $[Ca^{2+}]_0$ (see text). B. In late-pregnant myocyte, I_{TO} is seen only at small depolarizations (−10 mV), but is indistinct at more positive voltages. Main current is relatively smooth, rises gradually to maximum in ~30 ms, and decays only slightly over a few hundred milliseconds. It is insensitive to changes in $[Ca^{2+}]$ (see text).

noisy decaying current of a different nature. The I_{TO} had a highly negative voltage availability relation, with half-maximal inactivation at −77 mV (Wang et al. 1996). At the usual resting potential of −50 mV, only a small percentage of I_{TO} was available. It was half-maximally activated at 5 mV, and there was little overlap between the inactivation and activation curves. By analogy with observations in other tissues, a transient outward current has also been claimed on the basis of its susceptibility to 4-AP (see e.g. Miyoshi et al. 1991; Inoue et al. 1993). For the uterine myocyte, however, because the actions of 4-AP have never been fully studied, its specificity of action, and hence its value as a tool are unknown. Moreover, 4-AP was sometimes applied in combination with other nonspecific agents (such as TEA exceeding 10 mM), and the effects observed on a time scale too slow to show millisecond events. Thus, claims concerning I_{TO} in late-pregnant myocytes in which it is typically indistinct should be treated with reservation.

At least two other types of K^+ currents can be readily identified in uterine myocytes, a large-conductance Ca^{2+}-activated K^+ current and a Ca^{2+}-insensitive delayed rectifier–type current (Inoue et al. 1993; Wang et al. 1996). The former was most obvious in nonpregnant myocytes, where it was responsible for the noisiness of the current and for its facile responses to changes in $[Ca^{2+}]_o$. The latter was most obvious in late-pregnant myocytes, where it was the dominant current. Typically, it rose gradually to a maximum and maintained level in ~30 ms (at 22°C). It was unaffected by changes in $[Ca^{2+}]_o$ (Kao et al. 1989; Miyoshi et al. 1991).

These currents were probably present in mixed form in both nonpregnant and late-pregnant myocytes, but their relative contributions changed during pregnancy. The large-conductance Ca^{2+}-activated K^+ current in early pregnancy waned as pregnancy progressed, while the Ca^{2+}-insensitive delayed rectifier current assumed an increasing load of expressing the whole-cell I_K. One mechanism by which such a shift could occur is a change in the conditions of expressing the Ca^{2+}-activated K^+ current: in the nonpregnant myocyte, this current was half-maximally activated at 39 mV; in the late-pregnant myocyte, it was half-maximally activated at 63 mV. In support, at least two distinct single K^+ channels have been observed in detached membrane patches from late-pregnant myocytes: a 140-pS Ca^{2+}-sensitive channel and an ~15-pS channel (Wang et al. 1996). The half-maximal open-probability of the former occurred at 68 mV at pCa 7, in agreement with the half-maximal activation voltage in the whole-cell current. That ovarian steroids could alter the expression of the large-conductance Ca^{2+}-activated K^+ channel can be seen on myocytes from nonpregnant rabbits in different hormonal status (Yang et al. 1994). In detached membrane patches from hormone-naive (untreated) myocytes, the open-probability of the channel was half-maximal at 74 mV ($V_{0.5}$). In patches from estrogen-dominated myocytes, $V_{0.5}$ was 54 mV, whereas in patches from progesterone-dominated myocytes it was 78 mV.

Contributing to the manifested whole-cell currents are probably multiple concurrently active channels with similar activation properties, because selective blocking agents affected only part of the currents. For example, iberiotoxin, a peptidyl toxin from a scorpion venom that was highly selective for the large-conductance Ca^{2+}-activated K^+ channel (Galvez et al. 1990) even at 1 nM (forty-fold higher than the K_d), blocked some 80% of the noisy current but left about 20% residual current, which must be due to some other channel. Similarly, α-dendrotoxin from a mamba snake, which was selective for a small-conductance delayed rectifier channel (see Dreyer 1990) at several times K_d, blocked only part of the whole-cell I_K of late-pregnant myocytes.

The Ca^{2+}-activated K^+ currents discussed so far have been observed by different investigators in the presence of 0.5–1.0 mM EGTA in the pipette solution (Kao et al. 1989; Miyoshi et al. 1991; Wang et al. 1996). They are probably different from two Ca^{2+}-sensitive currents seen in the nonpregnant estrus rat uterine myocyte in the absence of any Ca^{2+}-buffering system in the pipette solution (Piedras-Rentaria et al. 1991). There, the Ca^{2+} activation was attributed to diffusion into the myocyte interior of 1.2 μM of Ca^{2+}, present as reagent contaminant in the pipette solution. Oddly, when 1 mM EGTA was incorporated into the pipette solution, one of the currents could be recorded for only a few minutes after membrane rupture, and the other not at all. A century after the concept of buffering was introduced, and four decades after divalent cation chelating agents became widely available, it is surprising that any serious work on Ca^{2+}-dependent cellular processes can still be done in the absence of Ca^{2+}-buffering systems.

A very slowly activating K^+ channel has been cloned from the estrogen-primed nonpregnant rat uterus, and expressed in the *Xenopus* oocyte (Boyle et al. 1987). Its functions are unknown, because a physiological counterpart with similarly slow kinetics has never been observed in any uterine myocytes.

3.3. Comparison of fresh myocytes with other model systems

No other visceral smooth muscle than the myometrium offers a better opportunity for assessing the usefulness of tissue-cultured material as a model for studying ionic channel functions, owing to the volume of publications on cultured myometrial cells. When a comparison of cultured and freshly dissociated uterine myocytes is made, so many significant differences are found in fundamental properties of ionic channels as to suggest that the channels pertain to different cells.

Some of the most important differences may be summarized as follows: (1) In freshly dissociated myocytes, a Na^+ current could be recorded readily in coexistence with a Ca^{2+} current (Kao et al. 1989; Ohya and Sperelakis 1989; Yoshino et al. 1990, 1997). The Na^+ current was present in nonpregnant myocytes from estrus rats (Wang and Kao 1993; Yoshino et al. 1997) and in myocytes from various stages of pregnancy (Inoue and Sperelakis 1991; Wang and Kao 1993; Yoshino et al. 1997). In cultured cells, with the exception of the human uterine myocyte (Young and Herndon-Smith 1991), none showed any Na^+ action potentials (Mollard et al. 1986) or any I_{Na}, all the inward current being attributed to I_{Ca} (Amedee et al. 1987; Rendt et al. 1992). Some cultured cells had no inward currents at all (Toro et al. 1990b). (2) In fresh myocytes, I_{Na} was elicited from holding potentials of -90 mV to -60 mV (Ohya and Sperelakis 1989; Wang and Kao 1993; Yoshino et al. 1990, 1997). The I_{Na} was blocked by TTX with a K_d of 27 nM (Ohya and Sperelakis 1989), characterizing the TTX receptor as a high-affinity type (see references in Kao and Levinson 1986). In cultured rat cells, I_{Na} was not elicited from -75 mV (Amedee et al. 1986; Rendt et al. 1992). It was expressed only after application of veratridine or sea anemone toxin, presumably following some interference with the Na^+-inactivation mechanism. Such an altered I_{Na} was then susceptible to TTX with an apparent K_d of 2 μM (Amedee et al. 1986), characterizing the receptor as a low-affinity type. The different TTX sensitivities are not trivial details, because they could reflect differences in amino acid compositions of the channel protein, such as those known to exist for sodium channels of nerves and cardiac myocytes (references in Kao and Levinson 1986). (3) In fresh (Yoshino et al. 1997) but not cultured (Rendt et al. 1992) nonpregnant myocytes, I_{Ca} could be blocked by nifedipine. (4) In fresh myocytes, half-maximal activation of the Ca^{2+} conductance in 3 mM Ca^{2+} was at -8 mV (Yoshino et al. 1997). In cultured cells, these properties were studied in 10 mM Ca^{2+}, which was known to cause a positive voltage shift by its screening effect on surface negative charges (Frankenhauser and Hodgkin 1957; Yamamoto et al. 1989a). Yet, the half-maximal activation in cultured cells was at −14 mV (Amedee et al. 1987) or at −7 mV (Rendt et al. 1992). (5) In fresh myocytes, I_{Ca} inactivated in two exponential phases with τs of 32 ms and 133 ms. In cultured cells the inactivation was appreciably faster with τs of 9 ms and 55 ms (Amedee et al. 1987). (6) In fresh myocytes, the steady-state I_{Ca} inactivation was almost all voltage-dependent, with half-maximal inactivation at −34 mV. In cultured cells, 10–13% of the I_{Ca} did not inactivate. Those that did had a half-maximal inactivation at −17 mV (Amedee et al. 1987) or −44 mV (Rendt et al. 1992). (7) In fresh myocytes, I_{Ca} reactivated with τs of 27 ms and 374 ms; in cultured cells, these values were 300 ms and 5,190 ms (Amedee et al. 1987).

Comparing tissue-cultured material with freshly dissociated myocytes is never easy, because of different practices in different laboratories, as is evident from the wide variations among cultured material just listed. Therefore, such a comparison is much more informative when it is based on work of the same investigators, using the same strain of animals and the same laboratory practices: in tissue-cultured myocytes from nonpregnant rat uterus, Toro et al. (1990b) found three K^+ currents – fast, intermediary, and slow I_Ks – with average activation τ of 0.7 ms, 6ms, and 15 ms, respectively. The fast I_K was associated with myocytes from animals in the estrus phase of the ovarian cycle, before ovulation and hence under estrogen domination. The intermediate I_K was associated with myocytes taken from diestrus animals, after ovulation and during secretion of progesterone. In freshly dissociated myocytes from estrus nonpregnant rat uterus, Piedras-Renteria et al. (1991) also described three K^+ currents on the basis of activation kinetics: a rapidly inactivating transient I_K with an activation τ of 1.3 ms, a fast I_K with an activation τ of 2.7 ms, and a slow I_K with an activation τ of 8.6 ms. Even accepting the transient I_K as new to the freshly dissociated myocyte, there is still little resemblance between the other currents in the two types of preparations. Neither the fast (τ = 0.7 ms) nor the slow (τ = 15 ms) I_Ks of the cultured cells were present in the fresh myocytes. The intermediate I_K of the cultured material (τ =6 ms), associated with progesterone of diestrus, had an activation rate closest to that of the slow I_K of the fresh myocyte, associated with the estrus state before progesterone was secreted. It thus appears that, by culturing, not only were kinetics altered but so was the hormonal state of the myocyte.

In the face of such glaring discrepancies, the inevitable conclusion is that the limited common qualitative features of fresh and cultured uterine myocytes (e.g., the presence of Ca^{2+} and K^+ currents) represent no more than some primitive functions expressed by a host of different cells possessing well-conserved homologous domains of channel proteins. The cultured myometrial cells possess none of the cell-specific functional ionic channels of the fresh uterine myocytes. Therefore, most cultured material used so far may consist of de-differentiating transitional forms, which are quite inappropriate as models for studying ionic channels of uterine myocytes.

Another model system used in studying myometrial ionic channels consists of incorporating isolated membrane fragments containing some ionic channels into artificial lipid bilayers (Toro, Ramos-Franco, and Stefani 1990a; Perez et al. 1993). In other instances, the methodology has been immensely useful in studying channel biology (see the references in Miller 1986). However, to be physiologically relevant, the first issue which needs to be clearly established is the origin of the microsomes: that is, whether or not they are derived from the plasma membrane. In the model studies, no unique surface markers were used (see e.g. Moczydlowski and Lattore 1983). Another issue of particular importance to uterine myocyte is the nature of the preparation. In the model studies, rat, pig, and human material were used – virtually interchangeably – without any information on hormonal status. The rationale that the properties of the large-conductance Ca^{2+}-activated K^+ channel were similar in the different species ignores the fact that the expression of that channel in the living myocyte changes markedly, depending on the hormonal status (Yang et al. 1994) and pregnancy status (Kahn et al. 1993; Wang et al. 1996). In the case of the channels from the human myometrium, disregard for physiological relevance was further demonstrated in the use of $[Ca^{2+}]_i$ considerably higher than could ever be attained in the living myocyte (Perez et al. 1993). Not surprisingly, the properties of the Ca^{2+}-activated K^+ channel in the model studies were therefore quite different from those in newly detached membrane patches from freshly dissociated uterine myocytes. For instance, for the rat myocyte, the open-probability (p_o) at 0 mV and pCa 6.8 was 0.18 in the model system (Toro et al. 1990a), but 0.003 in the fresh membrane (pCa 7). For the human myocyte, the ambient $[Ca^{2+}]_i$ in which the half-open probability occurred at 12 mV was 25.1 μM in the model system (Perez et al. 1993) but 0.5 μM in the fresh membrane (Kahn et al. 1993). For a $V_{0.5}$ of 56 mV, the ambient $[Ca^{2+}]_i$ was 7.2 μM in the model system, yet for a $V_{0.5}$ of 64 mV, the $[Ca^{2+}]_i$ was 0.05 μM in the fresh membrane. Hence the channels in the model system appear to be from 50 to over 100 times less sensitive to Ca^{2+} than those in the fresh membrane. The obvious questions are whether the channels are the same, and, even if they were, whether such large discrepancies can be ignored in studies of Ca^{2+}-sensitive processes.

4. Urinary tract myocytes

4.1. Ureteral myocytes

Judging by the shape of the action potentials, ureteral myocytes are among the most interesting of excitable cells. The action potential consists of a fast depolarizing phase followed by a rapid repolarization to a plateau (at around −15 to −20 mV) that lasts hundreds of milliseconds before final repolarization to the resting potential. The plateau is usually overlaid with irregular fast fluctuations as large as 10–15 mV (Kobayashi and Irisawa 1964; Washizu 1966; Kuriyama, Osa, and Toida 1967). Ionic currents have been studied in multicellular ureteral preparations (Shuba 1977), and more recently in freshly dissociated myocytes from the ureters of the guinea-pig (Imaizumi, Muraki, and Watanabe 1989, 1990; Lang 1989, 1990; Sui, Wang, and Kao 1990).

4.1.1. Inward currents

The inward current in the ureteral myocyte was carried mainly by Ca^{2+} through L-type channels. Its average density of 3.4 $\mu A/cm^2$ placed it at the low end of the spectrum of currents for visceral smooth myocytes. A standing controversy involves the role of Na^+ in the action potential. In some early studies, the action potential was unaffected either by replacing Na^+ with Li^+ (Washizu 1966) or by TTX (Kuriyama et al. 1967); in other studies, however, reducing extracellular Na^+ either prolonged or shortened the plateau, depending on the substituent (Kobayashi and Irisawa 1964). In double–sucrose gap studies, reducing $[Na^+]_o$ shortened the plateau, but Mn^{2+} (which blocked I_{Ca}) did not block the action potential (Shuba 1977). From these observations, both Na^+ and Ca^{2+} were thought to be involved in the action potential, possibly via a common channel (Shuba 1981).

In recent studies on dissociated ureteral myocytes, replacement of 87% of Na^+ with Li^+ shortened the I_{Ca} (Imaizumi et al. 1989), but the current carried by 1.5 mM Ba^{2+} through the Ca^{2+} channel was unaffected (Lang 1989). A fast, TTX-sensitive inward current that decayed rapidly was also observed (Lang 1990; Muraki, Imaizumi, and Watanabe 1991). On the other hand, Sui and Kao (1997a) failed to elicit any fast, TTX-sensitive inward currents even from a preconditioning voltage of −90 mV, nor did they find replacement of Na^+ by choline, TEA, or Tris to affect the amplitude or time course of the I_{Ca}, or of any outward K^+ current that might affect the inward current indirectly.

The I_{Ca} of ureteral myocyte differed from those of uterine or intestinal myocytes in its very slow rate of inactivation, with a τ_f of 217 ms and a τ_s of 2,456 ms. Associated with this slow decay was the absence or paucity of an inactivation mechanism dependent on $[Ca^{2+}]_i$, because both τ_f and τ_s remained constant in the voltage range of 0–30 mV (where I_{Ca} passed its maximum) and were also unaffected by Ba^{2+} (Sui 1993; Sui and Kao 1997b). Probably, the very slow inactivation is causally related to the lack of a $[Ca^{2+}]_i$-mediated Ca^{2+} inactivation.

In 3-mM $[Ca^{2+}]_o$, I_{Ca} was half-maximally inactivated at −16 mV and half-maximally activated at −1 mV, with an overlap region between −30 mV and 10 mV. The peak of this window current, at about −10 mV, reached ~25% of the maximum I_{Ca} (Sui 1993; Sui and Kao 1997b). Coupled with rather limited outward K^+ current (see Section 4.1.2), the long-lasting I_{Ca} would be responsible for the plateau of the action potential, usually located at about −20 mV. I_{Ca} reactivated via a two-exponential time course, with a τ_f of ~150 ms and a τ_s of ~2,000 ms. Even at 500 ms, less than half of the I_{Ca} was reactivated. Therefore, repetitive I_{Ca} is unlikely to cause the irregular voltage fluctuations on the plateau (Imaizumi et al. 1989).

4.1.2. Outward currents

Two outward K^+ currents have been recorded from the freshly dissociated ureteral myocyte. At hyperpolarized potentials, a transient current I_{TO}, was observed (Imaizumi et al. 1990; Sui 1993; Sui and Kao 1997c). It was half-maximally inactivated at −48 mV with a slope of 5 mV (Imaizumi et al. 1990). By this relation, at the usual resting potential of approximately −50 mV, almost half of this current was available to participate in physiological functions. However, since such an I_{TO} was not seen at a holding potential of −50 mV, its availability curve had to be more negative. Sui (1993) found the half-maximal inactivation at −76 mV, with a slope of 7 mV. Thus, at −50 mV, no more than 5% was available – probably insufficient to exert any significant physiological influence.

The other type of K^+ current was due to a large-conductance Ca^{2+}-activated

K^+ current, which was too noisy to be studied in detail in the whole-cell mode (Imaizumi et al. 1989; Lang 1989). Nonetheless, from the average current, it was small: 225 pA or 11 $\mu A/cm^2$ at 50 mV. From single-channel studies (Sui 1993; Sui and Kao 1997c), this channel had a unitary conductance of ~180 pS in asymmetrical K^+, was highly selective to K^+, and was susceptible to blockade by charybdotoxin (100 nM). It had a very low basal activity: at pCa 8, its half-open probability was at ~85 mV. It was highly sensitive to $[Ca^{2+}]_i$: at pCa 7, the half-open probability was shifted to ~25 mV. Ensemble reconstruction of such single-channel activities led to outward currents resembling those observed on whole-cell recordings, except for a more gradual activation. This difference is due to a lack of the surge in $[Ca^{2+}]_i$, attributable to I_{Ca}, that would have occurred in whole-cell recordings.

Thus, under usual physiological conditions, the long-lasting Ca^{2+} current and meager large-conductance Ca^{2+}-activated K^+ current contributed to a long plateau with occasional bursts of outward currents. The long action potential and its slow conduction velocity (Prosser, Smith, and Melton 1955) suggest that the active region of the ureter could be many millimeters in length. The contractile responses are likely to follow the same pattern, resulting in closure of long stretches of the ureter. In the absence of a sphincter at the ureterovesicular junction, such a closure could function as a valve to minimize reflux of urine, especially if intravesicular hydrostatic pressure were elevated.

4.2. Urinary bladder myocytes

Although action potentials of urinary bladder myocytes have been studied with impalement microelectrodes (Creed 1971; Mostwin 1986), ionic currents were not examined until Klöckner and Isenberg (1985a,b) worked on the freshly dissociated myocytes from the guinea-pig bladder. Until now, most investigations have been based on myocytes isolated from the detrusor portion of the guinea-pig bladder. The bladder myocyte had a resting potential of about −50 mV and a spikelike action potential, with an overshoot of about 20 mV, lasting ~35 ms (Klöckner and Isenberg 1989a; Sui 1993).

4.2.1. Inward currents

The inward current was entirely carried by Ca^{2+} through L-type channels. Among visceral smooth myocytes it was rather large, 12 $\mu A/cm^2$ (at ~22°C, Sui 1993) to 20 $\mu A/cm^2$ (at 35°C, Klöckner and Isenberg 1985b). Appreciable differences exist in the observations of these two groups, which may be partly due to differences in temperature. The maximum I_{Ca} occurred at −5 mV (Klöckner and Isenberg 1985b) or 15 mV (Sui 1993). At 35°C, I_{Ca} decayed by a three-exponential time course, with τs of 5, 40, and 220 milliseconds. The two faster components were strongly voltage-dependent, and the relation for the middle component was U-shaped with a minimum at −5 mV at which I_{Ca} was maximum. This feature, suggesting the presence of Ca^{2+}-mediated Ca^{2+} inactivation, was confirmed by direct measurement of $[Ca^{2+}]_i$ with Indo-1 (Ganitkevitch and Isenberg 1991). At 22°C, the decay of I_{Ca} followed a two-exponential time course with τs of 156 ms and 1,020 ms, both voltage-dependent. At 35°C, half-maximal inactivation occurred at −43 mV, with ~75% of I_{Ca} available at a resting potential of −50 mV. Half-maximal activation occurred at −14 mV. From −50 mV to −10 mV, a window current peaking at −30 mV to less than 10% of the maximum I_{Ca} could contribute to a slow "diastolic" depolarization and initiate spontaneous action potentials (Klöckner and Isenberg 1985b). At 22°C, half-maximal inactivation was at −21 mV and half-maximal activation at 12 mV (Sui 1993). Thus, the $V_{0.5}$ for activation and inactivation were both more positive by ~20 mV. A window current also shifted to between −30 mV and 10 mV, centering at −5 mV and reaching ~15% of the maximum I_{Ca}. At −50 mV, ~95% of the I_{Ca} was available, but spontaneous action potentials might be less likely to occur than in the myocytes studied by Klöckner and Isenberg (1985b).

4.2.2. *Outward currents*

The whole-cell outward current of bladder myocytes developed gradually even when elicited from a holding potential of −80 mV, indicating that no I_{TO} type of current was present (Sui 1993). The outward current was noisy, and was much attenuated when I_{Ca} was blocked (Klöckner and Isenberg 1985a; Sui 1993), indicating that a significant part of it was due to a large-conductance Ca^{2+}-activated K^+ current. This current has been examined via single-channel studies. In asymmetrical K^+ (135/5.4 mM), it had a unitary conductance of 162 pS at 0 mV. Its openings occurred in clusters with long closed periods in between. At pCa 8, $V_{0.5}$ was 56 mV; at pCa 7, it was shifted to 26 mV. The channel can be blocked by TEA from the outside surface, with a K_d of 255 nM. In 100 nM charybdotoxin outside, it was completely blocked.

Using the susceptibility to charybdotoxin as a tool, a second K^+ current was unmasked. As expected, the latter turned out to be insensitive to Ca^{2+}, outwardly rectifying, and decayed with a τ of 1–2 s. Its steady-state inactivation relation showed that 20% of the current did not inactivate and that 80% inactivated half-maximally at −73 mV. Thus, at a resting potential of −50 mV, ~30% of this current was available. Active single-channel openings with unitary conductances of ~27 pS were seen when elicited from a holding potential of −80 mV, but were rare from a holding potential of −30 mV. These properties suggest that this channel is probably a delayed rectifier–type channel.

In addition to these two types of K^+ currents, a voltage-insensitive K^+ channel that opened at low levels (less than 0.1 mM) but closed at higher levels (more than ~1 mM) of intracellular ATP (K_{ATP} channel) has been demonstrated under conditions when the large-conductance Ca^{2+}-activated K^+ current was completely suppressed (Bonev and Nelson 1993). Under normal physiological conditions, it is probably not sufficiently expressed to affect function.

Using an observed action potential as the voltage command and after selectively blocking the K^+ currents, the inward I_{Ca} had been isolated and integrated and the charge moved determined. The Ca^{2+} influx during the action potential was 73 aM (10^{-18} M). Into a mean morphometric cell volume of 5.3 pl, this amount of Ca^{2+} would raise $[Ca^{2+}]_i$ to 13.6 μM (Sui, Wang, and Kao 1993). This quantity is several times higher than what had been observed in various measurements of $[Ca^{2+}]_i$ by indicators. The difference may be due to a highly efficient intracellular Ca^{2+}-buffering system and to limitations of the photometric methods used. In all attempted determinations of $[Ca^{2+}]_i$, the voltage change applied was many times longer than the duration of a single action potential (~36 ms). Possibly, the Ca^{2+} influxed during an action potential was buffered rapidly and efficiently, and the average $[Ca^{2+}]_i$ might never be much higher than 1 μM. Nevertheless, the estimated rise of $[Ca^{2+}]_i$ to 13.6 μM simply demonstrates that the amount of Ca^{2+} entering the cell via the opened Ca^{2+} channel is more than enough to serve a myriad of physiological functions.

5. Interim assessment of current knowledge

Freshly dissociated smooth myocytes from some visceral organs, when properly isolated, possess many unique tissue-specific excitability properties, and are ideal preparations for studying cellular ionic channel functions. Until now, ionic channels in model systems, such as tissue-cultured material or channels incorporated into lipid bilayers, are too different from those in fresh myocytes to merit acceptance of those model systems as adequate substitutes. The practice of using Ba^{2+} to enhance inward currents is undesirable, because Ba^{2+} distorts some channel properties and renders interpretation of physiological phenomena difficult.

Whereas much has been learned about various ionic channels in smooth myocytes of the gastrointestinal tract, the uterus, and the lower urinary tract, the relation of cellular properties to physiological function is not always clear. For the lower urinary tract, fairly marked differences in ionic channels of ureteral and bladder myocytes need to be better integrated into their systemic functions. In the circular layer myocytes of the small intestine, no information is available to explain slow waves. In the uterine myocyte, much remains unclear on regulation of ionic channels by ovarian hormones and during pregnancy. In all visceral smooth myocytes, the problems of identifying, sorting, and allocating relative contributions of different K^+ channels remain unresolved. Although cloning provides powerful evidence on the identity of channels, it needs to be integrated with other approaches to help apportion the roles of individual types of channels for constructing a fuller understanding of the functions of smooth myocytes.

Acknowledgements

The work in my laboratory was supported in part by grants HD00378 and DK39371 from the National Institutes of Health.

References

Akbarali, H. I., and Giles, W. R. (1993). Ca^{2+} and Ca^{2+}-activated Cl^- currents in rabbit oesophageal smooth muscle. *J. Physiol. (Lond.)* 460: 117–33.

Akbarali, H. I., and Goyal, R. K. (1994). Effect of sodium nitroprusside on Ca^{2+} currents in opossum esophageal circular muscle cells. *Am. J. Physiol.* 266: G1036–G1042.

Akbarali, H. I., Hatakeyama, N., Wang, Q., and Goyal, R. K. (1995). Transient outward current in opossum esophageal circular muscle. *Am. J. Physiol.* 268: G975–G987.

Amedee, T., Mironneau, C., and Mironneau, J.. (1987). The calcium channel current of pregnant rat single myometrial cells in short-term primary culture. *J. Physiol. (Lond.)* 392: 253–72.

Amedee, C. T., Renaud, J. F., Jmari, K., Lombet, A., Mironneau, J., and Lazdunski, M. (1986). The presence of Na^+ channels in myometrial smooth muscle cells is revealed by specific neurotoxins. *Biochem. Biophys. Res. Commun.* 137: 675–81.

Anderson, N. C. (1969). Voltage-clamp studies on uterine smooth muscle. *J. Gen. Physiol.* 54: 145–65.

Anderson, N. C., Ramon, F., and Snyder, A. (1971). Studies in calcium and sodium in uterine smooth muscle excitation under current-clamp and voltage-clamp conditions. *J. Gen. Physiol.* 58: 322–39.

Bagby, R. M., Young, A. M., Dotson, R. S., Fisher, B. A., and McKinnon, K. (1971). Contractions of single smooth muscle cells from *Bufo marinus* stomach. *Nature* 234: 351–2.

Benham, C. D., and Bolton, T. B. (1986a). Calcium-activated potassium channels in single smooth muscle cells of rabbit jejunum and guinea-pig mesenteric artery. *J. Physiol. (Lond.)* 371: 45–67.

Benham, C. D., and Bolton, T. B. (1986b). Spontaneous transient outward currents in single visceral and vascular smooth muscle cells of the rabbit. *J. Physiol. (Lond.)* 381: 385–406.

Benham, C. D., Bolton, T. B., Lang, R. J. and Takewaki, T. (1985). The mechanism of action of Ba^{2+} and TEA on single Ca^{2+}-activated K^+ channels in arterial and intestinal smooth muscle cell membranes. *Pflügers Arch.* 403: 120–7.

Benham, C. D., Bolton, T. B., Lang, R. J., and Takewaki, T. (1986). Calcium-activated potassium channels in single smooth muscle cells of rabbit jejunum and guinea-pig mesenteric artery. *J. Physiol. (Lond.)* 371: 45–67.

Bielefeld, D. R., Hume, J. R., and Krier, J. (1990). Action potentials and membrane currents of isolated single smooth muscle cells of cat and rabbit colon. *Pflügers Arch.* 415: 678–87.

Bolton, T. B., Lang, R. J., Takewaki, T., and Benham, C. D. (1985). Patch and whole-cell voltage-clamp of single mammalian visceral and vascular smooth muscle cells. *Experentia* 41: 887–94.

Bolton, T. B., Tomita, T., and Vassort, G. (1981). Voltage clamp and measurement of ionic conductance in smooth muscle. In *Smooth Muscle: An Assessment of Current Knowledge.* (E. Bülbring, A. F. Brading, A. W. Jones, and T. Tomita, eds.). Austin: University of Texas Press, pp. 47–61.

Bonev, A. D., and Nelson, M. T. (1993). ATP-sensitive potassium channels in smooth muscle cells from guinea pig urinary bladder. *Am. J. Physiol.* 264: C1190–C1200.

Boyle, M. B., Azhderian, E. M., MacLusky, N. J., Naftolin, F., and Kaczmarek, L. K. (1987). *Xenopus* oocytes injected with rat uterine RNA express very slowly activating potassium currents. *Science* 235: 1221–1224.

Carl, A., McHale, N. G., Publicover, N. G., and Sanders, K. M. (1990). Participation of Ca^{2+}-activated K^+ channels in electrical activity of canine gastric smooth muscle. *J. Physiol. (Lond.)* 429: 205–21.

Carl, A., and Sanders, K. M. (1989). Ca^{2+}-activated K channels of canine colonic myocytes. *Am. J. Physiol.* 257: C470–C480.

Cole, K. S. (1968). *Membrane, Ions and Impulses.* Berkeley: University of California Press, pp. 267–9.

Cole, W. C., and Sanders, K. M. (1989). Characterization of macroscopic outward currents of canine colonic myocytes. *Am. J. Physiol.* 257: C461–C469.

Conner, J. A., Kreulen, D., Prosser, C. L., and Weigel, R. (1977). Interaction between longitudinal and circular muscle in intestine of cat. *J. Physiol. (Lond.)* 273: 665–89.

Creed, K. E. (1971). Effects of ions and drugs on smooth muscle membrane of the guinea-pig urinary bladder. *Pflügers Arch.* 338: 149–64.

Dreyer, F. (1990). Peptide toxins and potassium channels. *Rev. Physiol. Biochem. Pharmacol.* 115: 93–136.

Droogmans, G., and Callewart, G. (1986). Ca^{2+}-channel current and its modification by the dihydropyridine agonist BAY K 8644 in isolated smooth muscle cells. *Pflügers Arch.* 406: 259–65.

Fan, S. F., Wang, S. Y., and Kao, C. Y. (1993). The transduction system in the isoproterenol activation of the Ca^{2+}-activated K^+ channel in the guinea-pig taenia coli myocyte. *J. Gen. Physiol.* 102: 257–75.

Farrugia, G., Rae, J. L., Sarr, M. G., and Szurszewski, J. H. (1993a). Potassium current in circular smooth muscle of human jejunum activated by fenamates. *Am. J. Physiol.* 265: G873–G879.

Farrugia, G., Rae, J. L., and Szurszewski, J. H. (1993b). Characterization of an outward potassium current in canine jejunal circular smooth muscle and its activation by fenamates. *J. Physiol. (Lond.)* 468: 297–310

Farrugia, G., Rich, A., Rae, J. L., Sarr, M. G., and Szurszewski, J. H. (1995). Calcium currents in human and canine jejunal circular smooth muscle cells. *Gastroenterology* 109: 707–13.

Fay, F. S., and Delise, C. M. (1973). Contractions of isolated smooth muscle cells – Structural changes. *Proc. Nat. Acad. Sci. USA* 70: 641–3.

Felipe, A., Knittle, T. J., Doyle, K. L., and Tamkun, M. M. (1994). Primary structure and differential expression during development and pregnancy of a novel voltage-gated sodium channel in the mouse. *J. Biol. Chem.* 269: 30125–31.

Frankenhauser, B., and Hodgkin, A. L. (1957). The action of calcium on the electrical properties of squid axon. *J. Physiol. (Lond.)* 137: 218–44.

Galvez, A., Gimenez-Gallego, G., Reuben, J. P., Roy-Contacin, L., Feigenbaum, L., Kaczorowski, G. L., and Garcia, M. L. (1990). Purification and characterization of a unique potent peptidyl probe for the high conductance calcium-activated potassium channel from venom of the scorpion *Buthus tomulus. J. Biol. Chem.* 265: 11083–90.

Ganitkevitch, V. Ya., and Isenberg, G. (1991). Depolarization-mediated intracellular calcium transients in isolated smooth muscle cells of guinea-pig urinary bladder. *J. Physiol. (Lond.)* 437: 187–205.

Ganitkevitch, V. Ya., Shuba, M. F., and Smirnov, S. V. (1986). Potential-dependent calcium inward current in a single isolated smooth muscle cell of the guinea-pig taenia coli. *J. Physiol. (Lond.)* 380: 1–16.

Ganitkevich, V. Ya., Shuba, M. F., and Smirnov, S. V. (1987). Calcium-dependent inactivation of potential-dependent calcium inward current in an isolated guinea-pig smooth muscle cell. *J. Physiol. (Lond.)* 392: 431–49.

George, A. L., Knittle, T. J., and Tamkun, M. M. (1992). Molecular cloning of an atypical voltage-gated sodium channel expressed in human heart and uterus: Evidence for a distinct gene family. *Proc. Natl. Acad. Sci. USA.* 89: 4891–7.

Hamill, O. P., Marty, A., Neher, E., Sakmann, B., and Sigworth, F. J. (1981). Improved patch-clamp technique for high-resolution current recording from cells and cell-free membrane patches. *Pflügers Arch.* 391: 85–100.

Hart, P. J., Overturf, K. E., Russel, S. N., Carl, A., Hume, J. R., Sanders, K. M., and Horowitz, B. (1993). Cloning and expression of a K_v 1. 2 class delayed rectifier

K^+ channel from canine colonic smooth muscle. *Proc. Natl. Acad. Sci. USA* 90: 9659–63.

Hille, B. (1992). *Ionic Channels of Excitable Membranes*, 2nd ed. Sunderland, MA: Sinauer, pp. 524–44.

Hodgkin, A. L., Huxley, A. F., and Katz, B. (1952). Measurements of current-voltage relations in the membrane of the giant axon of *Loligo. J. Physiol. (Lond.)* 116: 424–48.

Hu, S. L., Yamamoto, Y., and Kao, C. Y. (1989a). The Ca^{2+}-activated K^+ channel and its functional roles in smooth muscle cells of the guinea pig taenia coli. *J. Gen. Physiol.* 94: 833–47.

Hu, S. L., Yamamoto, Y., and Kao, C. Y. (1989b). Permeation, selectivity and blockade of the Ca^{2+}-activated potassium channel of the guinea pig taenia coli myocyte. *J. Gen. Physiol.* 94: 849–62.

Imaizumi, Y., Muraki, Y., and Watanabe, M. (1989). Ionic currents in single smooth muscle cells from the ureter of the guinea-pig. *J. Physiol. (Lond.)* 411: 131–59.

Imaizumi, Y., Muraki, K., and Watanabe, M. (1990). Characteristics of transient outward current in single smooth muscle cells from the ureter of the guinea-pig. *J. Physiol. (Lond.)* 427: 301–24.

Inomata, H., and Kao, C. Y. (1976). Ionic currents in the guinea-pig taenia coli. *J. Physiol. (Lond.)* 255: 347–78.

Inomata, H., and Kao, C. Y. (1979). Ionic mechanisms of repolarization in the guinea-pig taenia coli as revealed by the actions of strontium. *J. Physiol. (Lond.)* 297: 443–62.

Inomata, H., and Kao, C. Y. (1984). Actions of Ba^{2+} on ionic currents of the guinea-pig taenia coli. *J. Pharmacol. Exp. Ther.* 233: 112–24.

Inoue, Y., Nakao, K., Okabe, K., Izumi, H., Kanda, S., Kitamura, K., and Kuriyama, H. (1990). Some electrical properties of human pregnant myometrium. *Am. J. Obstet. Gynecol.* 162: 1090–98.

Inoue, Y., Shimamura, K., and Sperelakis, N. (1993). Forskolin inhibition of K^+ current in pregnant rat uterine smooth muscle cells. *Eur. J. Pharmacol.* 240: 169–76.

Inoue, Y., and Sperelakis, N. (1991). Gestational change in Na^+ and Ca^{2+} channel current densities in rat myometrial smooth muscle cells. *Am. J. Physiol.* 29: C658–C663.

Kahn, R. N., Smith, S. K., Morrison, J. J., and Ashford, M. L. J. (1993). Properties of large-conductance K^+ channels in human myometrium during pregnancy and labour. *Proc. Roy. Soc. Lond. B* 251: 9–15.

Kao, C. Y. (1977). Recent experiments in voltage-clamp studies on smooth muscles.In: *Excitation–Contraction Coupling in Smooth Muscle.* (R. Casteel, T. Godfraind, and J. C. and Ruegg, eds.). Amsterdam, Elsevier, pp. 91–6.

Kao, C. Y. (1978). A calcium current in the rat myometrium. *Jpn. J. Smooth Muscle Res.* (suppl.) 14: 9–10.

Kao, C. Y. (1989). Electrophysiological properties of uterine smooth muscle. In *Biology of the Uterus.* (T. M. Wynn and W. P. Jollie, eds.). New York: Plenum, pp. 403–54.

Kao, C. Y., and Levinson, S. R. (1986). Tetrodotoxin, saxitoxin and the molecular biology of the sodium channel. *Ann. N.Y. Acad. Sci.* 479.

Kao, C. Y., and McCullough, J. R. . (1975). Ionic currents in the uterine smooth muscle. *J. Physiol. (Lond.)* 246: 1–36.

Kao, C. Y., Wakui, M., Wang, S. Y., and Yoshino, M. (1989). The outward current of the isolated rat myometrium. *J. Physiol. (Lond.)* 418: 20p.

Kao, C. Y., and Wang, S. Y. (1994). Effects of lidocaine on rat myometrial sodium channels and implications for the management of preterm labor. *Am. J. Obstet. Gynecol.* 171: 446–54.

Katzka, D. A., and Morad, M. (1989). Properties of calcium channels in guinea-pig gastric myocytes. *J. Physiol. (Lond.)* 413: 175–97.

Klöckner, U., and Isenberg, G. (1985a). Action potentials and net membrane currents of isolated smooth muscle cells (urinary bladder of the guinea-pig). *Pflügers Arch.* 405: 329–39.

Klöckner, U., and Isenberg, G. (1985b). Calcium currents of cesium loaded isolated smooth muscle cells (urinary bladder of the guinea pig). *Pflügers Arch.* 405: 340–8.

Kobayashi, M., and Irisawa, H. (1964). Effect of sodium deficiency on the action potential of the smooth muscle of ureter. *Am. J. Physiol.* 296: 205–20.

Kuriyama, H., Osa, T., and Toida, N. (1967). Membrane properties of the smooth muscle of guinea-pig ureter. *J. Physiol. (Lond.)* 191: 225–38.

Lang, R. J. (1989). Identification of the major membrane currents in freshly dispersed single smooth muscle cells of guinea-pig ureter. *J. Physiol. (Lond.)* 412: 375–95.

Lang, R. J. (1990). The whole-cell Ca^{2+} channel current in single smooth muscle cells of the guinea-pig ureter. *J. Physiol. (Lond.)* 423: 453–73.

Langton, P. D., Burke, E. P., and Sanders, K. M. (1989). Participation of Ca currents in colonic electrical activity. *Am. J. Physiol.* 257: C451–C460.

Lieberman, M., Hauschka, S. D., Hall, Z. W., Eisenberg, B. R., Horn, R., Walsh, J. V., Tsien, R. W., Jones, A. W., Walker, J. L., Poenie, M., Fay, F., Fabiato, F., and Ashley, C. C. (1987). Isolated muscle cells as a physiological model. *Am. J. Physiol.* 253: C349–C363.

Miller, C. (1986). *Ion Channel Reconstitution*. New York: Plenum.

Mironneau, J. (1974). Voltage clamp analysis of the ionic currents in uterine smooth muscle using the double sucrose gap method. *Pflügers Arch.* 352: 107–210.

Mironneau, J., and Savineau, J. P. (1980). Effects of calcium ions on outward membrane currents in rat uterine smooth muscle. *J. Physiol. (Lond.)* 302: 411–25.

Mitra R., and Morad, M. (1985). Ca^{2+} and Ca^{2+}-activated K^+ currents in mammalian gastric smooth muscle cells. *Science.* 229: 269–72.

Miyoshi, H., Urabe, T., and Fujiwara, A. (1991). Electrophysiological properties of membrane currents in single myometrial cells isolated from pregnant rats. *Pflügers Arch.* 419: 386–93.

Moczydlowski, E. G., and Lattore, R. (1983). Saxitoxin and quabain binding activity of isolated skeletal muscle membrane as indicators of surface origin and purity. *Biochim. Biophys. Acta* 732: 412–30.

Mollard, P., Mironneau, J., Amedee, T., and Mironneau, C. (1986). Electrophysiological characterization of single pregnant myometrial cells in short-term primary culture. *Am. J. Physiol.* 250: C47–C54.

Momose, K., and Gomi, Y. (1980). Studies on isolated smooth muscle cells. IV. Dispersion procedures for acetylcholine-sensitive smooth muscle cells of guinea pig. *Jpn. J. Smooth Muscle Res.* 16: 29–36.

Mostwin, J. L. (1986). The action potential of guinea pig bladder smooth muscle. *J. Urol.* 135: 1299–1303.

Muraki, K., Imaizumi, Y., and Watanabe, M. (1991). Sodium currents in smooth muscle cells freshly isolated from stomach fundus of the rat and ureter of the guinea-pig. *J. Physiol. (Lond.)* 442: 351–75.

Nakai, Y., and Kao, C. Y. (1983). Changing proportions of Na^+ and Ca^{2+} components of the early inward current in rat myometrium during pregnancy. *Fed. Proc.* 42: 313.

Ohya, Y., and Sperelakis, N. (1989). Fast Na^+ and slow Ca^{2+} channels in single uterine muscle cells from pregnant rats. *Am. J. Physiol.* 257: C408–C412.
Ohya, Y., Terada, K., Kitamura, K., and Kuriyama, H. (1986). Membrane currents recorded from a fragment of rabbit intestinal smooth muscle cell. *Am. J. Physiol.* 251: C335–C346.
Osa, T., and Katase, T. (1975). Physiological comparison of the longitudinal and circular muscles of the pregnant rat uterus. *Jpn. J. Physiol.* 25: 153–64.
Overturf, K. E., Russel, S. N., Carl, A., Vogalis, F., Hart, P. J., Hume, J. R., Sanders, K. M., and Horowitz, B. (1994). Cloning and characterization of a K_v1. 5 delayed rectifier K^+ channel from vascular and visceral smooth muscles. *Am. J. Physiol.* 267: C1231–1238.
Parkington, H. C., and Coleman, H. A. (1990). The role of membrane potential in the control of uterine activity. In *Uterine Function*. (M. E. Carsten and J. D. Miller, eds.). New York: Plenum, pp. 195–248.
Perez, G. J., Toro, L., Erulkar, S. D., and Stefani, E. (1993). Characterization of large-conductance calcium-activated potassium channels from human myometrium. *Am. J. Obstet. Gynecol.* 158: 653–60.
Piedras-Renteria, E., Toro, L., and Stefani, E. (1991). Potassium currents in freshly dispersed myometrial cells. *Am. J. Physiol.* 251: C278–C284.
Piper, H. M., and Isenberg, G. (1989). *Isolated Adult Cardiomyocytes*. Boca Raton, FL: CRC Press.
Prosser, C. L., Smith, C. F., and Melton, C. E. (1955). Conduction of action potentials in the ureter of the rat. *Am. J. Physiol.* 181: 651–60.
Rendt, J. M., Toro, L., Stefani, E., and Erulkar, S. D. (1992). Progesterone increases Ca^{2+} currents in myometrial cells from immature and non-pregnant adult rats. *Am. J. Physiol.* 31: C293–C301.
Sanders, K. M. (1989). Electrophysiology of dissociated gastrointestinal muscle cells. In: *Handbook of Physiology – The Gastrointestinal System I*. Bethesda, MD: American Physiological Society, pp. 1163–85.
Sanders, K. M., and Publicover, N. G. (1989). Electrophysiology of the gastric musculature. In: *Handbook of Physiology – The Gastrointestinal System I*. Bethesda, MD: American Physiological Society, pp. 187–216.
Shuba, M. F. (1977). The effect of sodium-free and potassium-free solutions, ionic current inhibitors and ouabain on electrophysiological properties of smooth muscle of guinea-pig ureter. *J. Physiol. (Lond.)* 264: 837–51.
Shuba, M. F. (1981). Smooth muscle of the ureter: The nature of excitation and the mechanisms of action of catecholamines and histamine. In: *Smooth Muscle: An Assessment of Current Knowledge* (E. Bülbring, A. F. Brading, A. W. Jones, and T. Tomita, eds.). Austin: University of Texas Press, pp. 377–548.
Sims, S. M. (1992). Calcium and potassium currents in canine gastric smooth muscle cells. *Am. J. Physiol.* 262: G859–G867.
Sims, S. M., Vivaudou, M. B., Hillemeier, C., Biancani, P., Walsh, J. V., and Singer, J. J. (1990). Membrane currents and cholinergic regulation of K^+ current in esophageal smooth muscle cells. *Am. J. Physiol.* 258: G794–G802.
Smirnov, S. V., Zholos, A. V., and Shuba, M. F. (1992). Potential-dependent inward currents in single isolated smooth muscle cells of the rat ileum. *J. Physiol. (Lond.)* 454: 549–71.
Sui, J. L. (1993). Ionic channel functions in smooth myocytes of lower urinary tract of guinea pigs. Ph.D. thesis, Health Science Center at Brooklyn, State University of New York.
Sui, J. L., and Kao, C. Y. (1997a). The roles of Ca^{2+} and Na^+ in the inward current

and action potentials of guinea-pig ureteral myocytes. *Am. J. Physiol.* 272: C535–42.

Sui, J. L. and Kao, C. Y. (1997b). Properties of inward calcium current in guinea-pig ureteral myocytes. *Am. J. Physiol.* 272: C543–49.

Sui, J. L. and Kao, C. Y. (1997c). Role of outward potassium currents in the action potentials of guinea-pig ureteral myocytes. *Am. J. Physiol.* (in press).

Sui, J. L., Wang, S. Y., and Kao, C. Y. (1990). Ionic currents in single myocytes of guinea pig ureter. *Biophys. J.* 57: 156a.

Sui, J. L., Wang, S. Y., and Kao, C. Y. (1993). Ca^{2+} influx during action potentials in smooth myocytes of guinea pig urinary bladder. *Biophys. J.* 64: A364.

Thornbury, K. D., Ward, S. M., and Sanders, K. M. (1992). Participation of fast-activating, voltage-dependent K currents in electrical slow waves of colonic circular muscle. *Am. J. Physiol.* 263: C226–C236.

Toro, L., Ramos-Franco, J., and Stefani, E. (1990a). GTP-dependent regulation of myometrial K_{Ca} channels incorporated into lipid bilayers. *J. Gen. Physiol.* 96: 373–94.

Toro, L., Stefani, E., and Erulkar, S. D. (1990b). Hormonal regulation of potassium currents in single myometrial cells. *Proc. Nat. Acad. Sci. USA.* 87: 2892–5.

Vassort, G. (1975). Voltage-clamp analysis of transmembrane ionic currents in guinea-pig myometrium: Evidence for an initial potassium activation triggered by calcium entry. *J. Physiol. (Lond.)* 252: 713–34.

Vogalis, F., Lang, R. J., Bywater, R. A. R., and Taylor, G. S. (1993). Voltage-gated ionic currents in smooth muscle cells of guinea pig proximal colon. *Am. J. Physiol.* 264: C527–C536.

Vogalis, F., Publicover, N. G., and Sanders, K. M. (1992). Regulation of calcium current by voltage and cytoplasmic calcium in canine gastric smooth muscle. *Am. J. Physiol.* 262: C691–C700.

Wakui, M., and Kao, C. Y. (1981). The late current in the rat myometrium studied with the aid of Co^{2+}. *Fed. Proc.* 40: 552.

Walsh, J. V., and Singer, J. J. (1980). Calcium action potentials in single freshly dissociated smooth muscle cells. *Am. J. Physiol.* 239: C162–C174.

Wang, S. Y., and Kao, C. Y. (1993). Ionic currents in the uterine myocyte during pregnancy. *Biophys. J.* 64: A366.

Wang, S. Y., Yoshino, M., Sui, J. L., and Kao, C. Y. (1996). Pregnancy and K^+ currents of freshly dissociated rat uterine myocytes. *Biophys. J.* 70: A396.

Washizu, Y. 1966. Grouped discharges in ureter muscle. *Comp. Biochem. Physiol.* 19: 713–28.

Weigel, R. J., Conner, J. A., and Prosser, C. L. (1979). Two roles of calcium during the spike in circular muscle of small intestine in cat. *Am. J. Physiol.* 237: C247–C256.

Yamamoto, Y., Hu, S. L. and Kao, C. Y. (1989a). Inward current in single cells of the guinea pig taenia coli. *J. Gen. Physiol.* 93: 521–550.

Yamamoto, Y., Hu, S. L., and Kao, C. Y. (1989b). Outward current in single smooth muscle cells of the guinea pig taenia coli. *J. Gen. Physiol.* 93: 551–64.

Yang, L., Wang, S. Y., and Kao, C. Y. (1994). Regulation by ovarian hormones of the large-conductance Ca^{2+}-activated K^+ channels of freshly dissociated rabbit uterine myocytes. *Biophys. J.* 66: A436.

Yoshino, M., Someya, T., Nishio, A., Yazawa, K., Usuki, T., and Yabu, H. (1989). Multiple types of voltage-dependent Ca channels in mammalian intestinal smooth muscle cells. *Pflügers Arch.* 414: 401–9.

Yoshino, M., Wang, S. Y. and Kao, C. Y. (1989). Ionic currents in smooth myocytes of the pregnant rat uterus. *J. Gen. Physiol.* 94: 38a.

Yoshino, M., Wang, S. Y. and Kao, C. Y. (1990). Inward current in freshly dispersed rat myometrial cells. *Biophys. J.* 57: 164A.

Yoshino, M., Wang, S. L., and Kao, C. Y. (1997). Sodium and calcium inward currents in freshly dissociated smooth myocytes of rat uterus. *J. Gen. Physiol.* (in press).

Young, R. C., and Herndon-Smith, L. (1991). Characterization of sodium channels in cultured human uterine smooth muscle cells. *Am. J. Obstet. Gynecol.* 164: 175–81.

Young, R. C., Herndon-Smith, L., and Anderson, N. C. (1991). Passive membrane properties and inward calcium current of human smooth muscle cells. *Am. J. Obstet. Gynecol.* 164: 1132–9.

4
Muscarinic Regulation of Ion Channels in Smooth Muscle

LUKE J. JANSSEN AND STEPHEN M. SIMS

1. Introduction

Acetylcholine (ACh) acts on muscarinic receptors to cause excitation of a wide range of smooth muscles. Early studies on multicellular smooth muscle preparations led to the view that muscarinic stimulation caused depolarization by activating channels permeable to Na^+, K^+, and Ca^{2+} (Bolton 1981). This proposed action of ACh in smooth muscle was similar to that described for neuromuscular transmission in skeletal muscle (Takeuchi and Takeuchi 1960; Katz 1966), where ACh binds to nicotinic ACh receptors which also function as nonselective cation channels. Support for this proposal was elusive in early studies of smooth muscles because it was difficult to apply the voltage-clamp technique – the method of choice for determining membrane conductance changes – to multicellular preparations, in large part because of difficulties in obtaining adequate spatial and temporal control of membrane potential (see the references in Singer and Walsh 1980a; or in Bolton et al. 1981). For this reason, the primary effect of ACh on membrane channels could not be distinguished from secondary changes accompanying depolarization. For example, ACh causes depolarization which in turn activates voltage-dependent Ca^{2+} channels (see Figure 1). Without voltage clamp, it would be impossible to distinguish between ACh activating Ca^{2+} channels directly or through an indirect mechanism involving depolarization.

Several important breakthroughs permitted clear and decisive studies of voltage- and transmitter-activated channels in smooth muscle. The first was the development of techniques for dissociating smooth muscle tissues into single cells suitable for physiological studies (Bagby et al. 1971; Fay and Delise 1973; reviewed in Sanders 1989). The second was the application of microelectrode voltage-clamp techniques to these cells for recording membrane potential and whole-cell currents (Singer and Walsh 1980a,b; Walsh and Singer 1980a,b). The third breakthrough was the development of patch-clamp techniques for high resolution recording both of whole-cell and single-channel currents (Hamill et al. 1981).

The purpose of this chapter is to review the excitatory effects of cholinergic agonists on smooth muscle, focusing on electrophysiological studies of single cells. Responses of intact muscles have been reviewed by others (e.g. Burnstock et al. 1963; Bolton 1979, 1981) and will not be considered here. Similarly, the types of ion channels expressed by smooth muscles have been reviewed elsewhere (e.g. Bolton 1989; Sanders 1989). We will first provide an overview of cholinergic excitation in smooth muscle to provide background for the main objective of this review: the ways that ACh regulates activity of ion channels to cause depolarization. We will also consider the signaling mechanisms mediating actions of ACh. A recurring theme will be the effects of cholinergic agonists on Ca^{2+} homeostasis, since Ca^{2+} plays a key role in excitation and contraction of smooth muscle (reviewed in Somlyo and Somlyo 1994). The ionic mechanisms underlying cholinergic excitation have been described elsewhere (Bolton 1989; Sanders 1989; Inoue and Shan 1993; Sims and Janssen 1993).

1.1. Cholinergic excitation: overview

Parasympathetic cholinergic nerve fibers project from the central nervous system via the vagus and sacral spinal segments, terminating on smooth muscle tissues throughout the body (Gillis et al. 1986; Gabella 1987). The density of the innervation and the proximity of nerve endings to the smooth muscle cells vary widely among tissues. As a result, the nature of cholinergic control in the different smooth muscle tissues can vary considerably. For example, a sparse innervation and large separation between nerve endings and smooth muscle cells in certain vascular smooth muscles (Somlyo and Somlyo 1968) and tracheal muscle (Daniel et al. 1986; Gabella 1987) account for the diffuse response to cholinergic input; in these tissues, the muscle acts as a syncytium. On the other hand, in tissues from the gastrointestinal tract, the high density and proximity of the cholinergic nerve endings to the muscle cells allow for a very localized and specific control over their activity (Burnstock 1979; Bolton and Large 1986); in these cases, one region of the tissue can be excited while adjacent regions are quiescent.

The target tissues of the cholinergic innervation are widespread, including the vasculature (arterial and venous), the tracheobronchial tree, the gastrointestinal tract, the sphincter pupillae, and the uterus (Somlyo and Somlyo 1968; Burnstock 1979; Bolton 1979, 1981; Bolton and Large 1986; Gabella 1987). Thus, cholinergic control of smooth muscle is of major importance to homeostasis in the organism and plays a central role in such diseases as asthma and related breathing disorders, hypertension, stroke, migraine, ileitis, and colitis.

Many (though not all) smooth muscles generate action potentials (Burnstock et al. 1963; Somlyo and Somlyo 1968; Bolton 1979; Burnstock 1979; Bolton and Large 1986). The action potential in most cases is due to activation of

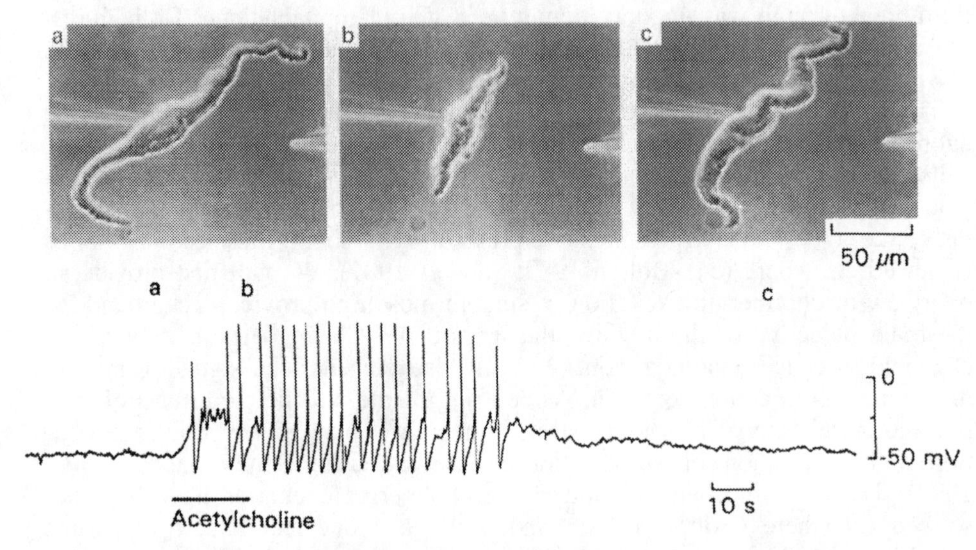

Figure 1. Acetylcholine (ACh) causes excitation and contraction of gastric smooth muscle cell from *Bufo marinus*. Membrane potential was recorded using intracellular microelectrode, seen at left in video frames. Contraction was simultaneously monitored on video, shown above. ACh (applied from micropipette at right, 50 μM) caused depolarization and action potentials, accompanied by reversible contraction. Timing of video frames is indicated by letters above voltage trace. Reproduced with permission from Sims and Janssen (1993).

voltage-dependent Ca^{2+} channels (Walsh and Singer 1980a,b; see also Bean 1989a, Bolton 1989, and Sanders 1989 for review). The resulting influx of Ca^{2+} causes further depolarization, elevation of cytosolic free Ca^{2+} concentration ($[Ca^{2+}]_i$) (Becker et al. 1989; Ganitkevich and Isenberg 1991), and contraction. In tissues generating action potentials, cholinergic agonists can depolarize the membrane to threshold, initiating action potentials. This is illustrated in Figure 1, where ACh caused depolarization and a burst of action potentials in a muscle cell from the stomach of the toad *Bufo marinus*. Simultaneous video recording revealed reversible contraction of the cell accompanying the burst of action potentials. Cholinergic stimulation can also modulate existing action potential activity, increasing the frequency of bursts, the number of action potentials per burst, and/or the amplitude and duration of the plateau potential. Lastly, as illustrated in Figure 2 for ileal muscle, ACh can cause prolonged depolarization associated with a cessation of action potentials.

In smooth muscles that do not produce action potentials, cholinergic agonists elicit depolarization with a much slower time course and smaller amplitude than is typical of action potentials (Bolton and Large 1986; Coburn and Baron 1990). Usually, the depolarization elicited is less than 10–20 mV and persists for several seconds or even minutes. As with regenerative action potentials, the depolarization may activate voltage-dependent Ca^{2+} channels, resulting in Ca^{2+} influx and contraction.

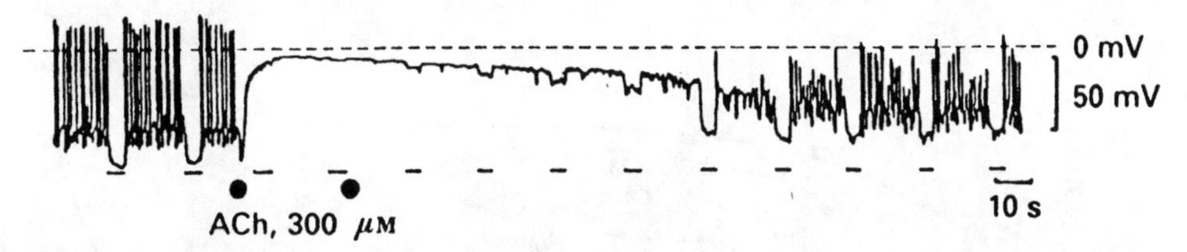

Figure 2. ACh-induced depolarization in guinea-pig ileal smooth muscle cell. Membrane potential was recorded in whole-cell configuration. Hyperpolarizing currents of −10 pA (4 s) were injected at times indicated by the bars below voltage trace. ACh (300 μM) was applied for period between dots below voltage trace. ACh caused depolarization and cessation of action potential activity, followed by recovery (at right). Amplitude of hyperpolarizations owing to injected current was reduced following ACh, indicating a membrane conductance increase accompanied depolarization. Reproduced with permission from Inoue and Isenberg (1990a).

2. Mechanisms of muscarinic excitation–contraction coupling

$[Ca^{2+}]_i$ is of central importance in controlling contraction of smooth muscles, since Ca^{2+}/calmodulin-activated myosin light chain kinase (MLCK) mediates phosphorylation of myosin, thereby initiating and/or maintaining contraction (Kamm and Stull 1989; Gerthoffer 1991; Somlyo and Somlyo 1994). In many smooth muscles $[Ca^{2+}]_i$ is maintained at ~100 nM (e.g., Williams et al. 1985, 1987; Yagi et al. 1988; Becker et al. 1989; Ganitkevich and Isenberg 1991; Pacaud and Bolton 1991; Vogalis et al. 1991) by mechanisms that extrude Ca^{2+} from the cell and sequester it into stores. The $[Ca^{2+}]$ in internal stores is estimated to be in the millimolar range (Iino 1989), so at rest there is a large gradient tending to move Ca^{2+} into the cytosol.

Muscarinic stimulation of smooth muscle elevates $[Ca^{2+}]_i$ to micromolar concentrations (toad gastric muscle, Williams et al. 1987; guinea-pig small intestine, Himpens and Somlyo 1988; guinea-pig jejunal muscle, Pacaud and Bolton 1991). Therefore, control of $[Ca^{2+}]_i$ is central to regulation of smooth muscle function. The mechanisms underlying Ca^{2+} extrusion and sequestration are beyond the scope of this review, and have been described by others (van Breemen and Saida 1989; Somlyo and Himpens 1989). Mechanisms for elevation of $[Ca^{2+}]_i$ include entry into the cell across the plasma membrane or release from intracellular stores, as summarized in Figure 3.

2.1 Influx of Ca^{2+}

2.1.1. Voltage-operated Ca^{2+} channels

During the earliest studies of excitation–contraction coupling in smooth muscles, influx of Ca^{2+} through voltage-dependent Ca^{2+} channels (Figure 3) was identified as being essential in many tissues (Somlyo and Somlyo 1968; Bolton 1979; Bolton and Large 1986). This view was based on the findings that cholinergic

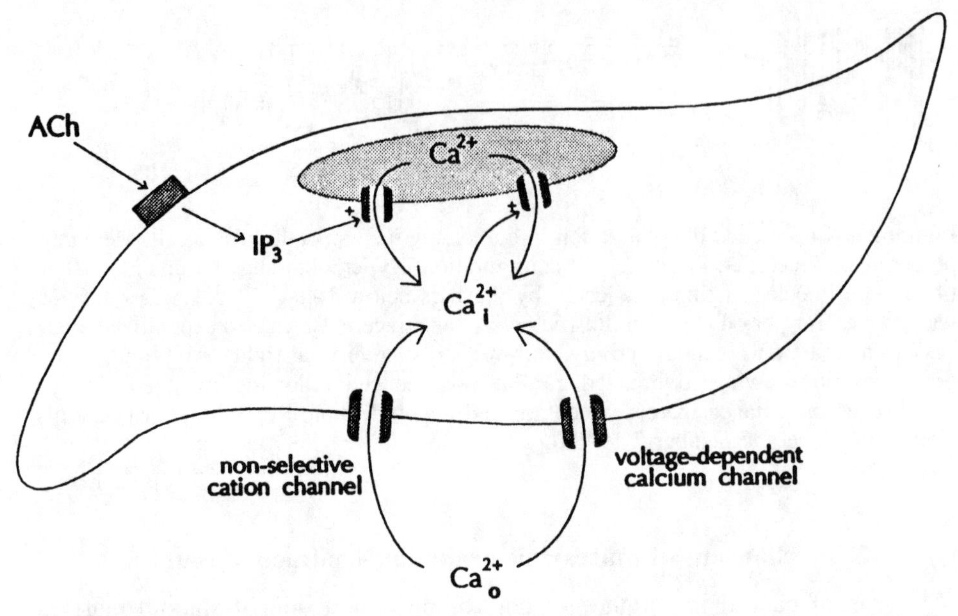

Figure 3. Schematic presentation of how ACh can cause elevation of $[Ca^{2+}]_i$ and contraction of smooth muscle. $[Ca^{2+}]_i$ can increase via influx of Ca^{2+} through voltage-dependent Ca^{2+} channels and possibly nonselective cation channels. Ca^{2+} is also released from intracellular stores (sarcoplasmic reticulum) through IP_3- and Ca^{2+}-activated channels.

agonist-induced contractions are accompanied by membrane depolarization, and that depolarizing the smooth muscle membrane by electrical stimulation produces contraction. On the other hand, membrane hyperpolarization or removal of external Ca^{2+} prevents these agonists from eliciting contractile responses, or elicits relaxations in precontracted tissues. In addition, these contractions are antagonized by blockers of voltage-dependent Ca^{2+} channels. Voltage-dependent Ca^{2+} channels have been identified in virtually all smooth muscles (Bean 1989a; Bolton 1989; Sanders 1989), although the classifications are beyond the scope of this review.

Opening of Ca^{2+} channels results in influx that is estimated to elevate $[Ca^{2+}]_i$ to 8–100 μM, assuming the Ca^{2+} distributes uniformly throughout the cytosol (e.g. Williams et al. 1987; Yamamoto et al. 1989; Vogalis et al. 1991). However, measurements of $[Ca^{2+}]_i$ using fluorescent indicator dyes indicate that the elevation of $[Ca^{2+}]_i$ is markedly smaller than predicted to arise from influx, with peak $[Ca^{2+}]_i$ of only ~1 μM (toad stomach, Becker et al. 1989; dog stomach, Vogalis et al. 1991; guinea-pig bladder, Ganitkevich and Isenberg 1991). The discrepancy between the estimated and the actual increases likely indicates that much of the Ca^{2+} entering the cell is rapidly buffered, sequestered into intracellular compartments, and/or extruded across the membrane. Fluorescence measurements from whole cells may be limited by spatial and temporal changes of $[Ca^{2+}]_i$, emphasizing changes in bulk cytosol. However, regional changes of $[Ca^{2+}]_i$, for example in submembrane compartments, may greatly influence the activity of ion channels. The ubiquity of voltage-dependent Ca^{2+} channels attests to the physiological importance of mechanisms underlying membrane depolarization.

2.1.2. Other Ca^{2+} entry pathways

In some tissues, excitation–contraction coupling involves influx of Ca^{2+} by a voltage-independent mechanism (Somlyo and Somlyo 1968; Bolton 1979, 1981; Bolton and Large 1986). For example, contractile responses are greatly reduced or abolished by removal of external Ca^{2+}, but are relatively unaffected by membrane hyperpolarization or classical voltage-dependent Ca^{2+} channel blockers (Farley and Miles 1977, 1978; Bolton and Large 1986). Similarly, certain agonists can elicit contractions with little or no depolarization (Farley and Miles 1977; Bolton and Large 1986; Janssen and Sims 1992; Sims 1992). Such observations led to the suggestion that there are voltage-insensitive "receptor-operated" channels through which Ca^{2+} enters (Bolton 1979; van Breemen et al. 1979). With only a few exceptions (e.g., ATP-activated channels, Benham and Tsien 1987; Benham 1989), such receptor-operated channels in smooth muscle have not been identified.

Other channels may also serve as Ca^{2+} entry pathways, such as the ACh-activated nonselective cation channels (Figure 3; see also Section 3.1). It is estimated from studies of ileal smooth muscle that ~10% of the inward current elicited by ACh is carried by Ca^{2+}, with the remainder carried by Na^+ (see Inoue and Shan 1993). Other nonselective cation channels, such as inwardly rectifying channels in gastric cells, are activated by stretch (Kirber et al. 1988). In the absence of other cations, these channels pass Ca^{2+}, although the channel conductance was reduced compared to physiological solutions containing Na^+. Stretch-activated channels may represent a mechanism for initiating stretch-induced contraction of smooth muscle both by causing depolarization of the membrane and by allowing limited Ca^{2+} entry. Ca^{2+} influx is now recognized as being evoked in many cell types in response to depletion of internal stores (calcium release–activated current, or CRAC; see e.g. Hoth and Penner 1992). Although little information exists regarding CRAC channels in smooth muscle, Ca^{2+} entry pathways contributing to refilling of internal Ca^{2+} stores have been explored. When Ca^{2+} stores in canine tracheal smooth muscle cells are depleted using cyclopiazonic acid, an inhibitor of the SR Ca^{2+} ATPase, activation of L-type Ca^{2+} channels leads to partial refilling (Janssen and Sims 1993b). Ca^{2+} also enters through a separate pathway, which is not voltage- or ACh-activated and is insensitive to blockade by dihydropyridine Ca^{2+} channel blockers. This "passive" refilling pathway has not yet been characterized in smooth muscle with respect to ionic selectivity, voltage-dependence, or pharmacological regulation, and thus deserves further attention.

2.2. Release of Ca^{2+} from internal stores.

It has long been recognized that cholinergic agonists can elicit contractions of many types of smooth muscle in the absence of external Ca^{2+}, even though KCl-induced contractions are abolished under these conditions (Somlyo and Somlyo 1968; Bolton 1979; Bolton and Large 1986). More recently it has been demonstrated that cholinergic agonists elicit a transient increase in $[Ca^{2+}]_i$ due to release of Ca^{2+} from stores in many cell types (Berridge 1993), including smooth muscle (Figure 3). The sarcoplasmic reticulum (SR) is widely believed to be the most physiologically important organelle for storing Ca^{2+} in smooth muscle (Somlyo and Himpens 1989; van Breemen and Saida 1989). Ca^{2+} can exit the SR by two pathways (Figure 3). As in skeletal and cardiac muscle, Ca^{2+} can stimulate the release of Ca^{2+} (Iino, 1989), with the release sites being activated by caffeine or ryanodine and blocked by ruthenium red (Ashida et al. 1988; Petersen and Wakui 1990; Ehrlich et al. 1994). The channels that mediate this release, the ryanodine receptors, have a unitary conductance of ~100 pS and are maximally activated when $[Ca^{2+}]$ at the cytosolic face of the channel is elevated into the micromolar range. Ca^{2+} is also released from the SR by submicromolar concentrations of inositol 1,4,5-trisphosphate ($InsP_3$), resulting in contraction (Suematsu et al. 1984; Somlyo et al. 1985; Walker et al. 1987; Kitazawa et al. 1989; Komori and Bolton 1991a,b). The $InsP_3$-activated channels have a smaller unitary conductance (~10 pS), are insensitive to caffeine or ryanodine, and are antagonized by heparin.

2.3. *Coupling of muscarinic receptors to excitation*

It was originally proposed that muscarinic receptors are directly coupled to channels permeable to Na^+, K^+, and Ca^{2+} (Bolton 1979; van Breemen et al. 1979), similar to the mechanism operating at the motor end plate in skeletal muscles, where ACh binds to the nicotinic ACh receptor and directly activates nonselective cation channels (Hille 1992). In smooth muscles, evidence for the existence of receptor-operated channels has been obtained from studies of ATP in rabbit ear arterial smooth muscle, which binds to receptors directly coupled to Ca^{2+}-selective channels, without any apparent involvement of diffusible second messengers (Benham and Tsien 1987; Benham 1989). However, such direct coupling of receptors to channels does not seem to be the case for muscarinic receptor-mediated responses, since latencies are on the order of milliseconds in skeletal muscle but can be as long as several seconds in smooth muscle (Bolton 1979; Burnstock 1979). The long latency for activation of membrane currents supports the hypothesis that ACh operates through second-messenger pathways, which are considered in what follows. These characteristics of muscarinic responses in smooth muscle are inconsistent with the hypothesis that ACh directly activates receptor/channel complexes.

Molecular and biochemical studies have revealed the structure and function of muscarinic receptors and the nature of their coupling to ion channels. Muscarinic receptors are members of a large family of proteins that includes the β-adrenergic and neurokinin receptors as well as rhodopsin (Hosey 1992; Brann et al. 1993). The amino acid sequence leads to a prediction of seven membrane-spanning regions, a long C-terminal hydrophilic sequence on the intracellular side of the membrane, and the N terminal on the extracellular side of the membrane. The second cytoplasmic loop between transmembrane segments I and

II is suggested to represent a site of interaction with heterotrimeric G proteins. Muscarinic receptor subtypes m1, m3, and m5 are coupled in most cases to $G_{\alpha q}$ or $G_{\alpha 11}$, resulting in activation of phospholipase C-β1 (PLC-β) and generation of 1,2-diacylglycerol (DAG) and $InsP_3$ (see Chapter 2 in this volume). DAG activates protein kinase C (PKC), whereas $InsP_3$ releases Ca^{2+} from internal stores and thereby activates MLCK (Berridge 1993; Somlyo and Somlyo 1994). Stimulation of these kinases leads to phosphorylation of myosin light chain and activation of contractile machinery, as well as phosphorylation of a number of channel types, thereby affecting both excitation and contraction.

3. Effects of muscarinic agonists on ionic currents

The development of methods for isolating single smooth muscle cells and the application of voltage-clamp techniques has permitted a direct study of ACh-regulated conductances. Two distinct conductance changes can cause the initial depolarization elicited by ACh, as illustrated in Figures 1 and 2. At rest, inward current across the cell membrane is balanced by outward current. Depolarization occurs when inward exceeds outward current, resulting in accumulation of positive charge inside the cell. Thus, ACh can cause depolarization either by *opening* channels carrying inward current (an increase in membrane conductance) or by *closing* channels carrying outward current (a decrease of membrane conductance; see Sims and Janssen 1993, fig. 1). Both of these conductance changes are elicited by cholinergic stimulation in smooth muscle, as summarized in Table 1. We will review the evidence that ACh activates inward nonselective cation and chloride currents and suppresses outward K^+ currents. ACh also enhances voltage-activated Ca^{2+} currents. There have been several reviews dealing with neurotransmitter regulation of ion channels in smooth muscle (e.g. Bolton 1989; Sanders 1989; Inoue and Shan 1993; Sims and Janssen 1993; Large and Wang 1996).

The membrane current (I_{ion}) through a population of channels can be described by

$$I_{ion} = N \cdot p_o \cdot i_{ion},$$

where N is the number of channels available to open, p_o is the probability of channel opening, and i_{ion} is the single channel current amplitude. The p_o term can be influenced by factors such as membrane potential, internal $[Ca^{2+}]$, pH_i, and/or the phosphorylation state of the channel. Furthermore, i_{ion} is given by

$$i_{ion} = \gamma \cdot (V_m - E_{ion})$$

where γ is the unitary channel conductance, V_m is the membrane potential, and E_{ion} is the equilibrium potential for that ion. These relationships provide a means for characterizing the ionic conductance(s) regulated by an agonist, since the selectivity of the ion channels is reflected in the reversal potential of the agonist-elicited current. Substitution of ions will result in changes in the reversal potential of the agonist-elicited current (since E_{ion} is altered) and may also give rise to changes in the magnitude of the current (since i_{ion} and/or p_o may be altered).

mary of excitatory cholingergic conductance changes in smooth muscle

Tissue	Unitary channel conductance (pS)	Pharmacological blockers	References
num	—	—	Benham et al. 1985; Benham & Bolton, 1986
ileum	25	Cd^{2+}, Ni^{2+}	Inoue et al. 1987; Inoue & Isenberg 1990a,b,c; Inoue 1991; Komori et al. 1992
orus	30	Cs^{+}, TEA	Vogalis & Sanders 1990
vein	190	—	Loirand et al. 1991
tric corpus	—	—	Sims 1992
guinea-pig trachea	—	—	Janssen & Sims 1992
ach	—	Ba^{2+}	Sims et al. 1985, 1988, 1990b
on	200	nifedipine	Cole et al. 1989
num	—	TEA	Benham et al. 1985; Benham & Bolton 1986
gus	—	TEA	Sims et al. 1990a
m	—	TEA	Shimada et al. 1991
porcine trachea	144	TEA	Janssen & Sims 1992; Kume & Kotlikoff 1991
gastric fundus	—	TEA	Lammel et al. 1991
ach	—	—	Clapp et al. 1987; Vivaudou et al. 1988
	—	—	Bielefeld et al. 1990
onary artery	—	Cd^{2+}, nifedipine	Matsuda et al. 1990
i	16.8	Cd^{2+}, nisoldipine	Kamishima et al. 1992
man, feline airway	24.3	—	Tomasic et al. 1992
cygeus	—	—	Byrne & Large 1987b
guinea pig trachea	—	SITS, niflumic acid	Janssen & Sims 1992
onchi	—	niflumic acid	Janssen 1996

Patch-clamp recording techniques (Hamill et al. 1981) provide a means for recording membrane potential (using current-clamp mode) or ionic currents (using voltage-clamp mode) from smooth muscle cells. Access to the cytoplasm can be achieved by rupturing the patch of membrane beneath the recording electrode (whole-cell configuration), allowing one to introduce known concentrations of ions or drugs into the cytoplasm. In some cases, dialysis may be undesirable, since essential soluble cytosolic constituents may be lost from the cell or unwanted factors may be introduced into the cell (e.g., the Ca^{2+} chelator EGTA). These problems can be avoided by using the perforated patch technique (Horn and Marty 1988; Korn et al. 1991), where the polyene antibiotics nystatin or amphotericin B are included in the electrode solution. These agents partition into the membrane under the patch electrode and permeabilize the membrane to monovalent ions. In this way, multivalent ions such as Ca^{2+} and large molecules remain in the cytoplasm, and membrane currents can be recorded from the whole cell without disruption of second-messenger systems.

3.1. Activation of nonselective cation current

Benham et al. (1985) used whole-cell patch-clamp techniques to investigate the mechanism underlying ACh-evoked depolarization in rabbit jejunal smooth muscle cells. During current clamp, current pulses were used to evoke hyperpolarizations, the amplitude of which would be proportional to membrane resistance (Ohm's law). They found that, during the cholinergic depolarization, the amplitude of the evoked hyperpolarizations was reduced, suggesting membrane resistance had decreased (i.e., membrane conductance had increased). When the membrane was clamped at $E_K = -78$ mV to rule out any contribution of K^+-selective currents, ACh evoked inward current. The reversal potential (V_{rev}) of this current did not correspond to the equilibrium potential of any single ion species, suggesting that a mixed conductance was activated. The value of V_{rev} was displaced in the positive direction when extracellular K^+ was increased from 6 mM to 45 mM ($E_K = -78$ mV and -25 mV, respectively), consistent with K^+ being a permeant ion. Furthermore, the fact that V_{rev} was always much more positive than E_K suggested that some ion(s) with more positive equilibrium potentials must also contribute to the ACh-evoked current: the only cations with very positive equilibrium potentials were Na^+ and Ca^{2+}. On the basis of these findings, Benham et al. (1985) concluded that ACh activated a nonselective cation conductance in rabbit jejunal smooth muscle.

In the first of a comprehensive series of studies of the ACh-activated nonselective cation conductance in guinea-pig ileal smooth muscle cells, Inoue et al. (1987) used patch-clamp techniques and low concentrations of ACh (1–10 μM) to resolve single-channel currents in whole cells. Higher concentrations activated more channels, preventing analysis of the single-channel events. An example of the single-channel currents activated by ACh in ileal smooth muscle cells is illustrated in Figure 4.

Removal of Na^+ from the bathing solution abolished the ACh-induced inward current, providing evidence that the single channels activated by ACh represent nonselective cation channels. When external $[Na^+]$ was reduced from 134 mM

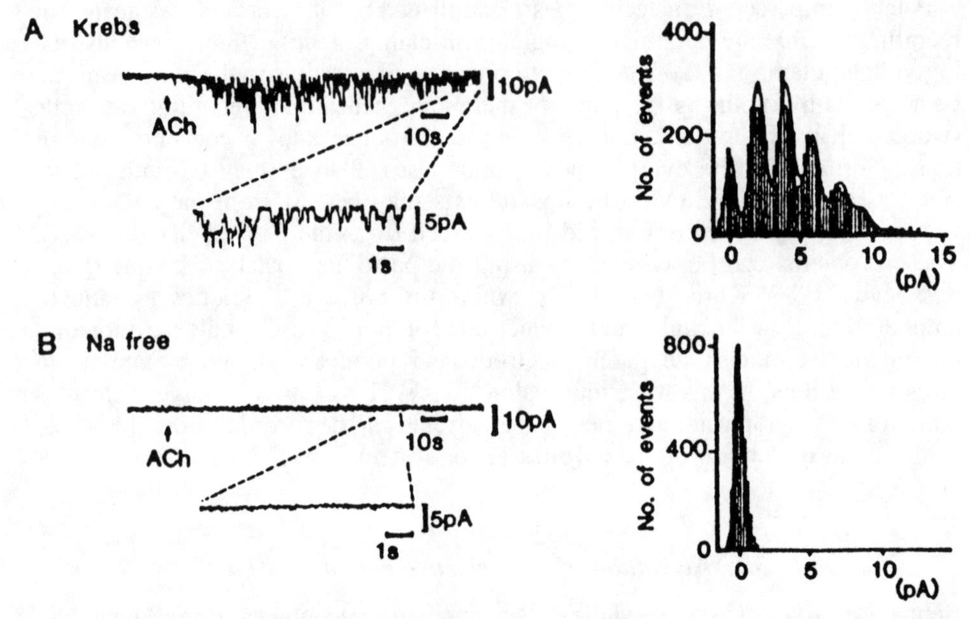

Figure 4. ACh activates single sodium-permeable channels in guinea-pig ileum studied in whole-cell configuration. ACh (30 μM) was applied in presence (A) and absence (B) of extracellular Na^+. In the presence of Na^+, ACh caused inward current and several channels were resolved. Removal of Na^+ abolished inward current. Expanded traces shown below. Amplitude histograms of current show multiple peaks fitted by sum of Gaussian distributions. Unitary channel amplitude was ~2 pA. In absence of Na^+, only a single peak was apparent, indicating baseline current. Reproduced with permission from Inoue et al. (1987).

to 67 mM, the slope of the single-channel *i–V* relation was reduced and V_{rev} of the *i–V* displaced 19 mV in the negative direction (from +23 mV in control solution to +2 mV in 67 mM Na^+). These data were used to calculate the relative permeabilities to Na^+ and to K^+ (i.e., P_K/P_{Na}), using the Goldman–Hodgkin–Katz equation, as follows:

$$\frac{P_K}{P_{Na}} = \frac{[Na]_o - [Na]_i \exp(FV_{rev}/RT)}{[K]_i(FV_{rev}/RT) - [K]_o},$$

where $[Na]_o$, $[Na]_i$, $[K]_o$, and $[K]_i$ are extracellular and intracellular concentrations of ions, respectively; *F*, *R*, and *T* denote Faraday's constant, the universal gas constant, and absolute temperature. From the shift in V_{rev} of 19 mV, and assuming that the effects of other cations (as well as anions) were negligible, P_K/P_{Na} was estimated to be ~4. In follow-up studies, Inoue and Isenberg (1990a) replaced external Na^+ with Li^+, K^+, or Cs^+ and found the reversal potential did not change, suggesting that the channels did not discriminate between these cations and thus confirming the nonselective cationic nature of this current. When external Na^+ was replaced with Ca^{2+} or Ba^{2+}, however, V_{rev} was displaced ~20 mV in the positive direction (i.e., toward E_{Ca}), suggesting that these channels were more permeable to divalent cations than to monovalent cations. Nonselective cation current in guinea-pig ileal smooth muscle was blocked potently by Cd^{2+} and less potently by Ni^{2+}, Mn^{2+}, Co^{2+}, and Mg^{2+} (Inoue 1991; Chen et al. 1993). Nonselective cation current in canine pyloric circular smooth muscle, however, was blocked by internal Cs^+ or by external TEA (Vogalis and Sanders 1990). The generally accepted strategy for identifying nonselective cation currents is ion substitution, as described previously. Using this approach, nonselective cation currents have been identified in cells of the canine pyloric circular smooth muscle (Vogalis and Sanders 1990), rat portal vein (Loirand et al. 1991), canine gastric smooth muscle (Sims 1992) and canine and guinea-pig tracheal smooth muscle (Janssen and Sims 1992). Properties of nonselective cation channels have been reviewed by Inoue and Shan (1993) and Isenberg (1993).

It is important to elaborate on the studies investigating the permeability of the nonselective cation channels to Ca^{2+}, as this issue is of great physiological relevance. As mentioned previously, Inoue and Isenberg (1990a) provided evidence that suggested Ca^{2+} could permeate the ACh-activated nonselective cation channels comparably or more readily than Na^+. Loirand et al. (1991) also found that, in the absence of all external Na^+, Ca^{2+} could contribute to the ACh-activated nonselective cation current in rat portal vein. These observations provide evidence that, in the absence of external Na^+, nonselective cation channels can provide entry pathways for Ca^{2+}. In Na^+-containing solution, Ca^{2+} has been estimated to contribute ~10% of the ACh-activated nonselective cation current (Inoue and Shan 1993). Combined electrophysiological and fluorescence studies of $[Ca^{2+}]_i$ led Pacaud and Bolton (1991) to conclude that Ca^{2+} entry through ACh-activated cation channels was very small under normal conditions. But definitive electrophysiological evidence for Ca^{2+} entry through these channels may be difficult to obtain, since $[Ca^{2+}]_i$ can modulate the current.

3.1.1. Gating properties

A common feature of cholinergic agonist-activated nonselective cation currents is that the current–voltage (*I–V*) relationships of the macroscopic currents are nonlinear (Benham et al. 1985; Inoue et al. 1987; Inoue and Isenberg 1990a; Nakazawa et al. 1990; Vogalis and Sanders 1990; Sims 1992). ACh-induced current is largest between approximately −50 mV and −20 mV, and decreases with depolarization and with hyperpolarization. An example of the ACh-activated cation current is illustrated in Figure 5, where nonselective cation current was recorded in a canine gastric corpus smooth muscle cell at steady-state voltages (Sims 1992). The *I–V* relationship of the peak ACh-induced current reverses direction close to 0 mV and so illustrates the nonlinear behavior of the macroscopic currents. The voltage dependence of gating of the ACh-activated cation current was investigated in guinea-pig ileum in voltage-jump experiments (Inoue and Isenberg 1990a). The steady-state conductance–voltage relationship is described by a Boltzmann distribution with half-maximal activation at −50 mV (Inoue and Isenberg 1990a). Hyperpolarization of the membrane during stimulation with cholinergic agonist causes an instantaneous increase in current am-

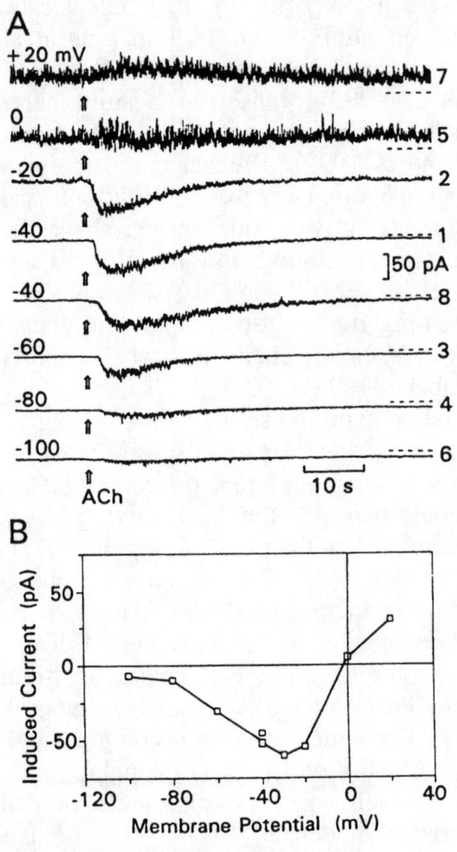

Figure 5. ACh activates nonselective cation current in canine gastric smooth muscle. Whole-cell currents were recorded using perforated-patch method. (A) Eight consecutive responses recorded from single cell held at potentials indicated to left of each trace. ACh was applied at arrows (1 s, 50 μM in application pipette). Sequence of measurements is indicated by numbers at right of traces. 5–7 minutes elapsed between each response. Two current responses are shown at −40 mV, representing the first and last responses recorded over a 55-minute period. At potentials negative to 0 mV, ACh evoked inward current accompanied by an increase in current noise. Outward current was evoked at 0 and +20 mV. Calibration bar applies to all traces except +20 mV, where it represents 100 pA. Dashed lines to right of current traces indicate zero current level. (B) *I–V* curve of ACh-induced current was nonlinear, with largest inward current between −40 mV and −20 mV, and smaller current at more negative potentials. Reproduced with permission from Sims (1992).

plitude (because of an increase in the driving force), followed by time-dependent relaxation of the current. The deactivation current relaxation is described by a single-exponential function with time constants dependent upon the membrane potential; these constants are ~28 ms at −180 mV and greater than 150 ms at potentials positive to −60 mV. On the basis of these observations, the authors proposed a two-state model (open ⇄ closed) for the voltage-dependent gating of the channel. In this model, the opening rate constant is largely unaffected by changes in V_m, whereas the closing rate constant is voltage-sensitive, decreasing at more positive potentials.

The whole-cell conductance activated by ACh in canine gastric smooth muscle cells also increases sigmoidally with depolarization (Sims 1992). In these cells, the half-activation potential of the ACh-activated conductance is -25 mV, with $\sim 500 \times 10^{-12}$ siemens (pS) activated near the resting potential of -60 mV and $\sim$2,500 pS activated at 0 mV. Therefore, at a typical resting membrane potential for these smooth muscle cells, approximately 1/4 to 1/2 of the maximum nonselective cation conductance is available to be activated by ACh. Since most smooth muscle cells have a resting membrane conductance of $\sim$1000 pS (assuming 1-GΩ input resistance), an increased conductance of 500 pS would strongly influence the membrane potential. Thus, as also concluded from studies of colonic muscle cells, activation of nonselective cation current strongly influences membrane potential of smooth muscle (Lee et al. 1993). Recent studies have revealed several additional ways that nonselective cation channels are regulated. Zholos and Bolton (1994) demonstrated that the voltage dependence of the nonselective cation current was influenced by the level of activation of G proteins. Inoue and co-workers (1995) found that extracellular protons also regulate gating of the channel. Further studies into such levels of control are needed to understand the general nature and physiological significance of these types of regulation.

The unitary *i–V* relationships for nonselective cation channels is linear, with slope conductance of 20–25 pS in guinea-pig ileum (Inoue et al. 1987) and rabbit portal vein (Inoue and Kuriyama 1993), 30 pS in canine pylorus (Vogalis and Sanders 1990), and around 200 pS in rat portal vein (Loirand et al. 1991). An example of the single-channel currents activated by ACh in ileal smooth muscle cells is given in Figure 4. Since the single-channel *i–V* relationship is linear, the decrease in macroscopic current at negative potentials must be due to a decrease in p_o rather than γ. Initial studies of the kinetics of single-channel currents at various potentials support this voltage dependence of opening (Inoue et al. 1987), although further studies of other cell types are required to more generally test the validity of this model.

3.1.2. *Does Ca^{2+} participate in activation of nonselective cation channels?*

Details of the coupling of muscarinic receptors to nonselective cation channels in smooth muscle are uncertain. G proteins are implicated, since GTPγS (which irreversibly activates G proteins) elicits inward currents with the same V_{rev} and ionic selectivity as those elicited by ACh (Inoue and Isenberg 1990c), whereas GDPβS prevents activation of cation current (Komori and Bolton 1990). The ACh-activated cation current is sensitvie to pertussis toxin (Inoue and Isenberg 1990c), suggesting involvement of $G_{\alpha i}$ and m2 or m4 subtypes of muscarinic receptor. While this view may be consistent with involvement of a membrane-delimited mechanism, evidence also exists for a diffusible regulator. Early studies discounted a role for $[Ca^{2+}]_i$ in regulation of this current (Benham et al. 1985; Inoue et al. 1987), but more recent evidence supported its involvement at some level. Increase of $[Ca^{2+}]_i$ through influx facilitates muscarinic cation current in ileal (Inoue and Isenberg 1990b), colonic (Lee et al. 1993), and gastric muscle (Sims, unpublished). Nonselective cation channels in vascular muscle are directly activated by Ca^{2+} (Loirand et al. 1991), like those in cardiac cells (Colquhoun et al. 1981) and neuroblastoma cells (Yellen 1982; Partridge and Swandula 1988; Isenberg 1993). The involvement of Ca^{2+} is further supported by a report showing that muscarinic cation current in jejunal cells is preceded by, and is dependent on, an increase in $[Ca^{2+}]_i$ (Pacaud and Bolton 1991). These authors also demonstrated that cholinergic elevation of $[Ca^{2+}]_i$ occurred in the absence of external Ca^{2+} and was antagonized by heparin, consistent with $InsP_3$ causing release of Ca^{2+} from SR. Whereas facilitation of muscarinic cation current by Ca^{2+} is well accepted, the consensus is that receptor occupation and G-protein activation are necessary for channel opening in visceral muscle. Increase of $[Ca^{2+}]_i$ alone is not sufficient to open channels in many muscles (Ganitkevich and Isenberg 1991; Lee et al. 1993), although nonselective cation current in canine gastric cells was elicited by caffeine, which activated the ryanodine receptor to release Ca^{2+} from the SR (Sims 1992). Evidence that Ca^{2+} contributes to activation of muscarinic cation current comes from studies demonstrating that removal of $Ca^{2+}{}_o$ depleted stores and eliminated contraction and current (Sims 1992). Furthermore, caffeine abolished an initial phase of the current (Lee et al. 1993). Thus, muscarinic receptor-mediated channel opening appears to involve both membrane-delimited and cytosolic signaling components.

Facilitation of the cation current in ileal cells is estimated to be approximately half-maximal between 100 nM and 300 nM $[Ca^{2+}]$, with maximal facilitation at greater than 1 μM $[Ca^{2+}]_i$ (Inoue and Isenberg 1990b). The mechanism of Ca^{2+}-mediated facilitation is uncertain (Lim and Bolton 1988). Facilitation is specific for Ca^{2+}, since it is not mimicked by Ba^{2+} or Sr^{2+} (Inoue and Isenberg 1990b). These authors also showed that facilitation was blocked by Ca^{2+} channel blockers D-600, nitrendipine (Figure 6) or Cd^{2+}. However, Cd^{2+}, Ni^{2+}, Mn^{2+}, Co^{2+}, and Mg^{2+} also rapidly suppressed nonselective cation current, possibly through direct interaction with the channel protein.

3.1.3. *Protein tyrosine kinases*

Many growth factors and some hormones exert their actions on cells by binding to cell surface receptors with intrinsic protein tyrosine kinase activity. Receptors for some growth factors share a common topology, consisting of an extracellular ligand-binding domain, a single membrane-spanning region, and a cytoplasmic tyrosine kinase domain. Ligand binding initiates receptor dimerization, allowing autophosphorylation of residues in the cytosolic domain, which in turn permits the binding and phosphorylation of substrates (Berridge 1993; Fantl et al. 1993). Effector molecules include PLC-$_{\gamma 1}$, phosphatidylinositol 3-kinase, and GTPase activating protein. The cascade of events initiated by receptor tyrosine kinases may involve various nonreceptor tyrosine kinases, and influences diverse processes such as Ca^{2+} homeostasis and gene transcription.

Growth factors have long been known to act on the gastrointestinal tract, not only regulating growth and developmental aspects of many cell types, but also

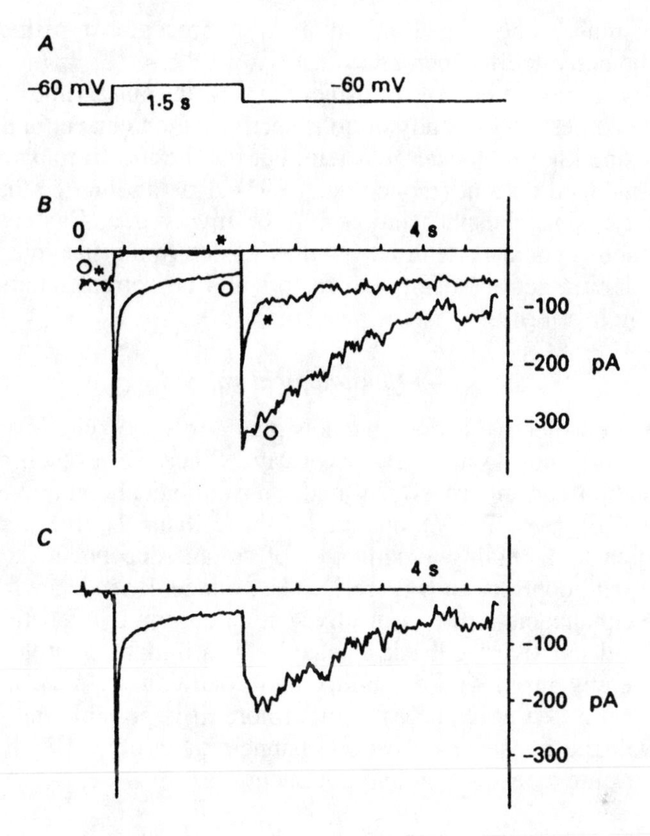

Figure 6. Ca^{2+}-mediated facilitation of ACh-induced nonselective cation current in guinea-pig ileal smooth muscle. Cell was held under voltage clamp in whole-cell configuration, with cesium aspartate in electrode to block K^+ currents. **(A)** During stimulation with ACh, cell was stepped from −60 mV to 0 mV for 1.5 s to elicit Ca^{2+} current. **(B)** Net currents during stimulation with ACh (300 μM) in absence (open circle) or presence (asterisk) of 10 μM nitrendipine. ACh elicited inward current seen at left as inward current below the baseline. This current was not affected by nitrendipine—see open circle and asterisk at left of **(B)**. Following repolarization to the holding potential, ACh-induced inward current was augmented, seen as large inward tail current. Nitrendipine blocked Ca^{2+} current and reduced tail current. **(C)** Digital subtraction of current traces in **(B)** reveals nitrendipine-sensitive current. Nitrendipine blocked voltage-dependent Ca^{2+} current as well as facilitation of nonselective cation current. This provides evidence that influx of Ca^{2+} through voltage-activated Ca^{2+} channels facilitates nonselective cation current. Reproduced with permission from Inoue and Isenberg (1990b).

acutely regulating smooth muscle contractility (Hollenberg 1994). Recent studies reveal that tyrosine kinases influence ion channels (Siegelbaum 1994). A combination of molecular and electrophysiological methods was used by Huang and co-workers (1993) to demonstrate that muscarinic suppression of a delayed rectifier K^+ channel was mediated both by Ca^{2+} and DAG. Surprisingly, however, tyrosine phosphorylation of the channels also occurred, leading the authors to conclude that muscarinic signaling involved nonreceptor tyrosine kinase(s), which might be activated by increased $[Ca^{2+}]_i$.

Evidence for involvement of tyrosine kinases in muscarinic signaling in smooth muscle comes from a study of nonselective cation current in ileal muscle, where the tyrosine kinase blocker genistein, but not the inactive analog daidzein, reduced ACh-induced current (Inoue et al. 1994). This finding led the authors to conclude that tyrosine phosphorylation may be involved in Ca^{2+} mobilization, thereby influencing channels. Further studies to determine the role of tyrosine kinases in mediating actions of growth factors and neurotransmitters in smooth muscle are a high priority.

3.1.4. Physiological roles

ACh-regulated nonselective cation channels play a role in excitation–contraction coupling in smooth muscle in at least two ways. Since the reversal potential for nonselective cation currents is ~0 mV under physiological conditions (with $E_{Na} \approx +60$ mV and $E_K \approx -87$ mV), opening of these channels gives rise to inward (depolarizing) current, leading to activation of voltage-dependent Ca^{2+} channels, influx of Ca^{2+}, and contraction (Figure 2). The voltage dependence for activation as well as the enhancement of nonselective cation current by elevation of $[Ca^{2+}]_i$ provide powerful positive feedback mechanisms to initiate or prolong depolarization, and thereby promote Ca^{2+} influx and contraction (Inoue and Isenberg 1990b; Sims 1992; Lee et al. 1993). Furthermore, it is possible that Ca^{2+} enters directly through the nonselective cation channels themselves (Figure 3), which may also contribute to excitation and contraction.

3.1.5. Overview

ACh activates nonselective cation conductance in a variety of smooth muscles. Activation of this conductance is apparent in current clamp by depolarization accompanied by an increase of membrane conductance. Under voltage clamp at the resting membrane potential, this conductance gives rise to an inward current. The nonselective cation channels discriminate poorly between monovalent cations and may allow Ca^{2+} entry. A common finding is that ACh-activated nonselective cation conductance exhibits voltage dependence, with increased activation at more depolarized potentials. This voltage dependence would give rise to positive feedback, since depolarization increases the conductance available to be activated by ACh.

The molecular mechanisms coupling muscarinic receptors to opening of nonselective cation channels require further clarification. A membrane-delimited interaction of receptor to channel is possible, but a number of findings point to the contribution of diffusible cytosolic messengers. A role for Ca^{2+} in activation of this cation channel is controversial. In some cells, elevation of $[Ca^{2+}]_i$ clearly facilitates the cation current, although ACh receptor activation is essential for channel opening. In other smooth muscles, evidence is consistent with cytosolic Ca^{2+} *directly* activating the channel. Heterogeneity of nonselective cation channels may account for these reported differences, but further investigation of the role for $[Ca^{2+}]_i$ is required. It remains to be determined whether these channels permit significant Ca^{2+} entry under physiological conditions.

3.2. Suppression of K^+ currents

3.2.1. M-current

In contrast to the activation of nonselective cation channels, ACh actually decreases the activity of K^+ channels in many cells. ACh was first shown to cause depolarization associated with a *decrease* of membrane conductance in gastric myocytes from the toad *Bufo marinus* (Sims et al. 1985). These observations suggested that ACh suppressed a K^+ conductance, because this was the only ion with an equilibrium potential negative to the resting potential. Closure of K^+ channels would reduce outward current, causing depolarization. Further support that ACh acted on K^+ channels came from the observation that the reversal potential for the current suppressed by ACh corresponded to E_K over a range of $[K^+]_o$. In the course of characterizing the K^+ current in gastric smooth muscle cells, two identifying features became apparent. The K^+ current was sensitive to blockade by muscarinic agonists and the underlying conductance exhibited time dependence and voltage dependence. A novel K^+ current with similar features had previously been identified in bullfrog sympathetic neurons by Adams, Brown and Constanti (1982a,b), and had been termed ''M current'' because of its sensitivity to muscarinic agonists. Based on these similarities, the ACh-sensitive K^+ current in smooth muscle cells was referred to as M current (Sims et al. 1985). Voltage-clamp traces in Figure 7 illustrate the general features of M current and the inward current elicited at -36 mV arising from its suppression. The conductance underlying M current in this smooth muscle is voltage-dependent, being almost completely deactivated at potentials more negative than -70 mV, increasing sigmoidally with depolarization, and being half-maximally activated at approximately -50 mV (Sims et al. 1985). This voltage dependence closely resembles that of M current in other cell types, such as sympathetic neurons (Adams et al. 1982a,b). Such voltage dependence also resembles that described for the nonselective cation conductance.

Further studies have revealed that, in addition to being suppressed by muscarinic agonists, M current can be *induced* in toad gastric smooth muscle cells by the β-adrenergic agonist isoproterenol (Sims et al. 1988, 1990a). Isoproterenol and cyclic adenosine 3',5'-monophosphate (cAMP) analogues increase the maximum conductance, slow the deactivation current relaxations, and shift the activation range of the M conductance to more negative potentials (Sims et al. 1990a). When considered in the framework of a two-state model of M-channel gating proposed by Adams et al. (1982a), these changes provide evidence that the activation of M current by β-adrenergic agents is due to a decrease in the closing rate constant for M channels, giving rise to an increase in mean open time (Sims et al. 1990a). In all cases, M current is suppressed by ACh or muscarine, acting at a locus downstream from regulation of adenylate cyclase and phosphodiesterase. Thus, M current in smooth muscle is under dual, antagonistic control by separate neurotransmitter systems. Activation of M current by the β-adrenergic system involves a cAMP-sensitive protein kinase pathway, since forskolin or phosphodiesterase-resistant analogs of cAMP activate M current (Sims et al. 1990a). Suppression of M current by muscarinic agonists appears to involve the diacylglycerol/protein kinase C pathway, since

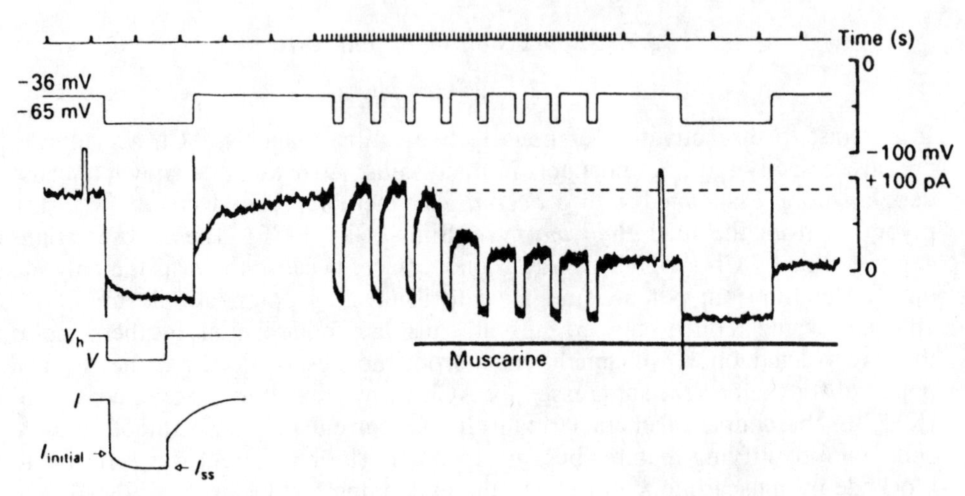

Figure 7. Cholinergic suppression of M current in gastric smooth muscle cell from toad. Cell was held under voltage clamp using intracellular microelectrode. Membrane potential was periodically hyperpolarized for 2 s as indicated in voltage trace at top. Membrane currents are shown in lower traces. Muscarine (500 μM in application pipette) caused net inward current and decrease in magnitude of current excursions elicited by hyperpolarizing commands, indicating a conductance decrease. Voltage jumps at beginning and end of record are shown on faster time scale, as indicated by timing trace at top. Upon hyperpolarization there was initial rapid current jump ($I_{initial}$) followed by slow inward current relaxation to steady-state level (I_{SS}), which represents outward K^+ current turning off. At end of 2-s pulse, there was smaller rapid current jump, followed by outward current relaxation to baseline, which is K^+ current turning back on. Expanded trace at right shows that current relaxations were abolished and initial rapid jump was smaller, the latter indicating reduction of chord conductance. 100 pA, 100 ms calibration pulses precede voltage jumps and were blanked from the slow portions of the current record for clarity. Reproduced with permission from Sims et al. (1985).

extracellular application of sn-1,2-dioctanoylglycerol (diC_8), a synthetic diacylglycerol and potent activator of protein kinase C, reversibly suppresses M current (Clapp et al. 1992). In contrast, 1,2-dioctanoyl-3-thioglycerol, a diacylglycerol analog that does *not* activate protein kinase C, fails to inhibit M current. Both ACh and di-C_8 suppress endogenous and isoproterenol-induced M current without altering the time course of M-current deactivation (Clapp et al. 1992). These results suggest that muscarinic suppression of M current mediated by diacylglycerol may be due to a reduction of *N*, the number of channels available to open, since the kinetics of the current – and hence the channel open probability – did not change.

Similar to cholinergic regulation of current in amphibian gastric muscle cells, ACh suppresses a steadily activated, TEA-sensitive K^+ current in guinea-pig gastric fundus smooth muscle cells (Lammel et al. 1991). ACh suppresses this current at potentials at which M current would not be active (see the foregoing and Sims et al. 1985), leading Lammel and coworkers to suggest that the two currents were distinct. Lammel et al. (1991) also reported that ACh suppressed a small Na^+ conductance in guinea-pig gastric fundus, but the significance of this to cholinergic excitation was uncertain.

3.2.2. Ca^{2+}-activated K^+ current

ACh suppresses Ca^{2+}-activated K^+ currents in several cell types (see Table 1). Cole, Carl, and Sanders (1989) showed that ACh reduced the time-dependent outward currents elicited in colonic smooth muscle cells by depolarizing voltage commands. This reduction was attributed to a 10–15-mV positive shift of the activation range of the current, which would result in suppression of the whole-cell current at all potentials. Nifedipine was shown to block $I_{K(Ca)}$, after which ACh caused no suppression in the outward K^+ current evoked by depolarization, suggesting that the current suppressed by ACh was indeed $I_{K(Ca)}$. In order to determine if ACh has a direct effect on Ca^{2+}-dependent K^+ channels or whether it acts indirectly by a reduction of voltage-activated Ca^{2+} current, Cole et al. (1989) also investigated the effects of ACh on large-conductance Ca^{2+}-activated K^+ channels using the cell-attached patch configuration. When ACh was added to both bath and pipette solutions, voltage-dependent activation of the K^+ channels was shifted positively compared to control patches, but the unitary channel conductance was not reduced.

More recently, cholinergic suppression of individual Ca^{2+}-activated K^+ channels has been investigated in excised outside-out patches from porcine tracheal smooth muscle (Kume and Kotlikoff 1991). Methacholine inhibits activity of Ca^{2+}-activated K^+ channels when $[Ca^{2+}]_i$ is in the physiologic range, without changing the single channel conductance. Similar to the conclusion reached by Cole et al. (1989), Kume and Kotlikoff interpreted their data as indicating that cholinergic stimulation shifted the voltage activation range in the positive direction. However, since the voltage activation curves did not reach saturation at positive potentials, it is also possible that their data could indicate that cholinergic stimulation simply reduced the number of channels available to open and/or the open probability, with no shift in activation parameters.

Receptor-channel coupling mechanisms have been investigated in several cases. Cholinergic suppression of K^+ current in colonic smooth muscle cells was irreversible when nonhydrolyzable GTP analogs (e.g., GTPγS or GppNHp) were included in the recording pipette (Cole and Sanders 1989). Furthermore, the effects of muscarinic agonists were abolished by pertussis toxin, leading Cole and Sanders to conclude that a pertussis toxin–sensitive G protein mediated suppression. Studies of excised Ca^{2+}-activated K^+ channels led Kume and Kotlikoff (1991) to a similar conclusion. Cholinergic suppression of channel activity was dependent on the presence of GTP at the cytosolic surface, was blocked by the inactive analogue GDPβS, and was abolished by treatment with pertussis toxin.

3.2.3. Spontaneous transient outward current

A variety of smooth muscles exhibit spontaneous transient outward currents (STOCs), which represent Ca^{2+}-activated K^+ channels opening in response to sporadic release of Ca^{2+} from intracellular stores (Benham et al. 1985; Benham and Bolton 1986; Bolton and Lim 1989; Nelson et al. 1995). The single-channel currents underlying STOCs have not been resolved, and thus it is not clear whether STOCs represent activation of the same Ca^{2+}-activated K^+ channels referred to previously. It has been suggested that these currents reflect Ca^{2+}-overloaded cells, and as such may reflect a pathophysiological condition (Benham and Bolton, 1986; Hume and Leblanc 1989). In any case, cholinergic agonists suppress STOCs in rabbit jejunum (Benham et al. 1985; Benham and Bolton 1986; Bolton and Lim 1989; Komori and Bolton 1990, 1991a), rabbit ear artery (Benham and Bolton 1986), feline esophagus (Sims et al. 1990b), rabbit ileum (Shimada et al. 1991), canine and guinea-pig trachea (Janssen and Sims 1992), and human bronchi (Janssen 1996).

The mechanism(s) whereby ACh suppresses STOCs is not known, but several possibilities exist. Cholinergic stimulation causes initial release of Ca^{2+} from stores, which is apparent from the activation of K^+ current seen both at the whole-cell (Benham et al. 1985; Benham and Bolton 1986; Komori and Bolton 1990; Sims et al. 1990b; Ganitkevich and Isenberg 1990a) and single-channel levels (Wade and Sims 1993). Following this release, there may be a period during which the stores are functionally depleted, accounting for suppressed channel activity. An alternative explanation is that ACh exerts its effect at the level of the channels, possibly involving phosphorylation of the channel protein (see e.g. Kume et al. 1989).

Evidence exists that the effects of cholinergic stimulation are exerted "upstream" from the channels, possibly at the level of Ca^{2+} release from stores. In addition to STOCs, spontaneous transient *inward* currents (STICs) occur in several muscles, representing Ca^{2+}-activated Cl^- currents (Janssen and Sims 1992; Wang et al. 1992). ACh suppresses both STOCs and STICs (Janssen and Sims 1992) and recovery of both occurs *pari passu.* The parallel regulation of these two separate conductances can most easily be explained by ACh acting primarily on the release of Ca^{2+} from stores, as opposed to acting downstream on the channels.

3.2.4. *Physiological roles*

It has been suggested (Cole et al. 1989) that decreased activity of K^+ channels could be advantageous for excitation of multicellular smooth muscle because the increase in membrane resistance would: (1) lead to an increase in the length constant of the tissue and the rate of impulse propagation; (2) conserve energy by reducing the need for inward current to sustain depolarization; and (3) provide a mechanism for delaying action potential repolarization. Indeed, cholinergic stimulation of gastrointestinal muscle causes prolongation of the action potential and an increase in amplitude of the plateau accompanied by large increase in the force generated by the muscle (Szurszewski 1987; Sanders 1989). Increases in the amplitude and duration of the action potential plateau phase have been shown to be key determinants of Ca^{2+} entry through voltage-dependent Ca^{2+} channels in colonic muscle cells. Small changes in membrane potential of the plateau phase give rise to large changes in $[Ca^{2+}]_i$ (Vogalis et al. 1991). Suppression of the K^+ currents responsible for repolarizing the action potential may contribute to the prolongation and enhancement of the action potential observed in tissues.

3.2.5. Overview

Based on studies of smooth muscle cells from a variety of species and organs, cholinergic suppression of K^+ currents appears to be of general significance. This mechanism has been identified using different cell isolation and recording techniques. Several K^+ currents are suppressed by cholinergic agonists, including M current and Ca^{2+}-activated K^+ currents. Therefore, cholinergic suppression of K^+ currents is a widespread and common mechanism contributing to cholinergic excitation of smooth muscles. Suppression of outward K^+ current gives rise to net inward current across the membrane, contributing to depolarization elicited by ACh. Indeed, cholinergic depolarization in many cells involves both suppression of K^+ channels and activation of others, such as nonselective cation and Cl^- channels.

Although muscarinic suppression of K^+ currents clearly involves G-protein–coupled processes, details of the receptor–channel coupling are just beginning to emerge. Studies of the large-conductance Ca^{2+}-activated K^+ channel provide preliminary indications that ACh suppresses channel activity by shifting the voltage dependence of channel opening to more positive potentials. Spontaneous transient outward K^+ currents are suppressed by cholinergic agonists in a variety of smooth muscles.

3.3. Activation of Cl^- current

Studies of vascular and visceral smooth muscles reveal that Ca^{2+}-activated chloride channels contribute to excitation in response to various agonists (reviewed by Large and Wang 1995). Early studies of rat anococcygeus muscle demonstrated that α-adrenoceptor stimulation caused activation of a voltage-dependent, Ca^{2+}-dependent Cl^- conductance (Byrne and Large 1987a). Muscarinic receptor stimulation elicited a current (I_{Cl}) with the same time course and reversal potential (Byrne and Large 1987b), leading to the suggestion that the two receptor types were coupled to the same Cl^- channels. However, interactions between the muscarinic and adrenergic agonists were not explicitly tested (e.g., additivity of the two responses and short-term heterologous desensitization of the response).

In canine and guinea-pig tracheal smooth muscle (Janssen and Sims 1992) and in human bronchial smooth muscle (Janssen 1996), ACh elicits inward current associated with an increase in membrane conductance. This current was typically apparent as a large inward current that peaked and declined during the continued presence of ACh (Figure 8A). In approximately 50% of guinea-pig cells and 10% of canine cells, the initial inward current was followed by current "oscillations" (Figure 8B). Such oscillations are suggestive of oscillations of $[Ca^{2+}]_i$ and Ca^{2+}-activated currents reported in a variety of other cell types (e.g. Petersen and Wakui 1990; Berridge 1993). The magnitude of the ACh-induced inward current in tracheal smooth muscle is reduced on removal of extracellular Na^+, but the inward current is still substantial, indicating that a portion of the inward component is not carried by Na^+. The inward current elicited by ACh in the absence of Na^+_o was attributed to Cl^- based on two findings (Janssen and Sims 1992). First, the reversal potential of the ACh-induced current is dependent on

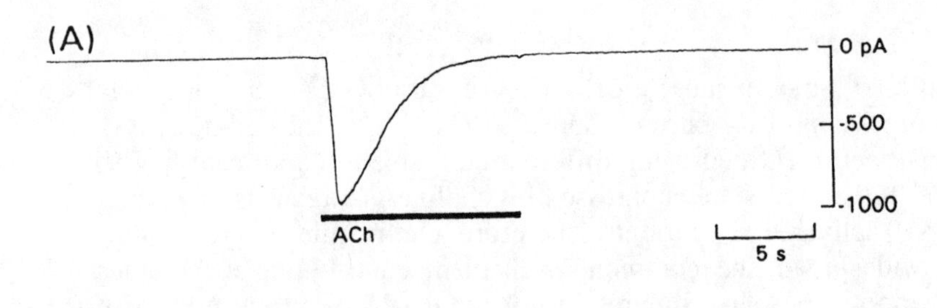

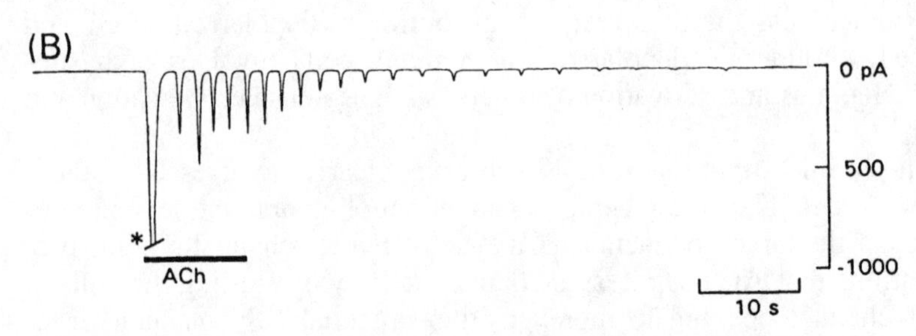

Figure 8. ACh elicits inward current oscillations in tracheal smooth muscle. Membrane currents recorded with nystatin perforated patch technique, with cells held at −60 mV. (A) In some cells ACh elicited brief inward current, which peaked and declined during continued presence of ACh (indicated by bar under current traces). (B) In other cells, ACh elicited inward current followed by current oscillations that persisted for ~1 min. Ion substitution studies revealed that inward current was due to nonselective cation and Cl^- currents. Cell in (A) from dog and cell in (B) from guinea pig. Reproduced with permission from Janssen and Sims (1992).

the Cl^- gradient. Second, the ACh-activated current is reversibly blocked by the Cl^- channel blockers SITS and niflumic acid, as illustrated in Figure 9.

Several observations support the conclusion that this Cl^- current is Ca^{2+}-activated. Cl^- current is accompanied by contraction, and depletion of stores (in Ca^{2+}-free solution, or using agents that inhibit the SR CaATPase, e.g., cyclopiazonic acid, CPA) abolishes the current (Janssen and Sims 1993b). Caffeine and ryanodine activate a current resembling the ACh-elicited current (reversal potential, latency, duration). It is noteworthy that receptor occupation is not required for channel opening, based on several observations. Spontaneous transient inward currents (STICs) are observed which reverse direction around E_{Cl} and which are coincident with STOCs (Wang et al. 1992; Janssen and Sims 1992, 1994a). Cl^- currents are also activated by Ca^{2+} influx (Janssen and Sims 1995). Finally, oscillations of current are observed (Figure 8B), consistent with oscillatory release of Ca^{2+} from intracellular stores.

Previous studies have revealed Cl^- conductances in several types of smooth muscle. For example, Ca^{2+}-activated Cl^- currents have been identified in rat portal vein (Pacaud et al. 1989) and rabbit ear artery (Amédée et al. 1990a, b), although in these latter cells the currents were elicited by noradrenaline. Daniel and coworkers (1988) investigated the mechanisms of action of ACh using microelectrode recording of membrane potential in intact tracheal smooth muscle strips.

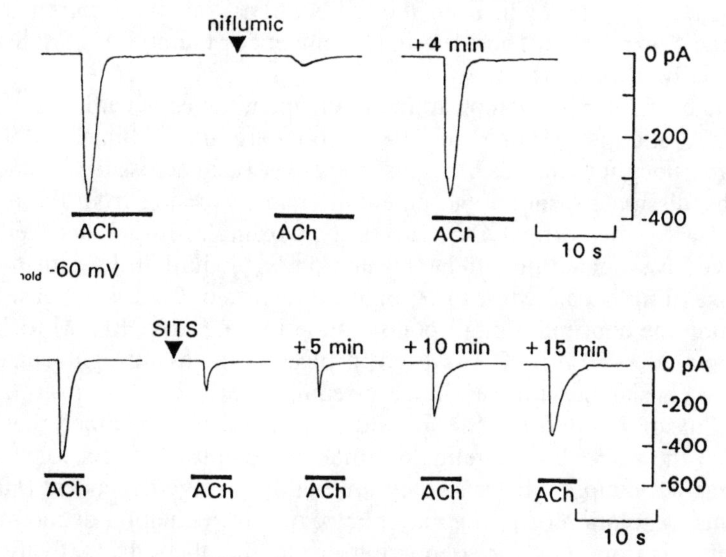

Figure 9. ACh-activated Cl^- current in canine tracheal smooth muscle is reduced by Cl^- channel blockers. Currents recorded at −60 mV using perforated-patch technique. Cells were in Na^+-free solution to eliminate contribution of nonselective cation current. ACh elicited brief inward Cl^- current, which peaked and declined during continued presence of ACh. Niflumic acid (100 μM in application pipette) or SITS (1 mM in application pipette) were applied for 30 s prior to second ACh stimulation, and largely blocked inward current. Recovery from blockade seen at various times after wash-out of blockers. 4–5 minutes elapsed between each application of ACh. Reproduced with permission from Janssen and Sims (1992).

Based on ion substitution experiments, these authors concluded that cholinergic agonists cause depolarization owing to an increase in Cl^- conductance of the cell membrane. In spite of inherent limitations of ion substitution in multicellular preparations, single-cell voltage-clamp data (Janssen and Sims 1992) confirm remarkably well the conclusions reached from studies of intact tissues (Daniel et al. 1988).

3.3.1. Physiological roles

I_{Cl} may play a role in excitation-contraction coupling in smooth muscle. The Cl^- equilibrium potential (E_{Cl}) in smooth muscles ranges from −33 mV to −8 mV (Aickin and Vermuë 1983; Gerstheimer et al. 1987; Aickin 1990). Activation of I_{Cl} at the resting potential would lead to an efflux of Cl^- from the cytosol and result in an inward, depolarizing current. Therefore, I_{Cl}, like the nonselective cation current described previously, may be part of a positive feedback loop, where Ca^{2+} released from internal stores causes inward current and depolarization, which in turn leads to influx of Ca^{2+} through voltage-dependent Ca^{2+} channels. However, ACh causes contraction in tracheal smooth muscle cells even when I_{Cl} is largely eliminated using Cl^- channel blockers (Janssen and Sims 1992). Thus, the Cl^- current per se seems to be unnecessary for contraction. An alternative role for I_{Cl} in tracheal tissues was proposed based on the observation that decreasing the electrode levels of $[Cl^-]$ reduces the fraction of cells that contract in response to ACh, even though ACh still elicits a current response (Janssen and Sims 1992). Thus, the $[Cl^-]_i$ influences the ability of cells to contract repeatedly in response to ACh.

Cl^- channels have been demonstrated in the membranes of cardiac and skeletal sarcoplasmic reticulum (Hals et al. 1989; Holmberg and Williams 1989). These channels are thought to allow Cl^- to distribute passively across the SR membrane and thereby dissipate osmotic and potential changes arising from the release or uptake of Ca^{2+}, which affect Ca^{2+} fluxes. For example, Joseph and Williamson (1988) have investigated the effects of altering $[K^+]$ and $[Cl^-]$ on $InsP_3$-induced Ca^{2+} release in rat hepatocytes. $InsP_3$-induced release of Ca^{2+} is increased ~20% by increasing the concentration of potassium gluconate from 40 mM to 120 mM, but is decreased ~50% if KCl is used in place of potassium gluconate. These findings are consistent with Ca^{2+} release requiring efflux of Cl^- from the SR as well as influx of K^+ into the SR. In addition, it has been claimed that the Cl^- channels on the sarcoplasmic reticulum membranes play a direct role in excitation–contraction coupling because they are highly voltage-dependent (Hals et al. 1989). Thus, regulation of $[Cl^-]_i$ may play a role in regulation of internal Ca^{2+}, which in turn is important for contraction. It may be, then, that activation of I_{Cl}, with concomitant efflux of Cl^-, may potentiate or prolong the release of Ca^{2+} from the SR.

Activation of I_{Cl} would lead to an efflux of Cl^- from the cytosol, since E_{Cl} is positive to the resting membrane potential. Assuming that the current elicited in tracheal smooth muscle cells by ACh in the absence of Na^+ is a pure Cl^- current, and that the decrease in $[Cl^-]_i$ associated with this Cl^- efflux is uniform throughout the cell volume, we estimated that ACh causes a decrease in $[Cl^-]_i$ of at least 3 mM (Janssen and Sims 1992). Close apposition of the SR under the plasma membrane may provide a diffusional barrier that would restrict the cytosolic volume from which Cl^- is lost, resulting in a more dramatic reduction of $[Cl^-]$ in a submembrane compartment.

I_{Cl} may also play a role in volume regulation during contraction. Based on morphometric analysis of light microscope images, Fay and Delise (1973) estimated that toad gastric smooth muscle cells decrease their volume by 30–40% during contraction initiated by electrical stimulation. If efflux of Cl^- was accompanied by K^+ (owing to concurrent activation of Ca^{2+}-dependent K^+ channels), the resulting loss of salt could contribute to reduction in cell volume.

3.4. Modulation of Ca^{2+} current

The changes in channel activity just described account for the initial depolarization elicited by ACh. Whether the depolarization is initiated by an activation of inward current or suppression of outward current, it can lead to opening of voltage-gated Ca^{2+} channels, allowing Ca^{2+} entry and contraction (see e.g. Figure 1). Ca^{2+} channels of cardiac muscle and neurons have been well described as being subject to regulation by neurotransmitters (Bean 1989a,b; McDonald et al. 1994); the same is true of Ca^{2+} channels in smooth muscle. ACh was first shown to increase the magnitude of voltage-activated Ca^{2+} current and slow its decay in gastric smooth muscle cells from the stomach of the toad *Bufo marinus* (Clapp et al. 1987). Further studies revealed that high voltage–activated (or L-type) Ca^{2+} channels alone are modulated by ACh in this tissue (Vivaudou et al. 1988). ACh enhancement of Ca^{2+} current in toad gastric cells was not accompanied by an apparent shift of the *I–V* relationship, possibly indicating that the voltage activation range of the channels was not altered by ACh. This leaves open the possibility of an increase in unitary channel current or an increase in the number of functional channels available to open. However, it is worth noting that studies of transmitter regulation of voltage-activated Ca^{2+} current in neuronal cells indicate that $I = V$ relationships do not provide sensitive measures of the voltage dependence with which Ca^{2+} channels open (Bean 1989b). Therefore, the lack of shift in the *I–V* relations during augmentation with ACh may not rule out a shift in the voltage activation range of the Ca^{2+} channels.

Other studies have documented cholinergic enhancement of voltage-activated Ca^{2+} current in mammalian smooth muscle cells, suggesting that this form of regulation may be of general significance in smooth muscles. In rabbit coronary artery smooth muscle, Matsuda and co-workers (1990) found that low concentrations of ACh (0.2 μM) increased the magnitude of depolarization-activated Ca^{2+} current by ~40% in 8 of 13 cells studied. This current was identified as an L-type Ca^{2+} current based on its voltage activation and inactivation properties as well as sensitivity to dihydropyridines. The effect of ACh required more than four minutes to occur, and was antagonized by atropine. Muscarinic stimulation also enhanced L-type Ca^{2+} current in circular smooth muscles of the cat colon (Bielefeld et al. 1990). This conclusion was based on the observation that depolarization-elicited inward Ca^{2+} current was barely detectable in control cells, but was visible approximately five minutes after stimulation with carbachol (1 μM).

Effects of cholinergic agonists have been studied on single Ca^{2+} channels in airway smooth muscle. Using the cell-attached patch technique, single-channel voltage-activated Ca^{2+} currents were recorded from rat bronchial smooth muscle cells, using 10 mM Ba^{2+} as the charge carrier and Bay K 8644 to increase L-type Ca^{2+} channel opening (Kamishima et al. 1992). Stimulation of cells with carbachol (10 μM) caused increased opening of 16.8-pS Ca^{2+} channels at depolarized potentials. Enhanced channel activity was shown to result from a hyperpolarizing shift of 9 mV in the activation range of Ca^{2+} channels, with no change in the unitary conductance or the number of available channels. Similar single-channel studies have been conducted on cultured human and freshly dispersed canine, ferret, and porcine tracheal muscle cells (Tomasic et al. 1992). Methacholine (50 μM or 1 mM) increased the open-state probability of a 24.3-pS Ca^{2+} channel. In one case illustrated, channel open probability at 0 mV increased from ~0.2 to 0.4. This effect was not due to membrane depolarization. It was also reported that, in some cases, currents arising from a smaller-conductance Ca^{2+} channel (9.5 pS using 80 mM Ba^{2+} as charge carrier) became apparent following stimulation with methacholine (Tomasic et al. 1992).

The mechanism whereby ACh enhances the Ca^{2+} current was investigated in toad gastric cells using synthetic analogs of DAG (Vivaudou et al. 1988). Cholinergic enhancement of the Ca^{2+} current was mimicked by the analog di-C_8 which activated PKC, but not by analogues which activated PKC only weakly (1,2-dioctanoyl-3-chloropropanediol) or not at all (1,2-dioctanoyl-3-thioglycerol). Accordingly, Vivaudou and co-workers proposed that DAG was a second messenger mediating the effects of ACh, possibly via PKC. Consistent with such a mechanism, Fish et al. (1988) demonstrated that voltage-activated Ca^{2+} current in vascular smooth muscle cells was enhanced by treatment with phorbol esters. Ca^{2+} channels are regulated by PKC in bag cell neurons of *Aplysia* (DeReimer et al. 1985) and by PKA in cardiac myocytes (McDonald et al. 1994).

Other pathways may also contribute to cholinergic enhancement of Ca^{2+} current in smooth muscle. McCarron et al. (1992) demonstrated a calcium-dependent enhancement of Ca^{2+} current in toad gastric smooth muscle, mediated by calmodulin-dependent kinase II, based on its inhibition by calmodulin inhibitory peptides. This mechanism for enhancement of Ca^{2+} current may be important in many cases where cholinergic agonists lead to depolarization and opening of voltage-dependent Ca^{2+}- channels (see e.g. Figure 1). This pathway could also be involved in modulation of Ca^{2+} currents in cases where ACh causes release of Ca^{2+} from intracellular stores.

L-type Ca^{2+} channels are also known to be inhibited by elevation of $[Ca^{2+}]_i$ (McDonald et al. 1994). Might excitation of smooth muscle and the accompanying increases of $[Ca^{2+}]_i$ result in direct modulation of Ca^{2+} channels? Indeed, muscarinic stimulation of visceral smooth muscles causes acute *suppression* of Ca^{2+} current in ileal, gastric, and tracheal muscles (Unno et al. 1995; Wade et al. 1996). Such acute suppression of Ca^{2+} current involves release of Ca^{2+} from stores, and is detected in single channels not exposed to agonist, indicating that a diffusible cytosolic factor (possibly Ca^{2+} itself) participates in such inhibition of Ca^{2+} channels.

3.4.1. Physiological roles

Cholinergic stimulation of smooth muscles causes depolarization via a number of different mechanisms. The depolarization activates voltage-dependent Ca^{2+} channels and contributes to contraction (Figure 3). Like the suppression of K^+ conductance, enhancement of voltage-activated Ca^{2+} currents would contribute to prolongation of the action potential and an increase in amplitude of the plateau as described, for example, in gastrointestinal smooth muscle (Szurszewski 1987; Sanders 1989).

Persistent Ca^{2+} current (window current) can contribute to physiologically relevant Ca^{2+} entry, as described in vascular and tracheal muscles (Ganitkevich and Isenberg 1990b; Fleischmann et al. 1994), which could serve to maintain contraction or contribute to refilling of stores. Following depletion by cyclopiazonic acid of internal Ca^{2+} stores in tracheal smooth muscle, activation of L-type Ca^{2+} channels led to partial refilling of stores (Janssen and Sims 1993b). This refilling occurred in the presence of cyclopiazonic acid, suggesting that the L-type Ca^{2+} channels might be in direct contact with the stores. This may represent an important process for recovery following stimulation with agents such as ACh, since the depolarization (caused in part by Ca^{2+} release from internal stores) would promote Ca^{2+} entry and refilling of the stores.

3.4.2. Overview

Cholinergic enhancement of voltage-activated Ca^{2+} currents appears to be widespread, having been described in vascular, gastrointestinal, and airway smooth muscles. In every case, this enhancement occurs in cells in which other conductance changes are also elicited by ACh, including activation of nonselective cation current, suppression of K^+ currents, and activation of Cl^- current. Therefore, although the primary depolarization is caused by these other mechanisms, enhancement of the depolarization-activated Ca^{2+} current will promote Ca^{2+} entry and excitation. The second-messenger pathways mediating the regulation of Ca^{2+} channels remain to be resolved: evidence has been obtained that this regulation may be mediated by diacylglycerol (possibly acting through activation of PKC) and/or calmodulin-dependent kinase II. Further studies are warranted to examine the role of increased $[Ca^{2+}]_i$ in mediating acute suppression of Ca^{2+} current.

3.5. Convergence and integration of signaling pathways

Electrophysiological studies of many cell types have revealed the convergence of different transmitter signaling pathways on ion channels (Hille 1992). Evidence for similar convergence exists in smooth muscle. Based on electrophysiological studies of intact tissues, it was suggested that cholinergic and histaminergic receptors mediate excitation by acting on the same ion channels (see e.g. Bolton et al. 1981). In studies where the effects of ACh and histamine were studied in the same cells, these agonists evoked the same membrane current responses. For example, nonselective cation current was activated and K^+ current suppressed by ACh as well as by histamine (Komori et al. 1992). In other studies where the effects of histamine alone was studied, histamine augmented a voltage-dependent Ca^{2+} conductance (Oike et al. 1992). In general, then, ACh and histamine act in a similar manner.

The convergence of histaminergic and cholinergic signaling pathways has been investigated in guinea-pig tracheal myocytes, where histamine and ACh evoke currents with similar latency, amplitude, duration, and ionic selectivity. Moreover, the responses are not additive and both agonists produce short-term heterologous desensitization (Janssen and Sims 1993a), providing evidence that histamine and ACh activate the same ion channels by the release of Ca^{2+} from intracellular stores. This is not surprising, since H_1 histamine receptors, like muscarinic receptors, are coupled via G proteins to PLC-β (Hill 1990; Berridge 1993). Similarly, neurokinin-activated and cholinergic signaling pathways may also converge, since neurokinin receptors are coupled to PLC-β (Mayer et al. 1992). Functional evidence for convergence of signaling pathways has been reported for several smooth muscles. For example, both substance P and ACh suppress K^+ current (M current) and augment L-type Ca^{2+} current in gastric smooth muscle cells from the toad (Sims et al. 1986; Clapp et al. 1989). In addition, ACh and substance P activate nonselective cation current in ileal smooth muscle (Nakazawa et al. 1990). Similar convergence occurs in tracheal muscle, where both agonists elicit Ca^{2+} release from stores and activate chloride current and contraction (Janssen and Sims 1994b). Although neurokinin agonists activate Cl^- channels in rabbit colonic smooth muscle cells, this is reported not to be mediated by elevation of $[Ca^{2+}]_i$ (Sun et al. 1992).

4. Summary

Electrophysiological studies of freshly isolated smooth muscle cells have revealed a variety of different membrane conductance changes that participate in muscarinic excitation. These include activation of nonspecific cation and Cl^- currents, the suppression of several types of outward K^+ currents, and the enhancement of voltage-activated Ca^{2+} current (Table 1). In most cases, several of these mechanisms have been found to operate together in the same cell. We now recognize that $[Ca^{2+}]_i$ plays a critical role not only in initiating contraction but in regulating membrane channel activity. For example, nonselective cation current in gastrointestinal muscle is facilitated by elevation of $[Ca^{2+}]_i$. Some nonselective cation and Cl^- channels are activated by elevation of $[Ca^{2+}]_i$, in many cases involving release from stores. These channels therefore participate in a positive feedback loop, where elevation of $[Ca^{2+}]_i$ by entry across the membrane or by release from internal stores causes or promotes depolarization, which in turn leads to further influx of Ca^{2+} through voltage-dependent Ca^{2+} channels. These features of cholinergic excitation of smooth muscle differ from established views, where depolarization is the primary event causing Ca^{2+} entry and contraction.

Acknowledgments

Studies carried out by the authors were supported by grants from the Medical Research Council of Canada and The Ontario Thoracic Society. L. J. Janssen is supported by a Scholarship from the Pharmaceutical Manufacturer's Association of Canada and the MRC. S. M. Sims is the recipient of a Scientist award from the MRC.

References

Adams, P. R., Brown, D. A., and Constanti, A. (1982a). M-currents and other potassium currents in bullfrog sympathetic neurones. *J. Physiol. (Lond.)* 330: 537–72.

Adams, P. R., Brown, D. A., and Constanti, A. (1982b). Pharmacological inhibition of the M-current. *J. Physiol. (Lond.)* 332: 223–62.

Aickin, C. C. (1990). Chloride transport across the sarcolemma of vertebrate smooth and skeletal muscle. In *Chloride Channels and Carriers in Nerve, Muscle, and Glial Cells* (F. J. Alvarez-Leefmans and J. M. Russell, eds.). New York: Plenum, pp. 209–49.

Aickin, C. C., and Vermuë, N. A. (1983). Microelectrode measurement of intracellular chloride activity in smooth muscle cells of guinea-pig ureter. *Pflügers Arch.* 397: 25–8.

Amédée, T., Benham, C. D., Bolton, T. B., Byrne, N. G., and Large, W. A. (1990a). Potassium, chloride and nonselective cation conductances opened by noradrenaline in rabbit ear artery cells. *J. Physiol. (Lond.)* 423: 551–68.

Amédée, T., Large, W. A., and Wang, Q. (1990b). Characteristics of chloride currents activated by noradrenaline in rabbit ear artery cells. *J. Physiol. (Lond.)* 428: 501–16.

Ashida, T., Schaeffer, J., Goldman, W. F., Wade, J. B., and Blaustein, M. P. (1988).

Role of sarcoplasmic reticulum in arterial contraction: Comparison of ryanodine's effect in a conduit and a muscular artery. *Circ. Res.* 62: 854–63.

Bagby, R. M., Young, A. M., Dotson, R. S., Fisher, B. A., and McKinnon, K. (1971). Contractions of single smooth muscle cells from *Bufo marinus* stomach. *Nature* 234: 351–3.

Bean, B. P. (1989a). Classes of calcium channels in vertebrate cells. *Annu. Rev. Physiol.* 51: 367–84.

Bean, B. P. (1989b). Neurotransmitter inhibition of neuronal calcium currents by changes in channel voltage dependence. *Nature* 340: 153–6.

Becker, P. L., Singer, J. J., Walsh, J. V., Jr., and Fay, F. S. (1989). Regulation of calcium concentration in voltage-clamped smooth muscle cells. *Science* 244: 211–14.

Benham, C. D. (1989). ATP-activated channels gate calcium entry in single smooth muscle cells dissociated from rabbit ear artery. *J. Physiol. (Lond.)* 419: 689–701.

Benham, C. D., and Bolton, T. B. (1986). Spontaneous transient outward currents in single visceral and vascular smooth muscle cells of the rabbit. *J. Physiol.(Lond.)* 381: 385–406.

Benham, C. D., Bolton, T. B., and Lang, R. J. (1985). Acetylcholine activates an inward current in single mammalian smooth muscle cells. *Nature* 316: 345–7.

Benham, C. D., and Tsien, R. W. (1987). A novel receptor-operated Ca^{2+}-permeable channel activated by ATP in smooth muscle. *Nature* 328: 275–8.

Berridge, M. J. (1993). Inositol trisphosphate and calcium signalling. *Nature* 361: 315–25.

Bielefeld, D. R., Hume, J. R., and Krier, J. (1990). Action potentials and membrane currents of isolated single smooth muscle cells of cat and rabbit colon. *Pflügers Arch.* 415: 678–87.

Bolton, T. B. (1979). Mechanisms of action of transmitters and other substances on smooth muscle. *Physiol. Rev.* 59: 606–718.

Bolton, T. B. (1981). Action of acetylcholine on the smooth muscle membrane. In: *Smooth Muscle: An Assessment of Current Knowledge*, (E. Bülbring, A. F. Brading, A. W. Jones, and T. Tomita, eds.). London: Arnold, pp. 199–217.

Bolton, T. B. (1989). Electrophysiology of the intestinal musculature. In: *Handbook of Physiology*, (S. G. Schultz, J. D. Wood, and B. B. Rauner, eds.). Bethesda, MD: American Physiological Society, pp. 217–50.

Bolton, T. B., Clark, J. P., Kitamura, K., and Lang, R. J. (1981). Evidence that histamine and carbachol may open the same ion channels in longitudinal smooth muscle of guinea-pig ileum. *J. Physiol.(Lond.)* 320: 363–79.

Bolton, T. B., and Large, W. A. (1986). Are junction potentials essential? Dual mechanism of smooth muscle cell activation by transmitter released from autonomic nerves. *Quart. J. Exp. Physiol.* 71: 1–28.

Bolton, T. B., and Lim, S. P. (1989). Properties of calcium stores and transient outward currents in single smooth muscle cells of rabbit intestine. *J. Physiol. (Lond.)* 409: 385–401.

Bolton, T. B., Tomita, T., and Vassort, G. (1981). Voltage clamp and the measurement of ionic conductances in smooth muscle. In: *Smooth Muscle: An Assessment of Current Knowledge*, (E. Bülbring, A. F. Brading, A. W. Jones, and T. Tomita, eds.). London: Arnold, pp. 47–63.

Brann, M. R., Jorgensen, H. B., Burstein, E. S., Spalding, T. A., Ellis, J., Penelope Jones, S. V., and Hill-Eubanks, D. (1993). Studies of the pharmacology, localization, and structure of muscarinic acetylcholine receptors. *Ann. N.Y. Acad. Sci.* 707: 225–36.

Burnstock, G. (1979). Autonomic innervation and transmission. *Br. Med. Bull.* 35: 255–62.

Burnstock, G., Holman, M. E., and Prosser, C. L. (1963). Electrophysiology of smooth muscle. *Physiol. Rev.* 43: 482–527.

Byrne, N. G., and Large, W. A. (1987a). Action of noradrenaline on single smooth muscle cells freshly dispersed from the rat anococcygeus muscle. *J. Physiol. (Lond.)* 389: 513–25.

Byrne, N. G., and Large, W. A. (1987b). Membrane mechanism associated with muscarinic receptor activation in single cells freshly dispersed from the rat anococcygeus muscle. *Br. J. Pharmacol.* 92: 371–9.

Chen, S., Inoue, R., and Ito, Y. (1993). Pharmacological characterization of muscarinic receptor-activated cation channels in guinea-pig ileum. *Br. J. Pharmacol.* 109: 793–801.

Clapp, L. H., Sims, S. M., Singer, J. J., and Walsh, J. V., Jr. (1992). A role for diacylglycerol in mediating the actions of acetylcholine on M-current in gastric smooth muscle cells. *Am. J. Physiol.* 263: C1274–C1281.

Clapp, L. H., Vivaudou, M. B., Singer, J. J., and Walsh, J. V., Jr. (1989). Substance P, like acetylcholine, augments one kind of Ca^{2+} current in isolated smooth muscle cells. *Pflügers Arch.* 413: 565–7.

Clapp, L. H., Vivaudou, M. B., Walsh, J. V., Jr., and Singer, J. J. (1987). Acetylcholine increases voltage-activated Ca^{2+} current in freshly dissociated smooth muscle cells. *Proc. Nat. Acad. Sci. USA* 84: 2092–6.

Coburn, R. F., and Baron, C. B. (1990). Coupling mechanisms in airway smooth muscle. *Am. J. Physiol.* 258: L119–L133.

Cole, W. C., and Sanders, K. M. (1989). G proteins mediate suppression of Ca^{2+}-activated K^+ current by acetylcholine in smooth muscle cells. *Am. J. Physiol.* 257: C596–C600.

Cole, W. C., Carl, A., and Sanders, K. M. (1989). Muscarinic suppression of Ca^{2+}-dependent K current in colonic smooth muscle. *Am. J. Physiol.* 257: C481–C487.

Colquhoun, D., Neher, E., Reuter, H., and Stevens, C. F. (1981). Inward current channels activated by intracellular Ca in cultured cardiac cells. *Nature* 294: 752–4.

Daniel, E. E., Kannan, M., Davis, C., and Posey-Daniel, V. (1986). Ultrastructural studies on the neuromuscular control of human tracheal and bronchial muscle. *Resp. Physiol.* 63: 109–28.

Daniel, E. E., Serio, R., Jury, J., Pashley, M., and O'Byrne, P. (1988). Effects of inflammatory mediators on neuromuscular transmission in canine trachea *in vitro*. In: *Mechanisms in Asthma: Pharmacology, Physiology, and Management*, New York: Alan Liss, pp. 167–76.

DeReimer, S. A., Strong, J. A., Albert, K. A., Greengard, P., and Kaczmarek, L. K. (1985). Enhancement of calcium current in *Aplysia* neurones by phorbol ester and protein kinase C. *Nature* 313: 313–16.

Ehrlich, B. E., Kaftan, E., Bezprozvannaya, S., and Bezprozvanny, I. (1994). The pharmacology of intracellular Ca^{2+}-release channels. *TIPS* 15: 145–49.

Fantl, W. J., Johnson, E. D., and Williams, L. T. (1993). Signalling by receptor tyrosine kinases. *Annu. Rev. Biochem.* 62: 453–81.

Farley, J. M., and Miles, P. R. (1977). Role of depolarization in acetylcholine-induced contractions of dog trachealis muscle. *J. Pharmacol. Exp. Ther.* 201: 199–205.

Farley, J. M., and Miles, P. R. (1978). The sources of calcium for acetylcholine-induced contractions in dog tracheal smooth muscle. *J. Pharmacol. Exp. Ther.* 207: 340–6.

Fay, F. S., and Delise, C. M. (1973). Contraction of isolated smooth muscle cells – Structural changes. *Proc. Nat. Acad. Sci. USA* 70: 641–5.

Fish, R. D., Sperti, G., Colucci, W. S., and Clapham, D. E. (1988). Phorbol ester increases the dihydropyridine-sensitive calcium conductance in a vascular smooth muscle cell line. *Circ. Res.* 62: 1049–54.

Fleischmann, B. K., Murray, R. K., and Kotlikoff, M. I. (1994). Voltage window for sustained elevation of cytosolic calcium in smooth muscle cells. *Proc. Nat. Acad. Sci. USA* 91: 11914–18.

Gabella, G. (1987). Innervation of airway smooth muscle: Fine structure. *Annu. Rev. Physiol.* 49: 583–94.

Ganitkevich, V., and Isenberg, G. (1990a). Isolated guinea-pig coronary smooth muscle cells. Acetylcholine induces hyperpolarization due to sarcoplasmic reticulum calcium release activating potassium channels. *Circ. Res.* 67: 525–8.

Ganitkevich, V. Y., and Isenberg, G. (1990b). Contribution of two types of calcium channels to membrane conductance of single myocytes from guinea-pig coronary artery. *J. Physiol. (Lond.)* 426: 19–42.

Ganitkevich, V. Y., and Isenberg, G. (1991). Depolarization-mediated intracellular calcium transients in isolated smooth muscle cells of guinea-pig urinary bladder. *J. Physiol. (Lond.)* 435: 187–205.

Gerstheimer, F. P., Mühleisen, M., Nehring, D., and Kreye, V. A. W. (1987). A chloride-bicarbonate exchanging anion carrier in vascular smooth muscle of the rabbit. *Pflügers Arch.* 409: 60–6.

Gerthoffer, W. T. (1991). Regulation of the contractile element of airway smooth muscle. *Am. J. Physiol.* 261: L15–L18.

Gillis, R. A., Quest, J. A., Pagani, F. D., and Norman, W. P. (1986). Control centers in the central nervous system for regulating gastrointestinal motility. In: *Handbook of Physiology – The Gastrointestinal System* (S. G. Schultz, J. D. Wood, and B. B. Rauner, eds.). Bethesda, MD: American Physiological Society, pp. 621–84.

Hals, G. D., Stein, P. G., and Palade, P. T. (1989). Single channel characteristics of a high conductance anion channel in "sarcoballs." *J. Gen. Physiol.* 93: 385–410.

Hamill, O. P., Marty, A., Neher, E., Sakmann, B., and Sigworth, F. J. (1981). Improved patch-clamp techniques for high-resolution current recording from cells and cell-free membrane patches. *Pflügers Arch.* 391: 85–100.

Hill, S. J. (1990). Distribution, properties, and functional characteristics of three classes of histamine receptor. *Pharmacol. Rev.* 42: 45–83.

Hille, B. (1992). *Ionic Channels of Excitable Membranes*, 2nd ed. Sunderland, MA: Sinauer Associates.

Himpens, B., and Somlyo, A. P. (1988). Free-calcium and force transients during depolarization and pharmacomechanical coupling in guinea-pig smooth muscle. *J. Physiol. (Lond.)* 395: 507–30.

Hollenberg, M.D. (1994). Tyrosine kinase pathways and the regulation of smooth muscle contractility. *TIPS* 15: 108–14.

Holmberg, S. R. M., and Williams, A. J. (1989). Single channel recordings from human cardiac sarcoplasmic reticulum. *Circ. Res.* 65: 1445–9.

Horn, R., and Marty, A. (1988). Muscarinic activation of ionic currents measured by a new whole-cell recording method. *J. Gen. Physiol.* 92: 145–59.

Hosey, M. M. (1992). Diversity of structure, signaling and regulation within the family of muscarinic cholinergic receptors. *FASEB J.* 6: 845–52.

Hoth, M., and Penner, R. (1992). Depletion of intracellular calcium stores activates a calcium current in mast cells. *Nature* 355: 353–6.

Huang, X. Y., Morielli, A. D., and Peralta, E. G. (1993). Tyrosine kinase-dependent suppression of a potassium channel by the G protein-coupled m1 muscarinic acetylcholine receptor. *Cell* 75: 1145–56.

Hume, J. R., and Leblanc, N. (1989). Macroscopic K^+ currents in single smooth muscle cells of the rabbit portal vein. *J. Physiol. (Lond.)* 413: 49–73.

Iino, M. (1989). Calcium-induced calcium release mechanism in guinea pig taenia caeci. *J. Gen. Physiol.* 94: 363–83.

Inoue, R. (1991). Effects of external Cd^{2+} and other divalent cations on carbachol-activated nonselective cation channels in guinea-pig ileum. *J. Physiol. (Lond.)* 442: 447–63.

Inoue, R., and Isenberg, G. (1990a). Effect of membrane potential on acetylcholine-induced inward current in guinea-pig ileum. *J. Physiol. (Lond.)* 424: 57–71.

Inoue, R., and Isenberg, G. (1990b). Intracellular calcium ions modulate acetylcholine-induced inward current in guinea-pig ileum. *J. Physiol. (Lond.)* 424: 73–92.

Inoue, R., and Isenberg, G. (1990c). Acetylcholine activates nonselective cation channels in guinea-pig ileum through a G protein. *Am. J. Physiol.* 258: C1173–C1178.

Inoue, R., Kitamura, K., and Kuriyama, H. (1987). Acetylcholine activates single sodium channels in smooth muscle cells. *Pflügers Arch.* 410: 69–74.

Inoue, R., and Kuriyama, H. (1993). Dual regulation of cation-selective channels by muscarinic and α_1-adrenergic receptors in the rabbit portal vein. *J. Physiol. (Lond.)* 465: 427–48.

Inoue, R., and Shan, C. (1993). Physiology of muscarinic receptor-operated nonselective cation channels in guinea-pig ileal smooth muscle. In: *Nonselective Cation Channels: Pharmacology, Physiology and Biophysics.* (D. Siemen and J. Hescheler, eds.). Verlag, Basel, Switzerland: Birkäuser.

Inoue, R., Waniishi, Y., and Ito, Y. (1995). Extracellular H^+ modulates acetylcholine-activated nonselective cation channels in guinea pig ileum. *Am. J. Physiol.* 268: C162–C170.

Inoue, R., Waniishi, Y., Yamada, K., and Ito, Y. (1994). A possible role of tyrosine kinases in the regulation of muscarinic receptor-activated cation channels in guinea pig ileum. *Biochem. Biophys. Res. Commun.* 203: 1392–97.

Isenberg, G. (1993). Nonselective cation channels in cardiac and smooth muscle cells. In: *Nonselective Cation Channels: Pharmacology, Physiology and Biophysics* (D. Siemen and J. Hescheler, eds.). Basel, Switzerland: Birkhäuser Verlag.

Janssen, L. J. (1996) Acetylcholine and caffeine activate Cl^- and suppress K^+ conductances in human bronchial smooth muscle. *Am. J. Physiol.* 270: L772–L781.

Janssen, L. J., and Sims, S. M. (1992). Acetylcholine activates nonselective cation and chloride conductances in canine and guinea-pig tracheal myocytes. *J. Physiol. (Lond.)* 453: 197–218.

Janssen, L. J., and Sims, S. M. (1993a). Histamine activates Cl^- and K^+ current in guinea-pig tracheal myocytes: Convergence with cholinergic signalling pathway. *J. Physiol. (Lond.)* 465: 661–77.

Janssen, L. J., and Sims, S. M. (1993b). Emptying and refilling of Ca^{2+} stores in tracheal myocytes as indicated by ACh-evoked currents and contraction. *Am. J. Physiol.* 265: C877–C886.

Janssen, L. J. and Sims, S. M. (1994a). Spontaneous transient inward currents and rhythmicity in canine and guinea pig tracheal smooth muscle cells. *Pflügers Arch.* 427: 473–80.

Janssen, L. J. and Sims, S. M. (1994b). Substance P activates Cl^- and K^+ conductances in guinea-pig tracheal smooth muscle cells. *Can. J. Physiol. Pharmacol.* 72: 705–10.

Janssen, L. J. and Sims, S. M. (1995). Ca^{2+}-dependent Cl^- current in canine tracheal smooth muscle cells. *Am. J. Physiol.* 269: C163–C169.

Joseph, S. K., and Williamson, J. R. (1988). Characteristics of inositol trisphosphate-mediated Ca^{2+} release from permeabilized hepatocytes. *J. Biol. Chem.* 261: 14658–64.

Kamishima, T., Nelson, M. T., and Patlak, J. P. (1992). Carbachol modulates voltage sensitivity of calcium channels in bronchial smooth muscle of rats. *Am. J. Physiol.* 263: C69–C77.

Kamm, K. E., and Stull, J. T. (1989). Regulation of smooth muscle contractile elements by second messengers. *Annu. Rev. Physiol.* 51: 299–313.

Katz, B. (1966). *Nerve, Muscle, and Synapse*. Toronto: McGraw-Hill.

Kirber, M. T., Walsh, J. V. Jr., and Singer, J. J. (1988). Stretch-activated ion channels in smooth muscle: A mechanism for the initiation of stretch-induced contraction. *Pflügers Arch.* 412: 339–45.

Kitazawa, T., Kobayashi, S., Horiuti, K., Somlyo, A. V., and Somlyo, A. P. (1989). Receptor-coupled, permeabilized smooth muscle: Role of the phosphoinositide cascade, G-proteins, and modulation of the contractile response to Ca^{2+}. *J. Biol. Chem.* 264: 5339–42.

Komori, S., and Bolton, T. B. (1990). Role of G-proteins in muscarinic receptor inward and outward currents in rabbit jejunal smooth muscle. *J. Physiol. (Lond.)* 427: 395–419.

Komori, S., and Bolton, T. B. (1991a). Calcium release induced by inositol 1,4,5-trisphosphate in single rabbit intestinal smooth muscle cells. *J. Physiol. (Lond.)* 433: 495–517.

Komori, S., and Bolton, T. B. (1991b). Inositol-trisphosphate releases stored calcium to block voltage-dependent calcium channels in single smooth muscle cells. *Pflügers Arch.* 418: 437–41.

Komori, S., Kawai, M., Takewaki, T., and Ohashi, H. (1992). GTP-binding protein involvement in membrane currents evoked by carbachol and histamine in guinea-pig ileal muscle. *J. Physiol. (Lond.)* 450: 105–26.

Korn, S. J., Marty, A., Connor, J. A., and Horn, R. (1991). Perforated patch recording. *Methods Neurosci.* 4: 364–73.

Kume, H., and Kotlikoff, M. I. (1991). Muscarinic inhibition of single K_{Ca} channels in smooth muscle cells by a pertussis-sensitive G-protein. *Am. J. Physiol.* 261: C1204–C1209.

Kume, H., Takai, A., Tokuno, H., and Tomita, T. (1989). Regulation of Ca^{2+}-dependent K^+-channel activity in tracheal myocytes by phosphorylation. *Nature* 341: 152–4.

Lammel, E., Dietmer, P., and Noack, T. (1991). Suppression of steady membrane currents by acetylcholine in single smooth muscle cells of the guinea-pig gastric fundus. *J. Physiol. (Lond.)* 432: 259–82.

Large, W.A., and Q. Wang. (1996). Characteristics and physiological roles of the calcium-activated chloride current in single smooth muscle. *Am. J. Physiol.* 271: C435–C454.

Lee, H. K., Bayguinov, O., and Sanders, K. M. (1993). Role of nonselective cation current in muscarinic responses of canine colonic muscle. *Am. J. Physiol.* 265: C1463–C1471.

Lim, S. P., and Bolton, T. B. (1988). A calcium-dependent rather than a G-protein mechanism is involved in the inward current evoked by muscarinic receptor stimulation in dialysed single smooth muscle cells of small intestine. *Br. J. Pharmacol.* 95: 325–7.

Loirand, G., Pacaud, P., Baron, A., Mironneau, C., and Mironneau, J. (1991). Large conductance calcium-activated nonselective cation channel in smooth muscle cells isolated from rat portal vein. *J. Physiol. (Lond.)* 437: 461–75.

Matsuda, J. J., Volk, K. A., and Shibata, E. F. (1990). Calcium currents in isolated rabbit coronary arterial smooth muscle myocytes. *J. Physiol. (Lond.)* 427: 657–80.
Mayer, E. A., Sun, X. P., and Willenbucher, R. F. (1992). Contraction coupling in colonic smooth muscle. *Annu. Rev. Physiol.* 54: 395–414.
McCarron, J. G., McGeown, J. G., Reardon, S., Ikebe, M., Fay, F. S., and Walsh, J. V., Jr. (1992). Calcium-dependent enhancement of calcium current in smooth muscle by calmodulin-dependent kinase II. *Nature* 357: 74–7.
McDonald, T. F, Pelzer, S., Trautwein, W., and Pelzer, D. J. (1994). Regulation and modulation of Ca^{2+} channels in cardiac, skeletal, and smooth muscle cells. *Physiol. Rev.* 74: 365–507.
Nakazawa, M., Inoue, K., Fujimori, K., and Takanaka, A. (1990). Difference between substance P- and acetylcholine-induced currents in mammalian smooth muscle cells. *Eur. J. Pharmacol.* 179: 453–6.
Nelson, M. T., Cheng, H., Rubart, M., Santana, L. F., Bonev, A. D., Knot, H. J., and Lederer, W. J. (1995). Relaxation of arterial smooth muscle by calcium sparks. *Science* 270: 633–7.
Oike, M., Kitamura, K., and Kuriyama, H. (1992). Histamine H_3-receptor activation augments voltage-dependent Ca^{2+} current via GTP hydrolysis in rabbit saphenous artery. *J. Physiol.* 448: 133–52.
Pacaud, P., and Bolton, T. B. (1991). Relation between muscarinic receptor cationic current and internal calcium in guinea-pig jejunal smooth muscle cells. *J. Physiol. (Lond.)* 441: 477–99.
Pacaud, P., Loirand, G., Lavie, J. L., Mironneau, C., and Mironneau, J. (1989). Calcium-activated chloride current in rat vascular smooth muscle cells in short-term primary culture. *Pflügers Arch.* 413: 629–36.
Partridge, L. D., and Swandula, D. (1988). Calcium-activated non-specific cation channels. *TINS* 11: 69–72.
Petersen, O. H., and Wakui, M. (1990). Oscillating intracellular Ca^{2+} signals evoked by activation of receptors linked to inositol lipid hydrolysis: mechanism of generation. *J. Membr. Biol.* 118: 93–105.
Sanders, K. M. (1989). Electrophysiology of dissociated gastrointestinal muscle cells. In: *Handbook of Physiology*, (S. G. Schultz, J. D. Wood, and B. B. Rauner, eds.). Bethesda, MD: American Physiological Society, pp. 163–86.
Shimada, T., Hamada, E., Terano, A., Sugimoto, T., and Kurachi, Y. (1991). β-agonists modulate ACh-inhibition of a K current in intestinal smooth muscle cells. *Biochem. Biophys. Res. Commun.* 179: 327–32.
Siegelbaum, S.A. (1994). Ion channel control by tyrosine phosphorylation. *Curr. Biol.* 4: 242–5.
Sims, S. M. (1992). Cholinergic activation of a nonselective cation current in canine gastric smooth muscle is associated with contraction. *J. Physiol. (Lond.)* 449: 377–398.
Sims, S. M., Clapp, L. H., Walsh, J. V., Jr., and Singer, J. J. (1990a). Dual regulation of M-current in smooth muscle cells: β-adrenergic-muscarinic antagonism. *Pflügers Arch.* 417: 291–302.
Sims, S. M., and Janssen, L. J. (1993). Cholinergic excitation of smooth muscle. *News in Physiol. Sci.* 8: 207–12.
Sims, S. M., Singer, J. J., and Walsh, J. V., Jr. (1985). Cholinergic agonists suppress a potassium current in freshly dissociated smooth muscle cells of the toad. *J. Physiol. (Lond.)* 367: 503–29.
Sims, S. M., Singer, J. J., and Walsh, J. V., Jr. (1988). Antagonistic adrenergic-muscarinic regulation of M-current in smooth muscle cells. *Science* 239: 190–3.
Sims, S. M., Vivaudou, M. B., Biancani, P., Hillemeier, C., Singer, J. J., and Walsh,

J. V., Jr. (1990b). Membrane currents and cholinergic regulation of K^+ current in esophageal smooth muscle cells. *American J. Physiol.* 258: G794–G802.

Sims, S. M., Walsh, J. V., Jr., and Singer, J. J. (1986). Substance P and acetylcholine both suppress the same K^+ current in dissociated smooth muscle cells. *Am. J. Physiol.* 251: C580–C587.

Singer, J. J., and Walsh, J. V., Jr. (1980a). Passive properties of the membrane of single freshly isolated smooth muscle cells. *Am. J. Physiol.* 239: C153–C161.

Singer, J. J., and Walsh, J. V., Jr. (1980b). Rectifying properties of the membrane of single freshly isolated smooth muscle cells. *Am. J. Physiol.* 239: C162–C174.

Somlyo, A. P., and Himpens, B. (1989). Cell calcium and its regulation in smooth muscle. *FASEB J.* 3: 2266–76.

Somlyo, A. P., and Somlyo, A. V. (1968). Vascular smooth muscle: I. Normal structure, pathology, biochemistry, and biophysics. *Pharmacol. Rev.* 20: 197–272.

Somlyo, A. P., and Somlyo, A.V. (1994). Signal transduction and regulation in smooth muscle. *Nature* 372: 231–6.

Somlyo, A. V., Bond, M., Somlyo, A. P., and Scarpa, A. (1985). Inositol trisphosphate-induced calcium release and contraction in vascular smooth muscle. *Proc. Nat. Acad. Sci. USA* 82: 5231–5.

Suematsu, E., Hirata, M., Hashimoto, T., and Kuriyama, H. (1984). Inositol 1,4,5-trisphosphate releases Ca^{2+} from intracellular store sites in skinned single cells of porcine coronary artery. *Biochem. Biophys. Res. Commun.* 120: 481–95.

Sun, X. P., Supplisson, S., Torres, R., Sachs, G., and Mayer, E. (1992). Characterization of large-conductance chloride channels in rabbit colonic smooth muscle. *J. Physiol. (Lond.)* 448: 355–82.

Szurszewski, J. H. (1987). Electrical basis for gastrointestinal motility. In: *Physiology of the Gastrointestinal Tract* (L. R. Johnson, ed.). New York: Raven, pp. 383–422.

Takeuchi, A., and Takeuchi, N. (1960). On the permeability of the end-plate membrane during the action of transmitter. *J. Physiol. (Lond.)* 154: 52–67.

Tomasic, M., Boyle, J. P., Worley, J. F., and Kotlikoff, M. I. (1992). Contractile agonists activate voltage-dependent calcium channels in airway smooth muscle cells. *Am. J. Physiol.* 263: C106–C113.

Unno, T., Komori, S., and Ohashi, H. (1995). Inhibitory effect of muscarinic receptor activation on Ca^{2+} channel current in smooth muscle cells of guinea pig ileum. *J. Physiol. (Lond.)* 484: 567–81.

van Breemen, C., Aaronson, P., and Loutzenhiser, R. (1979). Sodium-calcium interactions in mammalian smooth muscle. *Pharmacol. Rev.* 30: 167–208.

van Breemen, C., and Saida, K. (1989). Cellular mechanisms regulating $[Ca^{2+}]_i$ smooth muscle. *Ann. Rev. Physiol.* 51: 315–29.

Vivaudou, M. B., Clapp, L. H., Walsh, J. V., Jr., and Singer, J. J. (1988). Regulation of one type of Ca^{2+} current in smooth muscle cells by diacylglycerol and acetylcholine. *FASEB J.* 2: 2497–2504.

Vogalis, F., Publicover, N. G., Hume, J. P., and Sanders, K. M. (1991). Relationship between calcium current and cytosolic calcium in canine gastric smooth muscle cells. *Am. J. Physiol.* 260: C1012–C1018.

Vogalis, F., and Sanders, K. M. (1990). Cholinergic stimulation activates a nonselective cation current in canine pyloric circular muscle cells. *J. Physiol. (Lond.)* 429: 223–36.

Wade, G. R., Barbera, J., and Sims, S.M. (1996) Cholinergic inhibition of Ca^{2+} channels in guinea-pig gastric and tracheal smooth muscle. *J. Physiol. (Lond.)* 491: 307–19.

Wade, G. R., and Sims, S. M. (1993). Muscarinic stimulation of tracheal smooth muscle cells activates large-conductance Ca^{2+}-dependent K^+ channel. *Am. J. Physiol.* 265: C658–C665.

Walker, J. W., Somlyo, A. V., Goldman, Y. E., Somlyo, A. P., and Trentham, D. R. (1987). Kinetics of smooth and skeletal muscle activation by laser pulse photolysis of caged inositol 1,4,5-trisphosphate. *Nature* 327: 249–52.

Walsh, J. V., Jr., and Singer, J. J. (1980a). Calcium action potentials in single freshly isolated smooth muscle cells. *Am. J. Physiol.* 239: C162–C174.

Walsh, J. V., Jr., and Singer, J. J. (1980b). Penetration-induced hyperpolarization as evidence for Ca^{2+} activation of K^+ conductance in isolated smooth muscle cells. *Am. J. Physiol.* 239: C182–C189.

Wang, Q., Hogg, R. C., and Large, W. A. (1992). Properties of spontaneous inward currents recorded in smooth muscle cells isolated from rabbit portal vein. *J. Physiol. (Lond.)* 451: 525–37.

Williams, D. A., Becker, P. L., and Fay, F. S. (1987). Regional changes in calcium underlying contraction of single smooth muscle cells. *Science* 235: 1644–8.

Williams, D. A., Fogarty, K. E., Tsien, R. Y., and Fay, F. S. (1985). Calcium gradients in single smooth muscle cells revealed by the digital imaging microscope using Fura-2. *Nature* 318: 558–61.

Yagi, S., Becker, P. L., and Fay, F. S. (1988). Relationship between force and Ca^{2+} concentration in smooth muscle as revealed by measurements on single cells. *Proc. Nat. Acad. Sci. USA* 85: 4109–13.

Yamamoto, Y., Hu, S. L., and Kao, C. Y. (1989). Inward current in single smooth muscle cells of the guinea pig taenia coli. *J. Gen. Physiol.* 93: 521–50.

Yellen, G. (1982). Single Ca^{2+}-activated nonselective cation channels in neuroblastoma. *Nature* 296: 357–9.

Zholos, A. V., and Bolton, T. B. (1994). G-protein control of voltage dependence as well as gating of muscarinic metabotropic channels in guinea-pig ileum. *J. Physiol. (Lond.)* 478: 195–202.

5

Mechanics of Smooth Muscle Contraction

RICHARD A. MEISS

1. Smooth muscle mechanics in context

1.1. Smooth muscle mechanics and cellular function

An adequate understanding of the mechanical function of smooth muscle must integrate cellular events, extracellular interconnections and interactions, and subcellular and molecular events into a coordinated contraction of the tissue as a whole. At the center of this range of function is the smooth muscle cell, and its study has been approached from several directions. Approached from the tissue level, muscle function reveals behavior that requires a cellular explanation; approaching cellular function from the direction of the biochemical or physical interactions of isolated cellular constituents has provided some of these explanations. Attempts to understand smooth muscle mechanical function at all levels of organization are obviously necessary, and it is ultimately through the integration of all avenues of approach that an adequate picture will emerge. Mechanical studies of smooth muscle tissues inherently provide this kind of integration. This chapter will attempt to treat smooth muscle mechanics thoroughly enough to give the reader insight into both the adequacies and shortcomings of the kinds of mechanical measurements in widespread use in smooth muscle research.

The discussions in this chapter will be confined to vertebrate smooth muscle, mostly that of mammals, even though nonstriated muscles are found elsewhere in the animal kingdom, most notably among the mollusks (Prosser 1967; Ruegg 1971). Although much has been learned from invertebrate smooth muscle, the phylogenetic relationships between vertebrate and invertebrate muscles are unclear, and many aspects of their physiology and biochemistry are quite different. Although this unfortunately places them somewhat beyond the scope of this chapter, there are some good references available that can serve as a general introduction to invertebrate muscles and their many variations (Prosser 1973, 1982; Twarog et al. 1982).

Many of the topics of this chapter have been the subject of relatively recent reviews. The mechanical properties of vascular muscle have been reviewed by Murphy (1980), and those of gastrointestinal muscle mechanics by Meiss (1989). Both of these reviews, in addition to their special emphases, cover the general subject of smooth muscle mechanics and treat a number of basic areas not detailed in this chapter.

1.2. Comparisons with mechanical properties of skeletal muscle

Over the past several decades, research into smooth muscle mechanics has been strongly influenced by mechanical studies of skeletal muscle. These earlier studies have provided a framework that served well in providing critical conceptual links between skeletal muscle ultrastructure and mechanical properties, especially in the development of the sliding-filament/crossbridge hypothesis for skeletal muscle contraction. Since smooth muscle at first appeared to behave like skeletal muscle in most respects that were examined (see e.g. Bozler 1948; Csapo and Goodal 1954), a similarity of fundamental molecular mechanisms of contraction was assumed. To a great extent this situation has proven to be the case, especially in the basic area of myofilament interactions and biochemistry, where the fundamental relationships between actin and myosin are proving to be quite similar. The assumption of similar function has proven less successful (and perhaps misleading at times) in the study of the regulation of contraction, although it has served to point out the critical steps in the process. As an explanation for the effects of length on contractile mechanics of smooth muscle, the sliding-filament paradigm has provided little guidance, at least as far as our current incomplete knowledge of the ultrastructure of the contractile apparatus of smooth muscle allows us to judge. And finally, as a guide to understanding the lability and plasticity of smooth muscle function, skeletal muscle physiology has served more as an example of stability than as a source of information and insight.

2. Structural correlates of mechanical function

Even though the correlations are not as evident in smooth muscle as in skeletal muscle, the functional properties of smooth muscle are largely governed by the molecular structure and cellular and tissue organization of its constituent parts. Aspects of this organization have already been the subject of earlier chapters in this book, and additional information may be found in several excellent reviews of the ultrastructural organization of smooth muscle (Somlyo and Franzini-Armstrong 1985; Bagby 1990; Gabella 1990). An extensive review of the structural elements responsible for force transmission in smooth muscle tissues has been given by Gabella (1984). Although a consistent picture of the organization of the contractile apparatus of smooth muscle – both within cells and in its cell-to-cell extension – is now emerging, any close association between specific mechanical behaviors and structural details of the contractile apparatus, as has been true for skeletal muscle, is far from being realized in the case of smooth muscle.

Refinements of knowledge in both mechanics and structure will in time sharpen such associations.

3. Mechanical specializations of smooth muscle

3.1. Phasic and tonic muscle types

When classified on the basis of speed of activation and contraction, smooth muscles can be considered to be either *phasic* or *tonic* (Somlyo and Somlyo 1968, 1994). Phasic muscles contract and relax comparatively rapidly, and their contractions may be characterized by an initial transient peak that quickly declines to a plateau when activated by KCl depolarization (Somlyo and Somlyo 1968). They are generally not specialized to maintain large forces for long periods of time. Phasic muscles normally are activated by action potentials that may be spontaneous or brought about by neural activity. This category of muscle includes the uterus, urinary bladder, intestinal muscle, and some vascular muscles such as portal vein. Tonic muscles respond more slowly to stimulation by pharmacological or humoral agents; their course of contraction is much longer and they relax more slowly. They are specialized to maintain tension for long periods of time with low energy expenditure (the ''latch state''; see Section 7.2.1). Such muscles usually have cell membranes that do not respond with action potentials to electrical stimulation but are specialized for activation via membrane receptors for various agonists.

Although the principal difference between these two types is in their membrane activity, other factors also play a role. Phasic muscles have higher levels of regulatory enzymes (Fuglsang et al. 1993; Gong et al. 1992) for both activation and inactivation (relaxation). However, the differences persist even when activation is not a limiting factor, implying that the contractile proteins (more likely myosin than actin) have different catalytic properties, a factor that may be related to the isozymic variations in the regulatory light chain or in the myosin molecule itself. For example, Kelley et al. (1993) found a seven–amino acid insert near the ATP binding site on the head of the myosin molecules of chicken gizzard (a phasic muscle). This structural change was sufficient to confer upon this muscle a higher speed of shortening than found in aortic muscle (tonic) from the same animal, which lacked this short segment. The myosin light chains of both molecules were interchangeable without effect on the contraction kinetics. A further complication for this classification is that some muscles show either phasic or tonic activity, depending on the mode of stimulation. Using muscles generally considered to be phasic (rabbit urinary bladder and rat portal vein), Uvelius and Hellstrand (1980) produced ''tonic'' contractions by the use of KCl depolarization and ''phasic'' contractions with electrical stimulation. With stimuli adjusted to produce equal forces, they found that muscles activated tonically shortened more slowly when presented with a light isotonic load. When the chemical activation of living and chemically skinned guinea-pig taenia coli was of short duration, Hellstrand (1991) found that its shortening and resistance to stretch were typical of phasic muscle, using as a standard of comparison the same muscles subjected to long-term tonic stimulation. Compared with long-duration (''tonic'') activation, brief activation (''phasic'' stimulation) produced contractions showing more rapid crossbridge kinetics during recovery from a sudden length change.

3.2. Control of smooth muscle activity

As in skeletal muscle, calcium ions play a central role in initiating and controlling smooth muscle contraction. Cytosolic calcium may be increased by a variety of mechanisms (Somlyo and Himpens, 1989; van Bremen and Saida 1989), most of them quite different from those of skeletal muscle. Regardless of the means by which cytosolic calcium is controlled, its cellular function is to regulate actin–myosin interaction and thus control contraction, as will be summarized here very briefly and generally. As activation begins, calcium ions bind to the regulatory protein calmodulin, which is associated with the enzyme *myosin light chain kinase* (MLCK). This enzyme catalyzes the phosphorylation of a specific amino acid residue (serine-19) on the 20-kDa myosin (regulatory) light-chain, associated with each head of the myosin molecules that form the thick myofilaments. Myosin thus phosphorylated increases in its actin-activated Mg-ATPase activity by as much as a hundredfold, and its interaction with filamentous actin produces the mechanochemical phenomenon of muscle contraction. In competition with the activating processes just described, *myosin light chain phosphatase* (MLCP) continuously dephosphorylates the myosin light chains. During full activation, this process is overcome by the more rapid phosphorylation (forward) reactions; however, when the level of cytosolic calcium falls to a value too low to activate sufficient MLCK, relaxation ensues. The reduction in the concentration of cytosolic calcium is brought about by cellular mechanisms that sequester calcium internally, reduce its rate of entry, or actively remove it from the cell. It is characteristic of most smooth muscles that both intracellular calcium and the level of light chain phosphorylation are initially high and then decline as contractile force is maintained to some degree, although the specific temporal and biochemical details vary widely among various muscle types. Force maintenance, after its initial activation and development, is a complex process that is not fully understood. Most evidence indicates that myosin phosphorylation is a necessary component of smooth muscle activation, but much evidence also indicates that regulatory systems involving thin-filament control also play an important role in force maintenance (see e.g. Haeberle and Hemrich 1994).

4. Mechanical properties subject to experimental measurement

Smooth muscle contractile activity can be expressed in a variety of different ways. The muscle may develop force, it may shorten, and along with both of these modes it will undergo changes in stiffness. Figure 1 shows the basic contractile modes for smooth muscle. A variety of considerations determines the choice of contraction modes for experimental analysis. In some cases the mode will be chosen to represent as faithfully as possible the mechanical and biological conditions under which the tissue normally functions. In other cases artificial

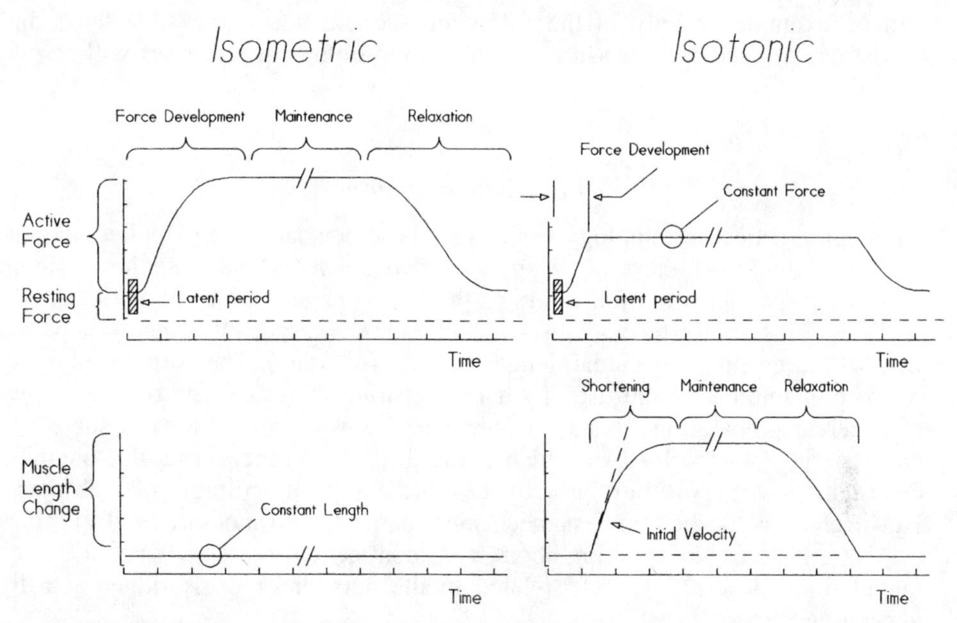

Figure 1. Terminology and relationships in muscle contraction. *Left panel: Isometric* contraction; length is held constant while force varies. Resting force is present before stimulation, and active force is developed after a latent period. Force output may be divided into the phases of force development, force maintenance, and relaxation. *Right panel: Isotonic* contraction; force is held constant while length varies. Muscle shortening begins as soon as the muscle has developed sufficient isometric force to lift an attached load. Length changes may be divided into three phases that are counterparts of the similar isometric phases. The time scale is arbitrary; depending on the muscle, it can range from seconds to many minutes. The duration of the maintenance phase depends on the duration of electrical or chemical stimulation. Note that in the isotonic contractions shown in Figures 3 and 5, the isometric force was allowed to reach a higher force than the ultimate load, and then the muscles were switched to lighter loads.

conditions of contraction may be set up to emphasize a particular mechanical attribute such as shortening velocity or resistance to forced extension. Often the limitations of the available experimental apparatus dictate the mode of analysis; this may be done at the risk of making measurements inappropriate to the physiological process being examined, because in many cases critical changes in mechanical performance will not vary in parallel with all of the mechanical functions. This is especially true in the case where the latch phenomenon is being investigated; over the course of a contraction, the relative capacities to shorten and to develop force may diverge dramatically (Dillon and Murphy 1982; Murphy 1994). In other cases, the relationship between force and stiffness may vary unexpectedly with different conditions of contraction (Meiss 1993, 1994; Gunst et al. 1995). In smooth muscle, to a much greater extent than in skeletal muscle, mechanical output is affected by the metabolic and regulatory state of the tissue and hence is subject to a wide variety of influences that may seriously modify mechanical performance. The most commonly used mechanical measurements will be examined briefly in the following section, where the advantages and limitations of each as a measure of internal physiological processes will be discussed.

4.1. Contractile force

Although practical terminology in this area is somewhat loosely applied, in general the term *force* refers to the pull exerted against a mechanoelectric force transducer. The term *tension* refers to the force per unit of cross-sectional area (F/CSA). This parameter may also be termed *stress*, especially when it is paired with measurements using small length changes (or strain). The normalized forms of force quantification are useful when comparing the force data from a variety of different preparations, but measurement of cross-sectional area is subject to uncertainties, not the least of which is the difficulty in measuring this area and determining what portion of it actually represents contractile muscle substance. However, force is the simplest mechanical parameter to measure, and in return it gives significant information about basic contractile processes, since the force output of a muscle is closely related to the number of crossbridges actually generating force.

4.2. Shortening

The extent and rate of change in length of a smooth muscle preparation can reveal much about the underlying function. Because shortening muscles perform physical work over some measurable span of time, the velocity of shortening is closely tied to the energy-producing reactions of the crossbridge cycle and thus to the enzymatic and mechanical functions of the crossbridges themselves. The extent and rate of shortening can also provide information about the level of activation and the regulatory state of the tissue, since the unloaded velocity of shortening appears closely matched to the maximal cycling rate of crossbridges (Hai and Murphy 1989). In addition, shortening measurements, especially when combined with measurements of stiffness, can shed light on the architecture of the contractile apparatus and the physical structure of the tissue.

4.3. Stiffness

The stiffness of contracting muscle under isometric conditions can provide an approximate measure of the number of crossbridges actually attached, whether or not they are actively generating force (Kamm and Stull 1986; Harris and Warshaw 1990). Stiffness of active muscle, sometimes called *dynamic stiffness*, is operationally defined as the incremental change in force (ΔF) in response to a change in length (ΔL), expressed as the ratio $\Delta F/\Delta L$. When ΔF is expressed as force/CSA (stress) and ΔL is expressed in relative terms as a fraction of the muscle length (strain, $\Delta L/L$), the stress/strain ratio is called the *elastic modulus*, or E. The elastic properties may also be expressed as the inverse of the stiffness; this is termed the *compliance* ($\Delta L/\Delta F$). The commonly used techniques for measuring stiffness (quick stretch, oscillatory length perturbations) induce an extra external strain in the crossbridge array and thus provide a straining force for those bridges that are not bearing force at the moment. This provides a means of detecting crossbridges that are attached but not generating force. The localization of muscle stiffness to skeletal muscle crossbridges has recently been shown to be only approximate, since mechanical and X-ray diffraction measurements have demonstrated that a significant portion of the overall compliance is associated with the thick and thin myofilaments (Goldman and Huxley 1994; Huxley et al. 1994; Wakabayashi et al. 1994). Such a caution must also hold for smooth muscle, but because of the complexity and less-regular organization of the cellular and tissue structure, a much smaller portion of the stiffness can be reliably associated with the crossbridges (Halpern et al. 1978; Mulvany and Warshaw 1981; Warshaw et al. 1988). It is reasonable, nevertheless, to assume that stiffness measurements can provide at least a temporal reflection of the state of the crossbridge array, especially if comparisons are made under closely comparable conditions of length and force.

5. Muscle preparations used for mechanical study

5.1. Whole-organ preparations

The choice of a suitable smooth muscle preparation for a particular study is obviously dictated by the experimental goals. If the aim is to keep conditions as nearly physiological as possible, then entire smooth muscle organs (such as gut segments, blood vessel segments, whole urinary bladders, etc.) may be employed. The disadvantage of using intact organs is that the smooth muscle is usually arranged in such a way that connective tissue and the special geometry of the organ distort the muscle activity and compromise its connections to the measuring apparatus. In some cases, however, the circular nature of a tissue can be turned into an advantage. Because most visceral smooth muscle does not have tendons in the usual sense, attachment to measuring apparatus can be difficult. Preparations of blood vessels, gut, and so on can be dissected in the form of rings which can then be borne on hooks attached to the apparatus. This avoids the problem of artificial attachment structures and further tissue damage. In other cases, the organ can be kept intact and mounted to an apparatus that measures luminal pressure as a reflection of the force generated by the smooth muscle component. Changes in the organ diameter can be used to calculate changes in muscle cell length.

5.2. Smooth muscle strips and isolated tissues

If the intent is to study the fundamental physiological processes of smooth muscle per se, many smooth muscle tissues may be dissected free of the organ in which they are found and may then be treated mechanically as muscle strips or ring preparations. The caudal artery of the rat can be cut in a helical manner, following the natural slow-pitch helix of the muscle cells, into a strip of sufficient length for studies of shortening velocity (Packer and Stephens 1985). A new method for stripping away the very stiff adventitial layer allows for the preparation of rings or strips of porcine carotid artery smooth muscle that can function without interference from a large parallel elastic component (Wingard et al. 1995). The taenia coli, a ribbonlike strip of longitudinal smooth muscle, can be dissected free from the surface of the intestine in guinea pigs or rabbits to form an intact, free-standing muscle strip (Gordon and Siegman 1971). Rings or strips of circular or longitudinal intestinal muscle, free of the underlying mucosa, may readily be prepared from cat duodenum (Meiss 1971), and similar preparations of rabbit uterine muscle have been used (Tanner et al. 1988). Strips of muscle with axial orientation of cells may be dissected from the urinary bladders of rats and rabbits (Uvelius and Gabella 1980). A few other muscles occur naturally as strips; the anococcygeus muscle of the rat may be dissected free from surrounding tissue with little damage (Gillespie 1971). Dog, pig, or rabbit trachealis muscle can readily be dissected free from the underlying connective tissue and from the ends of its cartilaginous supporting structures to form a well-ordered strip (Stephens et al. 1969; Gunst and Russell 1982). In female rabbits, the mesotubarium superius and the ovarian ligament muscles form the outer margin of supporting mesenteries in the reproductive system and can be dissected free with only a single cut along one edge. This produces a long and slender strip preparation with good axial orientation of cells and minimal tissue damage (Meiss 1975).

The lack of tendons in smooth muscle makes mechanically secure attachments to apparatus difficult. In some cases, very inextensible silk suture material or thread has been used to make knotted connections to the apparatus. Researchers have also used clips or small cylinders of aluminum foil to form artificial tendons (see e.g. Meiss 1984a; Pratusevich et al. 1995), and in many cases glues (such as cyanoacrylate cements, Tanner et al. 1988) have been used to make secure connections.

5.3. Isolated smooth muscle cells

Even more difficult problems are faced when the mechanical preparation, instead of being an intact tissue, is a single isolated cell. Much important information, however, has come from such measurements, which have been used to perform essentially the full range of classical mechanical measurements. The mechanical properties of such preparations are quite similar to those of the intact tissue, even though the cells are in a much different mechanical environment from cells within a tissue. Single-cell techniques have not become widespread because of the serious technical difficulties involved in the connection with and recording from the cells. Nonetheless, the results from such studies have been invaluable in understanding and defining the function of the contractile system when freed from interfering connective tissue structures. This work provides a basis for understanding both cellular mechanics and the cellular contribution to the collective mechanical properties of intact tissues; its importance must not be underemphasized, since it is ultimately within the inner workings of the single cell that the fundamental mechanical events of smooth muscle contraction must occur.

5.4. Isolated proteins and myofilaments

At the molecular level of organization and function, the use of the *in vitro motility assay* (Kron and Spudich 1986) has provided insight into the behavior of minimal assemblies of the functional mechanochemical components of muscle of all types. This has been quite useful in answering some long-standing questions in smooth muscle mechanics. The technique allows careful control of the type and specific composition (e.g., protein isoform) of the interacting actin, myosin, and putative regulatory molecules. In a typical application, a surface is coated with myosin heads of the desired type, and actin filaments coated with a fluorescent marker (rhodamine phalloidin) are allowed to interact with them in the overlying solution. From video analysis of the motion of the actin filaments, a number of tentative conclusions as to the nature of the interaction can be drawn.

At a further level of refinement, an individual actin filament (attached to two polystyrene beads coated with chemically modified myosin) can be held with ''optical tweezers'' (formed from intense laser beams; see Finer et al. 1994 and Simmons et al. 1993) and allowed to interact with a single captive myosin molecule. Alternatively, an actin filament can be attached to a flexible and calibrated microneedle and be pulled on by myosin molecules attached to a glass substrate (VanBuren et al. 1994). In these ways the force produced by the unitary event, a single crossbridge stroke, can be determined. In such isolated systems, however, interpretations must be made with caution because of the highly artificial circumstances. Nevertheless, such studies have confirmed, at the molecular level, some predictions that have arisen from experiments performed at a higher level of organization, and have given valuable insight into the unitary events of contraction.

6. Experimental approaches and methods

6.1. Mechanical measurements

Many of the techniques originally developed for the study of skeletal muscle have also been adapted to smooth muscle, and a number of specialized approaches tailored to the special requirements of smooth muscle have been devised. The slower contraction kinetics and greater compliance of smooth muscle generally place less stringent requirements on the speed and stiffness of measuring equipment used for smooth muscle (with some exceptions), although long-term baseline and calibration stability of measuring systems becomes correspondingly more important.

Most mechanical measurements on smooth muscle involve measuring the force of contraction in order to use this parameter as the experimental endpoint for some biochemical or pharmacological treatment (although this may not be the best indicator of contractile function). In such cases simple isometric measurements using a force transducer with its output displayed on a chart recorder have proven sufficient to monitor activity.

More complex protocols may call for simultaneous recording or control of both force and length. It is possible under special circumstances to use an isometric system to measure maximal shortening velocity. This involves the use of the "slack test," in which an isometrically contracting muscle is suddenly shortened (manually or automatically) and the time required to take up the resulting slack is measured (Edman 1979). This elapsed time can provide a measure of the maximal unloaded shortening velocity, but it cannot address intermediate velocities. If the parameters of interest are other velocities (or amounts) of shortening, then continuous recording of length is necessary. Although such measurements may be made using purely mechanical lever systems, often the low forces and slow movements involved make the use of feedback-controlled lever systems desirable (Hellstrand 1991; Meiss 1987, 1989, 1990, 1991, 1992, 1993, 1994). In many cases these systems also allow for computer-based data acquisition, a practice that allows for much more thorough analysis of data than is afforded by paper chart records.

A further refinement in measuring systems allows the determination of muscle stiffness, an important quantity in assessing some aspects of crossbridge function. Two principal methods of stiffness measurement are in current use. One method involves the sudden application of rapid increases or decreases in force (ΔF) or length (ΔL) and the faithful recording of the corresponding length or force adjustments, where the stiffness is defined as the ratio between these two quantities, $\Delta F/\Delta L$. This technique makes a direct measurement of the elastic properties of the muscle at a single instant; its proper use places stringent demands on the response-time capabilities of the transducers that measure force and length and deliver the sudden mechanical steps. The method can measure stiffness only at a single moment in time, and repeated contractions must be used to build up a complete picture of the muscle's elastic characteristics (see e.g. Peiper et al. 1978; Warshaw and Fay 1983). In the other important method of measuring stiffness, a continuous series of very small sinusoidal length changes (ΔL, or dL) are applied to the contracting muscle, which may be under either isotonic or isometric conditions. The response to this perturbation is a sinusoidal force oscillation (ΔF, or dF) that is proportional to the amplitude of the length perturbation and, more significantly, to the stiffness of the muscle. This method gives a continuous recording of the dynamic stiffness, dF/dL (Meiss 1978, 1993). Although there is not complete quantitative agreement between these two methods, both provide a proportional reflection of muscle stiffness and allow forming some conclusions regarding the state of the crossbridge array during various stages of the contraction.

6.2. Stimulation and activation

A large number of intact smooth muscle tissues are electrically excitable. These muscles are usually activated by electrical field stimulation, with the current (usually a train of pulses) delivered through the conductive bathing medium from wire or plate electrodes mounted on both sides of the tissue. In some cases the stimulating current directly causes depolarization or ion channel opening in the muscle cell membranes; in densely innervated tissue such as trachealis, the stimulus activates nerve elements within the tissue and muscle stimulation occurs as a result of neurotransmitter release. An advantage of electrical stimulation is that it can be varied in strength, duration, and frequency, and its effects are not subject to diffusional delays. For many tissues this allows a graded activation that can be exploited for experimental purposes, and it often forms a useful alternative to the slower activation provided by pharmacological stimulation or potassium depolarization.

Chemical activation is the method of choice for activating many smooth muscle tissues, especially those that respond to blood-borne activating substances. Vascular smooth muscle responds well to norepinephrine, histamine, et cetera, as well as to depolarization by KCl. With chemical agonists it is possible to modulate the force of contraction; in fact, in many cases smooth muscle contraction is used as the endpoint in the investigation of the efficacy of agonist drugs. From the mechanical standpoint, chemical activation produces steady and long-lasting contractions that can serve a variety of experimental purposes. On the other hand, diffusional delays are inherent in chemical activation, and rapid initial events cannot be followed faithfully.

A further elaboration of chemical stimulation can be applied to smooth muscle preparations whose cell membranes have been removed or rendered permeable to substances of low molecular weight, such as ATP, calcium ions, and small proteins such as calmodulin. Treatment with non-ionic detergents (e.g. Triton X-100, Gordon 1978; or saponin, Kargacin and Fay 1987) or by low-temperature glycerol treatment (Tanner et al. 1988) can produce selective dissolution of surface and/or internal membranes. Depending on the type and duration of treatment, however, important cellular constituents may be lost and require reintroduction from the bathing medium. Application of alpha toxin derived from staphylococcus bacteria can be used to form pores of controllable diameter in surface membranes (Kitazawa et al. 1989; Nishimura et al. 1988); this method has the advantage of keeping membrane-bound chemical receptors intact. Such permeabilized preparations (also called ''chemically skinned'') may be fully or partially activated by controlling the calcium concentration using a Ca^{2+}–EGTA buffer system. Once activated, these preparations can be used for a variety of mechanical experiments, although protocols must be modified in light of the relatively small number of times a preparation can be activated and relaxed.

One factor limiting the use of permeabilized preparations is the time lag in activation and relaxation caused by diffusional delays. In many cases this limitation has been overcome by using *caged compounds* (Somlyo and Somlyo 1990; Nishiye et al. 1993). These are modifications of ATP, ADP, IP_3, and other substances with attached side-chains that render them biologically inactive. When they are infused into a skinned preparation, they may be made suddenly available to the muscle interior by a brief and intense pulse of light of the appropriate wavelength. They are then present almost instantaneously and in high concentration at the site of their intended action.

7. Smooth muscle mechanics: experimental findings and their significance

The basic mechanical properties of muscle, regardless of the type, are by now well defined, and there is general agreement as to the ways in which the common functions of muscle can be described and quantified. All muscle shows a basic relationship between length and force, between shortening parameters and the applied load, and between the developed force and the stiffness. What follows is an attempt to place the specific attributes of smooth muscle into this overall context and then to concentrate on those aspects of smooth muscle function where special exceptions have been revealed. It is these exceptions, which arise from the many adaptations of smooth muscle to its special tasks, that have driven the progress of the study of smooth muscle and its mechanical function. This section will be biased toward areas in which a number of problems are currently being addressed experimentally by many of the methods outlined previously.

The sliding-filament/crossbridge hypothesis of skeletal muscle contraction has proven remarkably durable over the years (see e.g. Squire 1994; Cooke 1995) and, in spite of recent evidence that the notion of myofilament inextensibility will have to be revised (Goldman and Huxley 1994), it continues to be the guiding paradigm for skeletal muscle research. Extensions of this basic paradigm to the function of smooth muscle are now on a firm structural, biochemical, and phenomenological basis, and this state of affairs is not likely to change in any fundamental way. However, a growing body of recent work is beginning to show that structures such as the cytoskeleton and its components, and the architecture of the tissue and its extracellular matrix, impart to smooth muscles mechanical properties distinct from those associated with the crossbridge–myofilament system.

7.1. Activation of smooth muscle contraction

An important distinguishing feature of muscle as a tissue is its ability to undergo marked changes in its mechanical and chemical properties rapidly and in response to specific signals or stimuli. One of the significant characteristics distinguishing smooth muscle from skeletal muscle is the existence of multiple pathways for the activation (see e.g. Brozovich and Morgan 1989; Kamm and Stull 1989; Gunst et al. 1994) and the subsequent regulation of the contraction process. A detailed treatment of this complex subject is outside the scope of this chapter, and the reader is referred to a number of recent reviews (Kamm and Stull 1985; Hai and Murphy 1989; Stull et al. 1991; Allen and Walsh 1994; Murphy 1994), as well as to pertinent portions of the present volume.

It is important to emphasize at the outset that almost all experimental designs set up artificial conditions that may be quite unlike those encountered by a given smooth muscle in its physiological environment. The convenient approach of starting with a resting or quiescent preparation (often very difficult to manage with smooth muscle), activating it for study of mechanical properties, and then causing it to relax has been borrowed from skeletal muscle studies. While this approach can greatly simplify experimental design, it also places many muscles, especially those that are normally tonically active (such as vascular muscle), in a highly artificial situation and requires them to make mechanical and biochemical transitions that they would rarely encounter in normal function. Mechanical constraints are also likely to be rather artificial. Extreme shortening under unloaded conditions is rarely encountered in the functioning pulmonary or cardiovascular systems, and the isometric conditions imposed experimentally on isolated smooth muscles are not often found in realistic circumstances. Where possible, it is usually advisable to study the mechanical properties of a muscle in a fairly close approximation to the normal behavior to enhance the validity and appropriateness of the results. Some attempts have been made, for example, to provide shortening tracheal muscle with varying afterloads that may more closely approach its natural physiological working conditions (Ishida et al. 1990).

7.2. Isometric contraction

The most basic experimental expression of smooth muscle contractile activity is the isometric contraction, in which the muscle preparation does not change length during its activation and relaxation. For purposes of discussion it is useful to divide such a contraction into three parts (Figure 1, upper panel): the phase of *force development*, in which activation occurs and the biochemical and mechanical processes, beginning in the latent period, are set into operation; the phase of *force maintenance*, in which the initial internal events either persist or are modified in ways that favor steady-state force production; and the *relaxation* phase, in which the ongoing processes are terminated or reversed by specific chemical and mechanical processes. Because many, if not most, smooth muscles exhibit some tonic activity in the absence of stimulation, the phase of force development may begin against a background of ongoing low-level activity.

The rate at which force develops in a quiescent or resting preparation varies greatly with the type of muscle being observed and the mode of stimulation. Electrical stimulation of phasic smooth muscles (such as uterus, urinary bladder, and some gastrointestinal muscles) produces contractions that reach steady state in seconds or a few tens of seconds. Chemical stimulation of such muscles produces a slower rise of force, largely because of diffusional limitations. Tonic smooth muscles (such as many of those of the arterial system or the trachealis) are activated much more slowly, often over a period of minutes. For such muscles chemical stimulation is more practical than electrical activation (which may not be at all effective in producing activation). Muscles that may be activated by both of these methods may show mechanical properties that depend on the stimulation pathway and that involve different second-messenger systems.

Force development must begin with (actually, be preceded by, in the latent period) the onset of crossbridge interaction. This process is usually under the control of calcium ions, which regulate the phosphorylation of myosin light chains. Once this process has begun, actin–myosin interaction can take place. Often the first mechanical sign of this interaction is the development of increased stiffness before any external force is manifest (cf. Kamm and Stull 1986), indicating that crossbridges may initially attach in a low-force state (Warshaw et al. 1988; Chalovich et al. 1991). Further force development is accomplished by recruitment of additional crossbridges and by the cycling of activated crossbridges as they extend compliant elastic elements within the cells and tissue. Both of these processes lead to increased force and stiffness that approach a steady state as the activating mechanisms attain a level of saturation under the prevailing conditions. It is often unclear whether a given stimulation procedure has produced a maximal activation, although protocols that produce a saturating level of cytosolic calcium lead to the highest levels of myosin light chain phosphorylation.

The force maintenance phase continues or modifies the processes begun during force development. In many smooth muscles, especially tonic ones in which prolonged maintenance of force is the usual situation, the steady-state force is associated with a fall in internal (cytosolic) calcium levels and a progressive fall in the phosphorylation of myosin. These conditions lead to the development of the *latch* state, and force is maintained with a reduced expenditure of metabolic energy.

7.2.1. The latch state

It has long been known that smooth muscle is capable of maintaining contractile force for long periods of time without a high consumption of metabolic energy. This ability was initially considered to be the result of a tetanic contraction of a muscle with inherently slow kinetic processes (Bozler 1948; Ruegg 1971). The work of Murphy and collaborators (Dillon et al. 1981; Hai and Murphy 1992; Murphy 1994) showed that, in arterial muscle, the maintenance of isometric force after its initial onset was associated with a decreased shortening velocity capability, and that the velocity decline was in parallel with a fall in both myosin light chain phosphorylation and the internal Ca^{2+} concentration. This state of the muscle was termed the ''latch state'' (by way of analogy with the ''catch state'' – Ruegg 1971; Twarog et al. 1982 – of some molluscan muscles).

There is general agreement that an initially high cytosolic calcium concentration and level of light chain phosphorylation are necessary for the initial force development phase. However, despite the fact that the concept of the latch state has been under discussion and experimental analysis for a number of years, there is as yet no firm consensus as to its underlying molecular basis. The initial suggestion – that a latch bridge was an attached but dephosphorylated crossbridge that had formed in the normal way – has enjoyed significant experimental support over the years, but various objections have been raised against it on energetic grounds (Paul 1990) and on the growing evidence that thin-filament–based regulatory mechanisms (caldesmon, calponin) can contribute significantly to the economical maintenance of force (Hartshorne and Kawamura 1992; Haeberle and Hemric 1994; Walsh 1994). The four-state crossbridge model of Hai and Murphy (1988a) has provided a valuable framework for the testing of experimental ideas that relate to the identity and function of the putative latch bridges. It is beyond the scope of this chapter to resolve these difficulties, but other chapters will discuss various alternatives in the context of the regulation of contraction.

7.2.2. The length–tension relationship

Isometric contraction in all muscle types is a length-dependent phenomenon that usually shows optimal force development at some critical length. In skeletal muscle these dependencies can be related to the state of overlap of myofilaments within the sarcomeres, but in smooth muscle no such obvious correlation is apparent. The range of initial lengths over which smooth muscle can function is

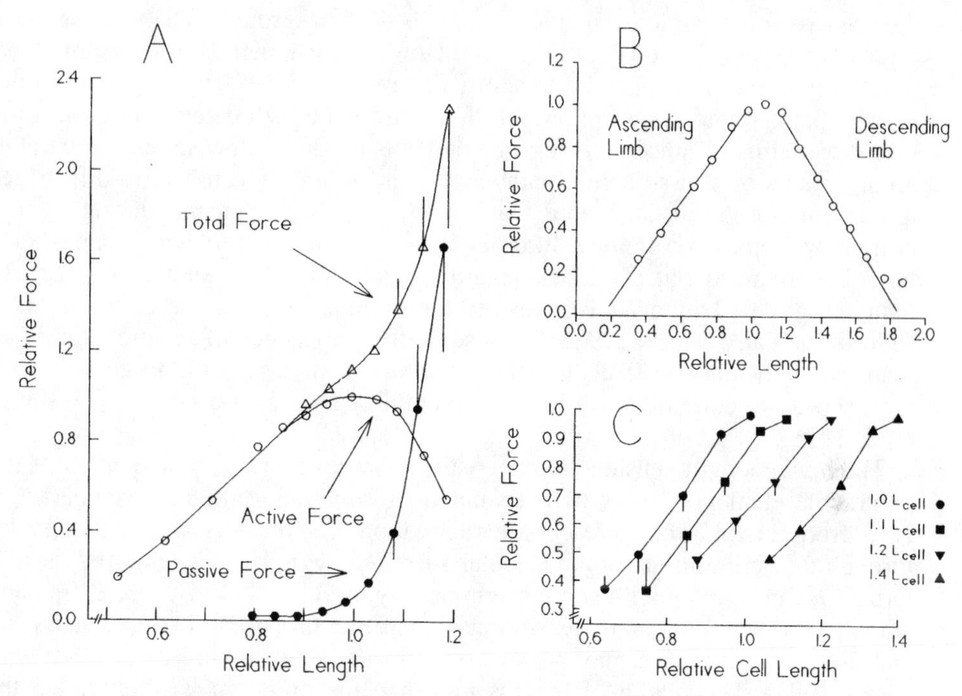

Figure 2. Length–tension curves of smooth muscle. *Left (A):* Typical length–tension curve for vascular (carotid) smooth muscle, showing the overall relationship. *Upper right (B):* Active length–tension curve of carotid artery muscle rings with the adventitial layer stripped off (see text). The ascending limb, length optimum, and descending limb are clearly shown. *Lower right (C):* Single-cell active length–tension curves from toad stomach. Cells were activated at lengths of 1.0, 1.1, 1.2, and 1.4 times L_o and quickly brought to the lengths indicated. The slopes of the curves were similar, although their positions along the length axis were shifted according to the starting length. This emphasizes the plasticity of the length–tension relationship in smooth muscle cells. [Redrawn, respectively, from Herlihy and Murphy 1973; Wingard et al. 1995; and modified from Harris and Warshaw 1991. Used with permission.]

very wide, although most physiological phenomena do not occur at the length extremes. The relationship between isometric length and force (tension) is usually described by the isometric length–tension diagram; the examples shown in Figure 2 also illustrate some features of the curve under special circumstances.

Unstimulated muscle shows a *passive* (or *resting*) length–tension characteristic. Stretching an unstimulated muscle (provided that the stretch does not activate it) usually leads to an initial increase in force followed by an exponential decline to some level higher than the prestretch force. This is the phenomenon of *stress-relaxation*, an expression of the viscoelastic (time-dependent) properties of the resting muscle. The steady-state poststretch force, when plotted against length over a wide range of isometric lengths, usually shows an exponential relationship between force and length, with the curve becoming quite steep at the greater lengths. This is an expression of the significant connective tissue component that is characteristic of most smooth muscle tissues. When the muscle is activated at lengths at which passive force is high, the extra force thus produced is added to the passive force, which remains present in the background. The sum of these forces is the total force (Figure 2, left). However, it is not safe to assume that, when active muscle is allowed to shorten from an extreme length, the fall in the passive background contribution will then follow the path determined from resting measurements, since activation of the cells produces new series and parallel arrangements of stress-bearing pathways. This ambiguity can lead to difficulties in determining the instantaneous load on the contractile elements during what is nominally isotonic shortening. In order to circumvent this problem, some studies have been carried out at ''slack length'' (also called ''initial length,'' *Li*). In many cases this length is also close to the optimal length as described in what follows (see also Meiss 1993). Because of the prevalence of resting ''tone'' in many smooth muscles (Ruegg 1971), it is not always possible to ensure completely passive conditions while also maintaining normal contractility (cf. Butler et al. 1976).

The *active* length–tension characteristic – that portion due to activation of the contractile elements – may be determined by subtracting the passive characteristics from the total force present after activation. The result is a curve that passes through an optimum at some characteristic tissue length. This length is usually termed L_o, in keeping with similar terminology used for skeletal muscle. In many smooth muscles, the length–tension curve is quite broad and flat, and statistical curve-fitting techniques may be necessary to define L_o (Wingard et al. 1995b; Meiss 1993). The region of lengths less than the optimum is often called the *ascending limb* of the length–tension curve; that above the optimum is termed the *descending limb.* Although the ascending limb is readily accessible to experimentation, the descending limb is complicated to study, because stretching the muscle to such lengths usually produces irreversible changes. Recent work in the laboratory of Murphy (Wingard et al. 1995a,b; see also part B of Figure 2) has led to a preparation of porcine carotid artery with its adventitial (connective tissue) and endothelial layers stripped off. This preparation can produce repeatable contractions at very great lengths, and upon return to shorter lengths the active force generation is not decreased. Such a preparation promises to be quite useful in the study of the energetics of the activation processes uncontaminated by the effects of active force production.

7.2.3. *Normalization of muscle force*

Since the cross-sectional area (CSA) of a muscle preparation varies as its length changes, so does the force/CSA. While corrections can be made for this effect, some uncertainty still exists. The cross-sectional area of actual contractile substance may vary at a rate different from that of the overall CSA because of changes in cell alignment and (perhaps) of varying myofilament overlap. This effect can be minimized by normalizing the force to the conditions at L_o for each muscle. Similar cautions are associated with measurements of isometric stiffness along the ascending limb. Force changes due to imposed length changes are subject to the same ambiguity as the developed force itself, and a length change of constant amplitude may represent a larger amount of internal shearing at the shorter muscle lengths. However, measurements of muscle stiffness in which both force and stiffness are normalized together by the same criteria at each muscle length have shown that the stiffness of some muscles appears to fall off with length less strongly than does the isometric force (Meiss 1978; Siegman et al. 1984). This implies that at the shorter lengths some crossbridges can attach without being able to contribute to the overall force. Measurements of active stiffness along the descending limb of the length–tension curve are made quite difficult by the large passive force and stiffness; however, the ''stripped'' carotid artery preparation mentioned previously may provide a means of addressing this problem in at least one important tissue.

In skeletal muscle, the length–tension curve for a given muscle preparation gives a stable and reproducible description of the muscle behavior. In some smooth muscles, however, there may be significant variability from time to time within the same preparation. Uvelius (1976) found that large amounts of passive stretch could reduce the active tension at any given length, an effect that could be reversed by subsequent contractile activity. Using single cells, Harris and Warshaw (1991) were able to produce several ascending-limb length–tension curves that had the same slope but were translated along the length axis (see Figure 2). Both of these results imply that the contractile apparatus of smooth muscle may be ''reset'' internally by external mechanical conditions.

7.2.4. Relationship between force and stiffness during isometric contraction

Isometric muscle stiffness, as measured by the force response to small length perturbations, shows a generally linear relationship between stiffness and force (Figure 3, lower right). This may reasonably be considered the consequence of two separate underlying situations, or some combination of both. In the first instance, if the tissue locus of active stiffness is the crossbridge array, and if the force generated by the crossbridges is proportional to their number, then the stiffness should be also. If the crossbridges and their cellular connections were completely rigid, then the extracellular connective tissue, with its typical exponential length–tension characteristics, should show a stiffness proportional to its force, a consequence of its exponential nature. Neither of these extremes is the actual case, but both probably contribute. Some idea of their relative contributions can be assessed by the analytical methods of Mulvany and Warshaw (1981), who were able to assign between 1/2 and 1/3 of the compliance of smooth muscle to cellular structures, with intercellular structures appearing to account for the remainder.

Of special interest are deviations from, and variations in, the general linear nature of the force–stiffness relationship (cf. Part E of Figure 3), because these are most likely to reflect special behavior of the crossbridge-specific portion of the isometric stiffness. For instance, early during the rise of isometric force, the ratio of stiffness to force (in arbitrary units) is very large (Gunst et al. 1995); this implies that crossbridges first attach in a low-force state before they begin active cycling (Kamm and Stull 1986; Warshaw et al. 1988). In some smooth muscle tissues, the relationship between force and stiffness is curved slightly downward; that is, as force increases, the stiffness increases also, but each successive incremental rise is smaller than the previous one (Gunst et al, 1995; Tanner et al. 1988). Such behavior could be due to a reduction in the force borne by each crossbridge as the activated population becomes larger.

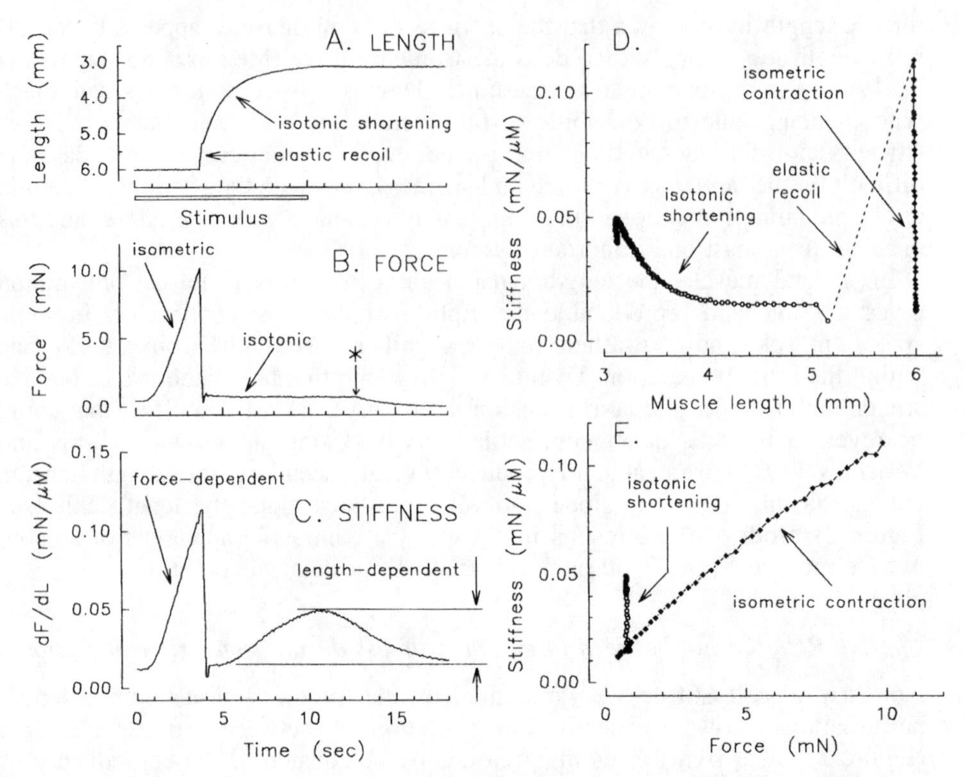

Figure 3. Stiffness dependencies in smooth muscle. *Left, from top to bottom:* (A) Muscle length; (B) contractile force, and (C) dynamic stiffness. Muscle develops isometric force, and the stiffness increases correspondingly. The muscle is then quickly switched to a low and constant isotonic afterload. During the early portion of shortening, stiffness is appropriate for the developed force. Stiffness then increases while force remains constant, reaching its maximum at the maximal extent of shortening. The stiffness can be divided into force-dependent and length-dependent portions as indicated. The asterisk marks the point at which potential isometric force fell to less than the afterload. *Upper right (D):* Stiffness plotted as a function of muscle length. At the extreme right, force-dependent stiffness increases at constant length. When the muscle shortens, note the increase in stiffness at short lengths, which may indicate an increase in crossbridge number that is not expressed as external force. Note also that this stiffness fell abruptly after stimulation ceased (horizontal line below length trace at left), while force and length remained essentially constant. This indicates a rapid reduction of an internal stress. *Lower right (E):* Muscle stiffness as a function of force. Here stiffness varies almost directly with isometric force, and during the isotonic shortening the stiffness (length-dependent) changes while force is constant. [Redrawn from Meiss 1992. Used with permission.]

7.2.5. Isometric relaxation

With cessation of stimulation, force declines along a nearly exponential time course that may be many times longer that the time course of activation. The decline of isometric force is associated with a decrease in the calcium bound to calmodulin, usually as a result of the resequestration of internal calcium or the active extrusion of cytosolic calcium. This in turn leads to a decline in the rate at which myosin is phosphorylated, and myosin light chain phosphatase activity predominates. If a contraction has been of any appreciable duration, both the calcium levels and the amount of myosin phosphorylation have already declined substantially. Crossbridge detachment also must accompany the reduction in activation, and may proceed along two or more distinct pathways. At least some of the attached crossbridges will be latch bridges (already dephosphorylated?), and the others will be normally cycling. This latter population can spontaneously detach as the proper point in their cycle is reached. It is quite probable that internal strain also plays a role in crossbridge detachment, since relaxation is most rapid when the force is highest, and a stretch applied during relaxation can hasten the process. Measurements of stiffness during relaxation show that it is higher at any given relaxation force than during force development (Meiss 1978, 1993). This indicates the presence of attached crossbridges that are not bearing force and implies that at least some of the bridges pass through a zero-force state on their way to detachment. It is also possible that the enhanced stiffness during and following relaxation (Meiss 1982, 1993; Stephens and Brutsaert 1982) is a reflection of some cytoskeletal involvement in the sustaining of isometric force.

7.3. Isotonic contraction

When an activated muscle shortens, the contractile apparatus faces a new set of mechanical conditions, and a new range of mechanical properties arises (see Figure 1, right). The velocity, load dependence, and extent of shortening become important variables. As muscle shortens, the crossbridges cycle, the interdigitation of myofilaments increases, and the geometry of the contractile apparatus continuously changes. One apparent consequence of these activities is the universal observation that smooth muscles shortening isotonically undergo continuous slowing as they shorten and finally reach a quasi-isometric equilibrium length (Arner and Hellstrand 1985; Harris and Warshaw 1990; Meiss 1994; cf. Figures 3 and 6). Since the external load has been kept constant, there is the possibility that shortening muscle faces an increasing length-dependent internal load that causes it to slow.

7.3.1. The question of internal loading

A number of experiments are best explained by the presence of an internal load (e.g., Harris and Warshaw 1990), although there has been no conclusive identification of the internal structure(s), real or virtual, that would provide such a load over a moderate range of lengths below L_o. It has been suggested, with some experimental support, that an internal load could arise from cytoskeletal connections that become fixed early during the course of activation and provide a resettable internal resistance to shortening. This aspect of the decrease in velocity during shortening was investigated by Gunst et al. (1994), who found that the velocity at any length during shortening depended upon the length of the tissue at the time of activation. Trachealis muscle activated at or near L_o was slower at any subsequent length, during that contraction, than when it was stimulated at shorter lengths (e.g., 0.7 times L_o). This effect was attributed to an internal load provided by cytoskeletal elements whose configuration was set at the time of stimulation, with greater internal loads being associated with the greater starting lengths.

Such internal loads appear to be at least in part a cell-based phenomenon. In single-cell experiments, evidence has been provided in support of the presence of a length-dependent (and time-independent) internal force that resists shortening and causes slowing during isotonic contraction (Harris and Warshaw 1990). This force also apparently causes re-extension while the cell relaxes (Kargacin and Fay 1987; Warshaw et al. 1987), accounting at least in part for the continuous slowing during isotonic shortening. Application of the F actin severing protein brevin to single cells and small strips of taenia coli muscle (Gailly et al. 1990, 1993) caused a twofold increase in unloaded shortening velocity and a concomitant fall in complex stiffness (Gailly et al. 1991). It has long been known that smooth muscle contains more actin than can be assumed to interact with myosin (Murphy et al. 1974; Ashton et al. 1975; Cohen and Murphy, 1978), and it has been suggested that two actin domains may exist (Small et al. 1986; Gailly et al. 1991), with contractile functions being localized within the actomyosin domain. The intermediate filament (cytoskeletal) domain contains actin filaments, along with the crosslinking protein filamin and the cytoplasmic dense bodies. The velocity increase and stiffness decrease, which occurred without any change in actomyosin ATPase activity, implied that crosslinked actin in the cytoskeletal domain had acted as an internal load. It has also been shown, under the even more simplified conditions of the *in vitro* motility assay (Warshaw and Trybus 1991), that crossbridges that are not phosphorylated (or that are inherently slower) can act as a load on the population of cycling crossbridges. In this regard, however, there is little evidence from energetic measurements on intact tissue to indicate an increased internal work production associated with the slowing typical of the latch phenomenon (Butler et al. 1987).

At extremes of shortening, however, there is evidence strongly suggesting that mechanical factors arising from distortion of the tissue structure provide a substantial internal load. Since the cells shorten at approximately constant volume, their increase in diameter at short lengths encounters opposition from the intercellular connective tissue that would have to yield to accommodate cellular expansion. Because this tissue component is comparatively stiff, it provides a force that opposes further shortening; this force is evident from the significant increase in longitudinal stiffness during extreme isotonic shortening (Meiss 1990, 1991, 1992; see Figures 3 and 6).

7.3.2. Shortening velocity

The force–velocity curve, which relates the initial velocity of shortening (see Figure 1, lower right) to the force exerted, provides the best comprehensive description of the isotonic behavior of muscle under isotonic conditions. Aside from the much lower shortening velocities, the basic form of the force–velocity curve for smooth muscle is similar to that of skeletal muscle, and it can be described by the same mathematical relationships, usually the Hill equation (Hill 1938). For many purposes, the most useful feature of the force–velocity curve is the initial rate of shortening at zero force, a value termed V_{max}, since this quantity is closely linked to the maximal cycling rate of the crossbridges. Because of its good agreement with conventional V_{max} values (Arner and Hellstrand 1985; Hai and Murphy 1989), in many studies the use of the "slack test" to estimate V_{max} can provide sufficient information about cycling rates to allow such measurements to substitute for complete determination of force–velocity curves.

All of the velocity effects discussed so far have been restricted to the relatively brief time scale of a single shortening. Over the long term, the shortening capability (as reflected in the shortening velocity) of many smooth muscles shows large variations with time. If an isometrically contracting muscle is released to a low isotonic load at various times during the steady-state force maintenance period (Dillon et al. 1981), it will shorten with an initial velocity that decreases as a function of the amount of time spent in isometric contraction. This reduction in velocity is often correlated with a decrease in myosin light chain phosphorylation and with a decrease in energy consumption, and has in fact become the hallmark of the latch-bridge phenomenon (see earlier discussion and references).

7.3.3. Isotonic relaxation

Although the mechanisms mentioned previously work to reduce crossbridge interaction, in isotonic relaxation they do so in the face of mechanical forces that tend to produce relative myofilament shearing as the muscle returns to its resting length. In the smooth muscles studied in this regard (Stephens et al. 1981; Stephens and Brutsaert 1982; Meiss, unpublished), the rate of isotonic relaxation was relatively independent of the loading conditions (see Figure 4), and relengthening occurs at a fairly constant rate over a wide range of afterloads. This type of relaxation has been termed "inactivation-dependent" (or "load-independent"), since nonmechanical factors associated with inactivation of myofilament interaction appear to dominate the rate of relaxation. In smooth (as compared with striated) muscle, the mechanisms of calcium resequestration are less aggressive, leading to a slower overall relaxation and to relative independence of the applied load. Such relaxation behavior could provide insight into the processes that control crossbridge detachment.

Although length-dependent internal forces that resist shortening have been postulated for smooth muscle, in most cases external force must be applied to return a contracted strip completely to its original length. Physiologically, this force is applied by the refilling of an organ, by the blood pressure, or by other similar means. This behavior is in contrast to the microscopically observed behavior of single isolated smooth muscle cells, which appear to relengthen quite well on their own (Bagby and Fisher 1973; Kargacin and Fay 1987).

7.4. Combinations of isotonic and isometric contraction

Most contractions of smooth muscle involve some degree of shortening as well as the development of significant force (Figures 1, 3, and 5). In cases of mixed contraction, the final mechanical result is highly dependent on the pathway by which it is reached (Uvelius 1976; Meiss 1993; see also Figure 5).

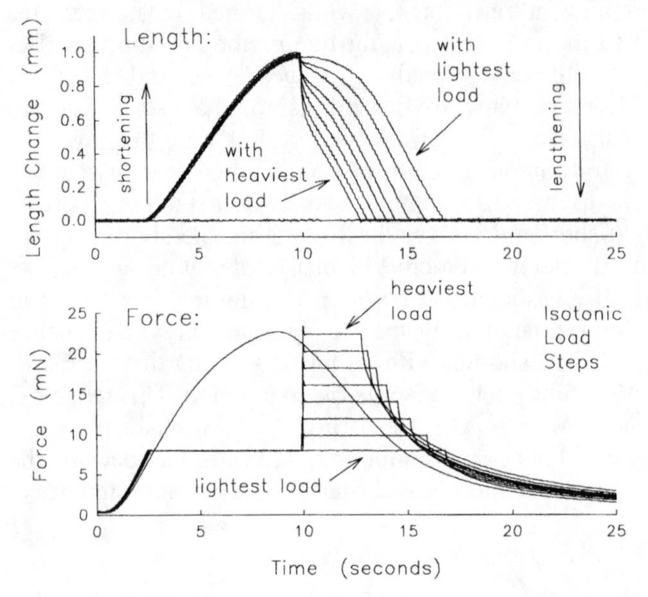

Figure 4. Isotonic relaxation in smooth muscle. In each of a series of eight contractions, all superimposed, an ovarian ligament muscle was allowed to shorten under a constant force (load). After 10 seconds the stimulus was turned off, and then step increases in force were applied. As indicated by the arrows, the lightest load was associated with the most delayed relaxation and the heaviest with the earliest, with the others falling between these extremes. In each case, after an initial elongation proportional to the force step (load-dependent relaxation), the muscle elongated at a rate independent of the applied force (inactivation-dependent relaxation). An isometric contraction is shown for comparison. [Meiss, unpublished data.]

7.4.1. *Effects on force and velocity*

The interactions between isotonic and isometric contraction are most strikingly demonstrated in the case of the length-tension relationship (see Figure 5, lower left). A series of isometric contractions, each made at a different starting length, defines the ascending limb of the conventional isometric length–tension curve (upper curve). At the length L_o (5 mm in this case), the maximal force is reached via path *AC*. If the muscle is then set to L_o and allowed to shorten isotonically as much as possible under a moderate afterload, it will come to a stop (via path *AA*') at a length significantly short of the isometrically defined curve. If the muscle is held at this length, allowed to relax, and is then held isometric at the new length, it will develop the expected amount of isometric force (via path *BB*'). Thus two different length–tension curves, generated along separate pathways, both characterize the same muscle. The right side of Figure 5 shows pooled data from eight muscles to emphasize the difference between these two approaches to length-dependent force. Similarly, if a muscle strip is allowed to shorten isotonically for some time and is then held isometric for the remainder of the contraction, the redeveloped isometric force will fall short of its expected value by an amount directly proportional to its previous shortening (Meiss 1993).

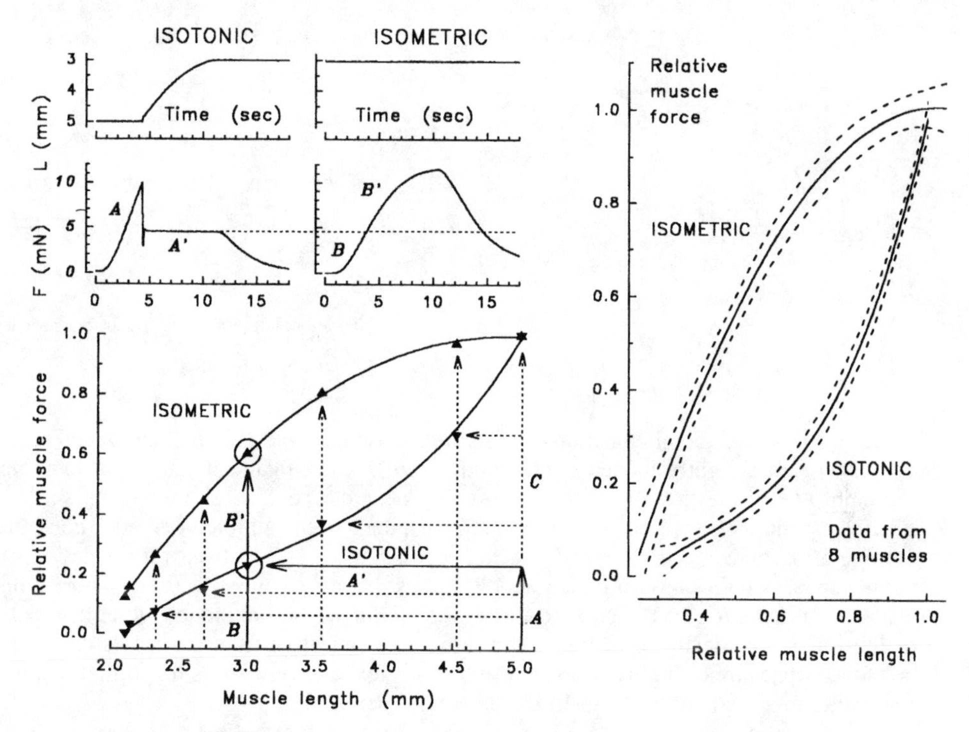

Figure 5. Isometric and isotonic approaches to the length–tension curve. *Upper left panels:* Typical contraction protocol, illustrating the approach. At *A, A'*, muscle was allowed to shorten as much as it could and then held at the new length during relaxation. In the next contraction and at this shorter length, muscle was restimulated (*B, B'*) to produce isometric tension, which far exceeded the isotonic force at this length. *Lower left:* Complete isotonic and isometric length–tension curves. The path shown by the solid arrows and designated *A, A'* corresponds to the isotonic contraction illustrated above and traces the isotonic path to maximum shortening (from 5 mm down to 3 mm). Note that force first rises at constant length (*A*) until the isotonic force is reached, and then the muscle shortens (*A'*). The solid arrows at 3 mm, marked *B, B'*, show the path of the subsequent contraction (as illustrated above) that was fully isometric at a length of 3 mm. The fully isometric contraction at 5 mm is shown by the path *A, C*. Overall, the curve marked "ISOMETRIC" was reached by contractions traced by the dotted vertical arrows, and the "ISOTONIC" curve was reached by contractions traced by the horizontal arrows, with each contraction beginning at a different developed force from a length of 5 mm. See text for further details. *Right:* Pooled isotonic and isometric length-tension curves, generated as shown at the left, from a number of ovarian ligament muscles. [Modifed from Meiss 1993. Used with permission.]

When the force–velocity characteristics of such muscles were determined during the period of force redevelopment, it was found that the maximal isometric forces were decreased as expected; however, the V_{max} values, reached when shortening under very light loads, were significantly enhanced. The mechanisms underlying these apparently related behaviors are not yet clear. Hai and co-workers (Hai 1991; Hai and Szeto 1992; Hai and Ma 1993) have shown that during unloaded shortening both the peak and the steady-state phosphorylation were greater than

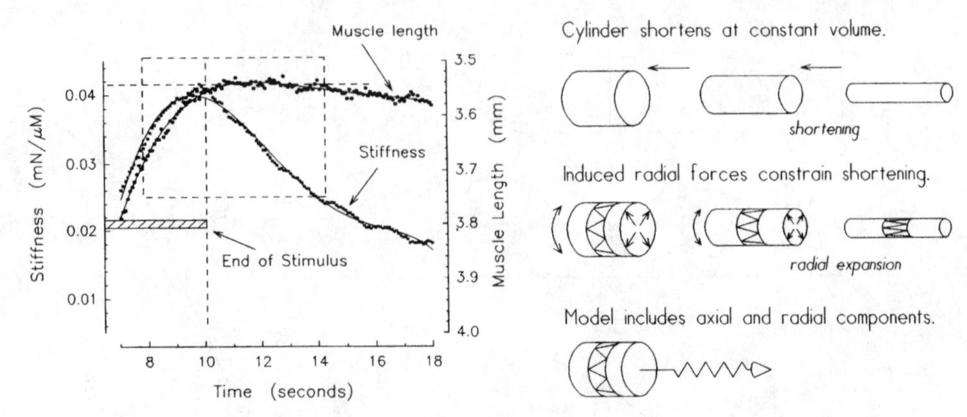

Figure 6. Effects of radial constraints on smooth muscle stiffness. *Left:* Length and stiffness data from a lightly loaded contraction of an ovarian ligament muscle at extreme shortening. Only the portion of the contraction near the peak of shortening is shown. When the stimulus was turned off (horizontal bar), the length and force remained constant for several seconds, while the stiffness fell significantly. This indicates a relaxation of internal stresses built up during contraction, forces that act to oppose extreme shortening in smooth muscle. *Right:* Essential features of the radial constraint model, which accounts for the forces demonstrated at the left in terms of connective tissue elements opposing the radial expansion of highly shortened tissue; see text and references for further details. [*Left* panel modified from Meiss 1992. Used with permission.]

during isometric contractions at short lengths, and that the length sensitivity appeared to be related to the coupling of membrane processes to the elevation of internal (cytosolic) Ca^{2+}. In recent experiments in which rapid length changes were made just prior to and during isometric contraction (Mehta et al. 1996), the measured cytosolic calcium and myosin light chain phosphorylation levels were not sufficient to account for the differences in redeveloped force. This suggests that crossbridge-based mechanisms were not the most important factor, opening up the possibility that cytoskeletal remodeling involving the contractile apparatus might form the basis of such behavior (Gunst et al. 1995).

7.5. Muscle stiffness during mixed contractions

Changes in the conditions of contraction that have specific effects on the muscle stiffness may be arranged into distinct categories (see Figure 3). During the rise of isometric force, the stiffness is force-dependent in a nearly linear manner. As shortening begins, stiffness enters a phase that is also length-dependent. Early in shortening, the magnitude of the stiffness is appropriate to the isotonic force level; during later stages of shortening, length-dependent stiffness may rise sharply even though force remains constant. Cessation of stimulation causes a fall in stiffness at constant force and length, giving evidence of activation-dependent stiffness (Figure 6, left). All of these stiffness components can be categorized as *axial* stiffness, since they are measured along the force-generating axis of the tissue.

Ample evidence suggests that the active stiffness resides in both intracellular (crossbridge and cytoskeleton) and extracellular structures. In all active stiffness the ''prime mover'' must be traced to crossbridge activity, because attached crossbridges ultimately form the anchors to which the other components are attached and through which their stiffness is measured. In addition, of those components mentioned, the mechanical state of the crossbridges is the one most readily changed, whether by stimulation or by applied mechanical changes (good information on the mechanical lability of the cytoskeletal contribution to muscle stiffness is currently not available). Smooth muscle tissues that have relatively small passive (parallel) elastic structures can be forcibly elongated while undergoing isometric contraction. If a controlled (constant-rate) stretch is applied, the muscle rapidly yields and elongates with only a modest rise in force (Gunst and Russell 1982; Meiss 1982; Singer et al. 1986). The stiffness, though it increases, rises less than its force-dependent values would predict, the difference being proportional to the rate of elongation (Meiss 1987). This mechanical situation has been modeled as an array of activated crossbridges that make transient connections as the myofilaments are forced to slide past one another. Such considerations as these suggest that acute changes in stiffness, under a variety of mechanical conditions, can serve as an indication of the status of the crossbridge array, even though on a static (unperturbed) basis only 1/3 to 1/2 of the stiffness can be localized to the crossbridges. This supports the assumption, invoked by a number of experimenters, that measurements of active muscle stiffness provide a valuable tool for rapidly assessing the status of the smooth muscle contractile system (Hellstrand 1991, 1994; Warshaw et al. 1988).

7.6. Mechanical properties of single cells

Several investigators have attempted to correlate the mechanical function of a tissue with that of its constituent cells (Driska et al. 1978; Halpern et al. 1978; Uvelius and Gabella 1980) and have found a close correspondence between the dimensional changes of isolated cells and the overall tissues. However, the most direct approach to this problem is the study of the mechanical properties of individual isolated cells, upon which essentially the full range of classical mechanical measurements have been performed. For example, studies of the length–tension relationship (Harris and Warshaw 1991, the force–velocity curve (Warshaw 1987), crossbridge elasticity and kinetics (Warshaw and Fay 1983; Warshaw et al. 1988; Brozovich and Yamakawa 1993), and passive elastic properties (Glerum et al. 1990) have all been carried out with an impressive degree of refinement. Tissue sources have included toad stomach (Fay 1977) and vascular tissues (Driska and Porter 1986; Brozovich and Yamakawa 1993). Cells are isolated intact and living by careful enzymatic digestion of the connective tissue that binds them into organized structures. By knotting the ends of cells isolated from toad stomach smooth muscle to microtools mounted to highly sensitive force transducers (Fay 1977) and displacement-generating devices, isometric and isotonic contractions can be recorded. A number of significant findings have emerged from these studies.

The contractile force of single cells has been shown to be length-dependent to the same extent as found in intact tissues (Harris and Warshaw 1991), with an important exception. The relative position of the ascending limb of the length–tension curve (the only portion observed) on the length axis depended on the length at which the cell was activated, although the slope of the curve was not affected (Figure 2, lower right). This implies an activation-dependent structural accommodation of the contractile apparatus that determines the subsequent degree of length-dependent myofilament overlap and/or crossbridge interaction.

Force–velocity curves obtained from single cells (Warshaw 1987) showed the familiar hyperbolic shape found in intact tissues, and V_{max} values from single cells showed a close correlation with the values obtained by the slack test. Such results, along with the transient responses to sudden length changes (Warshaw and Fay 1983) and measurements of the crossbridge compliance (at a value 1.5 times that of skeletal muscle, Warshaw et al. 1988) indicate that the structural constraints present in intact tissue do not have a qualitative effect on the measured kinetic properties of the crossbridge mechanisms of the cells. This adds confidence to similar measurements made on intact tissues.

The use of single cells represents a simplification that has permitted highly focused studies on mechanical properties, and the results achieved are a tribute to the experimental skills of the investigators. The results from these measurements have been in good general agreement with results from whole tissues, but one cost of this simplification has been a new kind of uncertainty. In single-cell experiments the cells are held only at the ends, and evidence suggests that a significant portion of the mechanical force generated within a single cell is normally transmitted to membrane-associated dense bodies located along its lateral surfaces for the whole length of the tissue (Fay et al., 1983; Kargacin et al. 1989). In fact, careful observations of shortening isolated cells bearing surface markers have revealed "corkscrewlike" shortening (Warshaw et al. 1987) that indicates the presence of relatively short, spirally arranged contractile units. Because smooth muscle cells in an intact tissue normally function as part of a distributed mechanical network in which their individual motions are constrained, the contractile mechanics of isolated cells may be qualitatively different from their function in a normal mechanical environment. Although this represents a caution, it does not diminish the importance of single-cell studies for illuminating important aspects of the contractile mechanism.

7.7. Mechanical studies of isolated myofilaments

The introduction of the *in vitro* motility assay (as briefly outlined in Section 5.4) into the study of smooth muscle has served to emphasize that smooth and skeletal muscle share a common basis for contraction. It has also allowed the identification and analysis of features unique to the smooth muscle contractile and regulatory systems. A number of variations of the *in vitro* motility assay have been applied to the study of basic smooth muscle mechanics (see e.g. Warshaw et al. 1990; Harris et al. 1994) to shed light on questions that have arisen from experiments on isolated cells and tissues. It is usually assumed, for example, that the maximal unloaded shortening velocity of a muscle represents the maximal cycling rate of its crossbridges. In terms of the *in vitro* motility assay, this implies that the motion of actin filaments on a surface coated with a single type of myosin likewise represents the condition of unloaded shortening and approximates the maximal velocity capability of the contractile system. Analysis of the effects of MgATP and its crossbridge-cycle metabolites has borne this out (Warshaw et al. 1991), provided that care is taken to ensure a constant and properly defined chemical environment (Haeberle and Hemric 1995). The time-dependent modulation of shortening velocity in intact muscle is associated with a decrease in myosin light chain phosphorylation; this is a partial description of the latch state, as put forth by Dillon et al. (1981), who suggested that dephosphorylated crossbridges could act as an impediment to shortening. In the context of the *in vitro* motility assay, if an actin filament were being propelled by normally phosphorylated myosin, then unphosphorylated myosin (which can bind weakly to actin but which catalyzes a very low rate of ATP hydrolysis) could retard the motion. This notion has been tested in a number of ways. When a combination of unphosphorylated and phosphorylated myosins were copolymerized into myosin filaments (Warshaw et al. 1990), the velocity of actin sliding fell rapidly when the copolymer was composed of greater that 75% unphosphorylated myosin. With copolymer filaments of skeletal muscle and either state of smooth muscle myosin, the velocity of actin sliding was very sensitive to the amount of inherently slower smooth muscle myosin. It was also emphasized that *un*phosphorylated and *de*phosphorylated myosin were equivalent; it was not necessary for the myosin first to have been phosphorylated, placed in the assay environment, and then dephosphorylated in order for it to exert its retarding effect.

These results were consistent with a model (Warshaw et al. 1990) explaining the effect as a mechanical interaction between slow crossbridges that undergo compression (i.e., are negatively strained) when driven by the faster crossbridges that are also interacting with the same actin filament. The model embodied several assumptions: that individual crossbridges have a hyperbolic force–velocity curve similar in shape to that of the muscle from which they are derived, and are described by the same ratio of the parameters a and p_o (Hill 1938); and that the slower crossbridges follow the same hyperbolic force–velocity curve for either positive or negative strain. Evidence for most of the assumptions (derived from studies of more nearly intact systems) has been provided; for example, negatively strained crossbridges (Somlyo et al. 1988) have been demonstrated in chemically skinned smooth muscle, and this behavior is a feature of Huxley's (1957) crossbridge model. Some aspects of the force–velocity behavior have also been independently supported (Edman 1979; Oiwa et al. 1990). The model predicts, with a fair degree of accuracy, the experimentally measured quantitative effects of myosin mixtures on actin in the *in vitro* motility assay, as well as its most important contention: that the velocity reduction is a mechanical effect transmitted along the actin filaments. The suggestion of Hai and Murphy (1988?) that the slowed velocity may reflect an average decrease in cycling rate (perhaps as communicated along the thick filaments) is not a feature of the model, and subsequent experiments (Harris et al. 1994) with a variety of monomeric myosin types showed that the integrity of a thick filament was not necessary for a mixture of myosins to have an interactive effect.

Studies with monomeric myosins from heart, smooth, and skeletal muscle (Harris et al. 1994) bear out the general notion that slower myosins can impede the action of the faster, and that such an interaction may explain the velocity-altering effects of mixtures of myosin isozymes (Somlyo 1993) as well as at least part of the latch phenomenon. Critical characteristics of the myosin molecules themselves have been analyzed in the motility assay; Kelley and coworkers (1993) were able to account for the different kinetic properties of intestinal and vascular muscle myosin on the basis of an insert of seven amino acids into a region of the myosin head near the ATP binding site. Motility studies also provide a further molecular explanation for the findings of Barany (1967), who found that the inherent actomyosin ATPase rate of a given muscle correlated closely with its maximal velocity, and that this agreement was a function of the type of myosin, not the actin, involved in the reactions. This lack of species specificity of actin in the *in vitro* motility assay has also been amply demonstrated; actins from many sources (including both smooth and skeletal muscle) can serve interchangeably in the motility assay (cf. Harris and Warshaw 1993a).

A question of long standing in the study of smooth muscle is that of the force capability of smooth muscle crossbridges (Murphy et al. 1974), since smooth muscle, on the basis of cross-sectional area, produces as much force as skeletal muscle does but with only 1/5 of the myosin content. A refinement of the *in vitro* motility assay is beginning to provide an answer (VanBuren et al. 1994, 1995). A single fluorescently labeled actin filament was attached to a highly compliant glass microneedle, and a known length of actin was allowed to interact with myosin heads (from either smooth or skeletal muscle) bound to the rigid substrate. Developed force was determined by measuring the deflection of the calibrated microneedle; the calculated force of a single smooth muscle myosin molecule was on the order of 0.8 piconewtons, in contrast with 0.2 pN for skeletal muscle. Although these figures are probably an underestimate because of the random orientation of the myosin heads (see e.g. VanBuren et al. 1994; Yanagida and Ishijima 1995), they do confirm that smooth muscle myosin produces approximately four times the force per crossbridge head of a skeletal muscle myosin molecule. Further explanation of the difference may lie in the relative duty cycles of smooth and skeletal muscle myosins; that is, the relative amount of time spent in the high-force state during the crossbridge cycle. Whereas both smooth and skeletal muscle show similar low duty cycles under the unloaded conditions of the motility assay (Harris and Warshaw 1993b), the situation may be different under loaded conditions, since the kinetics of some crossbridge transitions are sensitive to strain (Huxley 1957; Eisenberg et al. 1980). Experiments using laser-based optical traps (Finer et al. 1994) are being performed with smooth muscle contractile components in order to measure the kinetics of single actin–myosin interactions (VanBuren et al. 1995) and to focus further on this question and on the related question of the physical size of the crossbridge power stroke in smooth muscle; both of these parameters are important in a final answer to the question of the magnitude of the elementary force-generating reactions.

Many experiments have used the *in vitro* motility assay to test hypotheses about the molecular regulation of smooth muscle, especially by the myofilament-binding proteins caldesmon, calponin, and tropomyosin (see e.g. Haeberle and Hemric 1994; Kurumi and Chacko 1995; Jaworski et al. 1995). See Chapter 7 of this volume for further discussion.

7.8. Effects of tissue architecture on smooth muscle mechanics

Except for minute structures such as the smallest arterioles or precapillary sphincters, most smooth muscle tissues consist of a great number of muscle cells located in a matrix of connective tissue and ground substance. Locally the cells tend to share the same orientation, which results in a mechanically coordinated and directed contraction. This arrangement is part of a mechanical syncytium, and there is mechanical continuity between the intracellular contractile apparatus (myofilaments, crossbridges, and cytoskeleton) and the connective tissue of the extracellular matrix (or with other cells) via the focal adhesion plaques (see Gabella 1984). Lateral (radial) tissue integrity is maintained by connections that run at an angle (up to 90 degrees) to the force-transmitting axis of the tissue, and recent experiments suggest an experimental approach to developing a functional mechanical picture of smooth muscle tissues and how cells are interconnected.

In the developing ''radial constraint hypothesis,'' the significant rise in axial stiffness seen with extreme shortening occurs because the radial expansion of shortening cells is opposed by such radially oriented connective tissue. This connective tissue would also function as a resistance to shortening, imposing an additional physical load on the shortening cells. The concept is supported by several lines of evidence (Meiss 1992, 1993): forcing the tissue to swell or shrink by using osmotic challenges causes directionally appropriate change in length-dependent stiffness; and activation-dependent changes in axial stiffness at constant force and length (Figures 3 and 6) are consistent with this model. An idealized physical representation of the stiffness components, the basis for a simple mathematical model of the behavior, is also shown in Figure 6.

Other approaches have been taken to work out the cell/tissue problem. Cells have been teased out of tissue that has been chemically fixed at various lengths, and cell lengths correspond well to overall tissue lengths (see e.g. Driska et al. 1978). Microscopically visible markers, either artificially attached (Meiss 1984) or present within the tissue (Halpern et al. 1978), have been shown to change spacing in proportion to the overall changes in tissue dimensions. Finally, measured properties of individually isolated and characterized cells have been used to calculate properties of a whole tissue composed of such cells, and interexperiment agreement has been reasonable.

8. Muscle mechanics as an end-point indicator in experimental investigations

The principal physiological role of smooth muscle is to produce some sort of mechanical output important to the function of a particular organ or system. One may therefore assume that the cellular biochemical and biophysical reactions have evolved to optimize (or at least to enable) specific kinds of mechanical behavior. When smooth muscle mechanical behavior is viewed in this context, it becomes apparent that the choice of mechanical function to be observed must depend on the sort of internal process under investigation. It should also be kept in mind that the mechanical behavior measured in typical laboratory situations represents some sort of idealized condition that usually lies at an extreme. Although purely isotonic or isometric conditions are a biological rarity, their judicious investigational use can often allow a sharper focus on a particular problem. Some strengths and weaknesses of various kinds of mechanical measurements are considered next.

8.1. Force as a measure of crossbridge activity

An obvious manifestation of contractile activity is the generation of isometric force. It is the simplest contraction parameter to measure, and in many cases it may serve as an adequate endpoint. However, there are a number of factors to consider regarding its use. Unless very special circumstances obtain (such as thiophosphorylation of the myosin light chains in a skinned muscle), it is difficult to know what fraction of a muscle's full force-generating capacity is being manifested. For example, KCl depolarization of many smooth muscles, especially phasic ones, produces a peak of force that soon relaxes to some lower steady-state value that is likely to represent the activity of fewer crossbridges. The length sensitivity of force is also a consideration, especially in light of our incomplete knowledge of events determining the length–tension curve. These difficulties can often be overcome by the use of internal controls; for example, force can be normalized to a standard stimulus intensity (or agonist concentration) at L_o.

Force measurements per se can give inconsistent information about crossbridge kinetics because of delays in activation and/or agonist diffusion. However, some workers (Klemt et al. 1981; Meiss 1993) have used the rate of force *re*development (dF/dt) that occurs following the mechanical interruption of an isometric contraction. This bears a close, although complicated, relationship to the instantaneous cycling rate of crossbridges. It should also be kept in mind that the time-dependent development of the latch state means that the same force at two different times during a contraction can represent quite different internal states. In this case, additional mechanical parameters must be considered.

8.2. Shortening velocity as an index of crossbridge cycling rate

An important criterion of the latch state is the reduced velocity of shortening that can be measured during the period of force maintenance. For valid comparisons to be made, the conditions of measurement must be properly chosen, especially with regard to muscle length and afterload. For instance, tracking the shortening velocity capability during the course of an isometric contraction (by quick releases to a constant light load or by the slack test) can be done at a constant muscle length. However, tracking the maximal velocity capability of a muscle throughout the course of a moderately loaded isotonic contraction by the use of zero-load clamps (cf. Stephens et al. 1986) is complicated by the changing muscle length between measurements, because velocity is quite sensitive to the length at which the contraction begins (Gunst et al. 1993; Meiss 1993).

8.3. Isometric stiffness as an adjunct to force measurements

As has been noted previously, the dynamic stiffness of muscle is usually a simple, nearly linear function of the isometric force. However, the differences in stiffness at the same force during contraction and relaxation (Meiss 1978, 1993) and the progressive increase in stiffness during a maintained contraction (Gunst et al. 1995) emphasize that mechanical states that can appear to be similar may have underlying differences. Although the cellular and tissue loci of smooth muscle stiffness are not completely understood, determining stiffness in conjunction with force measurements can serve as an empirical tool that can sharpen the ability to detect changes in contractile parameters.

9. Problems remaining in smooth muscle mechanics

The array of investigative tools that has been brought to bear on the study of smooth muscle function is impressive indeed. The power of recent techniques such as the *in vitro* motility assay and its variations, the use of the methods of molecular biology to manipulate the constituents of the contractile system, and the use of advanced optical, electron microscopic, and immunological staining techniques has been shown in the rapid advance of knowledge of the contractile mechanism of smooth muscle. Many long-standing questions have either been answered or can now be put to experimental tests. The impressive gains made in many of these areas certainly show the importance of understanding the function of the elemental parts of a complex system in relative isolation from other factors. From one perspective, some of these gains may be considered as ends in themselves. On the other hand, these findings must be integrated into an understanding of smooth muscle at higher levels of organization and in the overall body economy. The study of muscle mechanics can provide a natural framework to direct this reintegration and to pose additional questions and means of approach to a number of important areas that are still not well understood. The following items (not an exhaustive list by any means) describe questions at several levels of complexity. They appear to be areas in which much remains to be done; with the array of tools now available, answers are likely to be forthcoming.

1. *Providing a better definition of the geometry of the contractile system.* Although recent work has made impressive progress in establishing relationships among dense bodies and the myofilament array in single cells (Bennet et al. 1988; Kargacin et al. 1989) and in the structural assembly of the myofilaments themselves (Cooke et al. 1989), the structures involved are still hard to visualize clearly enough to provide a structural basis for specific mechanical tests of their integrated function.

2. *Quantification of thick and thin filament overlap.* Despite a relative paucity of definitive evidence, the conviction remains that the length–tension behavior of smooth muscle can be explained in large part by varying overlap between thick and thin filaments. The fact that force is generated at all argues that overlap must occur, since the filament and crossbridge structures are rather well understood. What is not clear is the role that varying overlap might play in determining

the length dependence of force; certainly some of this phenomenon is due to length-dependent calcium handling, phosphorylation, and signal transduction mechanisms. More precise knowledge of the degree of filament overlap and lateral spacing as a function of length, as well as the development of some sort of real-time marker of the changes in these parameters, would allow the construction of a sliding-filament theory of smooth muscle contraction that is more than an analogy to the situation in skeletal muscle.

3. *The role of the cytoskeleton in organizing the mechanical output of the contractile system.* The superimposability of repeated smooth muscle contractions, especially those involving significant changes in length, argues that the smooth muscle contractile system is supported by structural elements that determine and preserve its geometric integrity. Experiments involving length changes at critical times (Meiss 1993; Gunst et al. 1995; Mehta et al. 1996) suggest that these structural elements may also be rapidly modified. Current research (cf. Pavalko et al. 1995) is seeking to determine the mechanism of these changes.

4. *Force transmission throughout a smooth muscle tissue: the role of cell attachments and the extracellular matrix.* Although the mechanical behavior of intact smooth muscle has been well characterized in a number of cases, and that of single cells has likewise been carefully investigated, significant questions arise as to the relationship between the whole and its parts. This is an area where study could prove quite valuable, because of the possibility of independent alteration (experimentally, physiologically, and pathologically) of cellular mechanics, the properties of the extracellular matrix, and the attachments of cells to the matrix and to each other.

5. *Smooth muscle "mechanical memory" and short-term modifications of the contractile system.* Mechanical conditions just prior to a contraction, and conditions during a contraction as well, can influence later events in a predictable, path-dependent way (Gunst et al. 1993, 1995; Pavalko et al. 1995). Such well-documented influences are a part of a short-term memory system that must have its basis in the structure and chemistry of smooth muscle cells. It is likely that the cytoskeleton and associated macromolecules in the cell provide a mechanism for this phenomenon, and it has also been suggested (Pratusevich et al. 1995) that short-term modifications to myosin filament structure may contribute to such phenomena. A start toward working out some possible mechanisms has been made, and further progress in this area would be valuable in providing a more comprehensive and general understanding of the overall contractile mechanism in smooth muscle cells.

10. Summary

The specialized mechanical properties of smooth muscle are the manifestation of its cellular and molecular activity. An important approach to understanding the internal function of smooth muscle is the careful measurement of its mechanical properties under a variety of externally imposed conditions. Current knowledge of mechanical properties has been obtained by experiments with whole organs, isolated tissues, individual cells, and on isolated proteins and myofilaments. Isometric specializations of smooth muscle include its ability to function over a wide range of lengths and to maintain force economically for long periods of time, while isotonic performance is strongly time- and length-dependent, with velocity decreasing as a function of duration of contraction and the degree of shortening. The stiffness of contracting muscle reveals additional mechanical properties not apparent with measurements of force and length alone. Contraction under mixed external conditions reveals significant interactions between these two modes.

The use of muscle mechanics as an endpoint in experimental design, an important practical matter, requires consideration of the underlying cellular, molecular, and tissue-level events of contraction. Much of this specialized information has come from experiments on isolated single cells and reconstituted systems of contractile proteins and myofilaments (the *in vitro* motility assay). Work in its early stages is focusing on the role of the cytoskeleton in determining cellular mechanical properties, and the mechanical interactions among cells in intact tissues are being studied in an effort to relate cellular mechanics to the mechanics of tissues as a whole.

References

Allen, B. G., and Walsh, M. P. (1994). The biochemical basis of the regulation of smooth-muscle contraction. *Trends Biochem. Sci.* 19: 362–68.

Arner, A., and Hellstrand, P. (1985). Effects of calcium and substrate on force–velocity relation and energy turnover in skinned smooth muscle of the guinea-pig. *J. Physiol. (Lond.)* 360: 347–65.

Ashton, F. T., Somlyo, A. V., and Somlyo, A. P. (1975). The contractile apparatus of vascular smooth muscle: Intermediate high voltage stereo electron microscopy. *J. Mol. Biol.* 98: 17–29.

Bagby, R. M. (1990). Ultrastructure, cytochemistry, and organization of myofilaments in vertebrate smooth muscle cells. In: *Ultrastructure of Smooth Muscle* (P. M. Motta, ed.). Boston: Kluwer, pp. 23–61.

Bagby, R. M., and Fisher, B. A. (1973). Graded contractions in muscle strips and single cells from *Bufo marinus* stomach. *Am. J. Physiol.* 225: 105–9.

Barany, M. (1967). ATPase activity of myosin correlated with speed of muscle shortening. *J. Gen. Physiol.* 50 (suppl.): 197–218.

Bennett, J. P., Cross, R. A., Kendrick-Jones, J., and Weeds, A. G. (1988). Spatial pattern of myosin phosphorylation in contracting smooth muscle cells: Evidence for contractile zones. *J. Cell Biol.* 107: 2623–9.

Bozler, E. (1948). Conduction, automaticity, and tonus of visceral muscles. *Experientia* 4: 213–18.

Brozovich, F. V., and Morgan, K. G. (1989). Stimulus-specific changes in mechanical properties of vascular smooth muscle. *Am. J. Physiol.* 257: H1573–H1580.

Brozovich, F. V., and Yamakawa, M. (1993). Agonist activation modulates cross-bridge states in single vascular smooth muscle cells. *Am. J. Physiol.* 264: C103–C108.

Butler, T. M., Siegman, M. J., and Davies, R. E. (1976). Rigor and resistance to stretch in vertebrate smooth muscle. *Am. J. Physiol.* 231: 1509–14.

Butler, T. M., Siegman, M. J., and Mooers, S. U. (1987). Slowing of crossbridge cycling rate in mammalian smooth muscle occurs without evidence of an increase in internal load. *Prog. Clin. Biol. Res.* 245: 289–301.

Chalovich, J. M., Yu, L. C., and Brenner, B. (1991). Involvement of weak binding crossbridges in force production in muscle. *J. Muscle Res. Cell Motil.* 12: 503–6.
Cohen, D. M., and Murphy, R. A. (1978). Differences in cellular contractile protein contents among porcine smooth muscles: Evidence for variation in the contractile system. *J. Gen. Physiol.* 72: 369–80.
Cooke, P. H., Fay, F. S., and Craig, R. (1989). Myosin filaments isolated from skinned amphibian smooth muscle cells are side-polar. *J. Muscle Res. Cell Motil.* 10: 206–20.
Cooke, R. (1995). The actomyosin engine. *FASEB J.* 9: 636–42.
Csapo, A., and Goodall, M. (1954). Excitability, length-tension relation and kinetics of uterine muscle contraction in relation to hormonal status. *J. Physiol. (Lond.)* 126: 384–95.
Dillon, P. F., Aksoy, M. O., Driska, S. P., and Murphy, R. A. (1981). Myosin phosphorylation and the crossridge cycle in arterial smooth muscle. *Science* 211: 495–7.
Dillon, P. F., and Murphy, R. A. (1982). Tonic force maintenance with reduced shortening velocity in arterial smooth muscle. *Am. J. Physiol.* 242: C102–C108.
Driska, S. P., Damon, D. N., and Murphy, R. A. (1978). Estimates of cellular mechanics in an arterial smooth muscle. *Biophys. J.* 24: 525–40.
Driska, S. P., and Porter, R. (1986). Isolation of smooth muscle cells from swine carotid artery by digestion with papain. *Am. J. Physiol.* 251: C474–C481.
Edman, K. A. (1979). The velocity of unloaded shortening and its relation to sarcomere length and isometric force in vertebrate muscle fibers. *J. Physiol. (Lond.)* 291: 143–59.
Eisenberg, E., Hill, T. L., and Chen, Y. (1980). Cross-bridge model of muscle contraction. Quantitative analysis. *Biophys. J.* 29: 195–227.
Fay, F. S. (1977). Isometric contractile properties of single isolated smooth muscle cells. *Nature* 265: 553–6.
Fay, F. S., Fujiwara, K., Rees, D. D., and Fogarty, K. E. (1983). Distribution of alpha-actinin in single isolated smooth muscle cells. *J. Cell Biol.* 96: 783–95.
Finer, J. T., Simmons, R. M., and Spudich, J. A. (1994). Single myosin molecule mechanics: Piconewton forces and nanometre steps. *Nature* 368: 113–19.
Fuglsang, A., Khromov, A., Torok, K., Somlyo, A. V., and Somlyo, A. P. (1993). Flash photolysis studies of relaxation and cross-bridge detachment: Higher sensitivity to tonic than phasic smooth muscle to MgADP. *J. Muscle Res. Cell Motil.* 14: 666–73.
Gabella, G. (1984). Structural apparatus for force transmission in smooth muscles. *Physiol. Rev.* 64: 455–77.
Gabella, G. (1990). General aspects of the fine structure of smooth muscles. In: *Ultrastructure of Smooth Muscle* (P. M. Motta, ed.). Boston: Kluwer, pp. 1–22.
Gailly, P., Gillis, J. M., and Capony, J. P. (1991). Complex stiffness of smooth muscle cytoplasm in the presence of Ca-activated brevin. *J. Muscle Res. Cell Motil.* 12: 333–9.
Gailly, P., Gillis, J. M., and Capony, J. P. (1993). Influence of Ca-activated brevin on the mechanical properties of skinned smooth muscle. *Adv. Exp. Med. Biol.* 332: 205–12.
Gailly, P., Lejeune, T., Capony, J. P., and Gillis, J. M. (1990). The action of brevin, an F-actin severing protein, on the mechanical properties and ATPase activity of skinned smooth muscle. *J. Muscle Res. Cell Motil.* 11: 293–301.
Gillespie, J. S. (1971). The rat anococcygeus; A new, densely innervated smooth muscle preparation. *Br. J. Pharmacol.* 43: 430P.
Glerum, J. J., Van Mastrigt, R., and Van Koeveringe, A. J. (1990). Mechanical proper-

ties of mammalian single smooth muscle cells. iii. passive properties of pig detrusor and human *a terme* uterus cells. *J. Muscle Res. Cell Motil.* 11: 453–62.

Goldman, Y. E., and Huxley, A. F. (1994). Actin compliance: Are you pulling my chain? *Biophys. J.* 67: 2131–36.

Gong, M. C., Cohen, P., Kitazawa, T., Ikebe, M., Somlyo, A. P., and Somlyo, A. V. (1992). Myosin light chain phosphatase activities and the effects of phosphatase inhibitors in tonic and phasic smooth muscle. *J. Biol. Chem.* 267: 14662–8.

Gordon, A. R. (1978). Contraction of detergent-treated smooth muscle. *Proc. Nat. Acad. Sci. Usa* 75: 3527–30.

Gordon, A. R., and Siegman, M. J. (1971). Mechanical properties of smooth muscle. I. Length-tension and force-velocity relations. *Am. J. Physiol.* 221: 1243–9.

Gunst, S. J., al-Hassani, M. H., and Adam, L. P. (1994). Regulation of isotonic shortening velocity by second messengers in tracheal smooth muscle. *Am. J. Physiol.* 266: C684–C691.

Gunst, S. J., Meiss, R. A., Wu, M., and Rowe, M. (1995). Mechanisms for the mechanical plasticity of tracheal smooth muscle. *Am. J. Physiol.* 268: C1267–C1276.

Gunst, S. J., and Russell, J. A. (1982). Contractile force of canine tracheal smooth muscle during continuous stretch. *J. Appl. Physiol.* 52: 655–63.

Gunst, S. J., Wu, M. F., and Smith, D. D. (1993). Contraction history modulates isotonic shortening velocity in smooth muscle. *Am. J. Physiol.* 265: C467–C476.

Haeberle, J. R., and Hemric, M. E. (1994). A model for the coregulation of smooth muscle actomyosin by caldesmon, calponin, tropomyosin, and the myosin regulatory light chain. *Can. J. Physiol. Pharmacol.* 72: 1400–9.

Haeberle, J. R., and Hemric, M. E. (1995). Are actin filaments moving under unloaded conditions in the *in vitro* motility assay? *Biophys. J.* 68: 306s–311s.

Hai, C. M. (1991). Length-dependent myosin phosphorylation and contraction of arterial smooth muscle. *Pflügers Arch.* 418: 564–71.

Hai, C., and Ma, C. B. B. (1993). Fluoroaluminate-and GTP gammaS-induced stress, shortening and myosin phosphorylation in airway smooth muscle. *Am. J. Physiol.* 265: L73–L79.

Hai, C., and Murphy, R. A. (1988a). Regulation of shortening velocity by cross-bridge phosphorylation in smooth muscle. *Am. J. Physiol.* 255: C86–C94.

Hai, C., and Murphy, R. A. (1988b). Cross-bridge phosphorylation and regulation of latch state in smooth muscle. *Am. J. Physiol.* 254: C99–C106.

Hai, C., and Murphy, R. A. (1989). Ca^{2+}, crossbridge phosphorylation, and contraction. *Ann. Rev. Physiol.* 51: 285–98.

Hai, C., and Murphy, R. A. (1992). Adenosine 5'-triphosphate consumption by smooth muscle as predicted by the coupled four-state crossbridge model. *Biophys. J.* 61: 530–41.

Hai, C., and Szeto, B. (1992). Agonist-induced myosin phosphorylation during isometric contraction and unloaded shortening in airway smooth muscle. *Am. J. Physiol.* 262: L53–L62.

Halpern, W., Mulvany, M. J., and Warshaw, D. M. (1978). Mechanical properties of smooth muscle cells in the walls of arterial resistance vessels. *J. Physiol. (Lond.)* 275: 85–101.

Harris, D. E., and Warshaw, D. M. (1990). Slowing of velocity during isotonic shortening in single isolated smooth muscle cells. Evidence for an internal load. *J. Gen. Physiol.* 96: 581–601.

Harris, D. E., and Warshaw, D. M. (1991). Length vs. active force relationship in single isolated smooth muscle cells. *Am. J. Physiol.* 260: C1104–C1112.

Harris, D. E., and Warshaw, D. M. (1993a). Smooth and skeletal muscle actin are mechanically indistinguishable in the *in vitro* motility assay. *Circ. Res.* 72: 219–24.

Harris, D. E., and Warshaw, D. M. (1993b). Smooth and skeletal muscle myosin both exhibit low duty cycles at zero load *in vitro*. *J. Biol. Chem.* 268: 14764–8.

Harris, D. E., Work, S. S., Wright, R. K., Alpert, N. R., and Warshaw, D. M. (1994). Smooth, cardiac and skeletal muscle myosin force and motion generation assessed by cross-bridge mechanical interactions *in vitro*. *J. Muscle Res. Cell Motil.* 15: 11–19.

Hartshorne, D. J., and Kawamura, T. (1992). Regulation of contraction–relaxation in smooth muscle. *News in Physiological Sciences* 7: 59–64.

Hellstrand, P. (1991). Mechanics of the crossbridge interaction in living and chemically skinned smooth muscle. *Adv. Exp. Med. Biol.* 304: 85–96.

Hellstrand, P. (1994). Cross-bridge kinetics and shortening in smooth muscle. *Can. J. Physiol. Pharmacol.* 72: 1334–7.

Herlihy, J. T., and Murphy, R. A. (1973). Length-tension relationship of smooth muscle of the hog carotid artery. *Circ. Res.* 33: 275–83.

Hill, A. V. (1938). The heat of shortening and the dynamic constants of muscle. *Proc. Roy. Soc. Lond. B.* 126: 136–95.

Huxley, A. F. (1957). Muscle structure and theories of contraction. *Prog. Bioph. Mol. Biol.* 7: 255–318.

Huxley, A. F., Stewart, A., Sosa, H., and Irving, T. (1994). X-ray diffraction measurements of the extensibility of actin and myosin filaments in contracting muscle. *Biophys. J.* 67: 2411–21.

Ishida, K., Pare, P. D., Blogg, T., and Schellenberg, R. R. (1990). Effects of elastic loading on porcine trachealis muscle mechanics. *J. Appl. Physiol.* 69: 1033–9.

Jaworski, A., Anderson, K. I., Arner, A., Engstrom, M., Gimona, M., Strasser, P., and Small, J. V. (1995). Calponin reduces shortening velocity in skinned taenia coli smooth muscle fibers. *FEBS Letts.* 365: 167–71.

Kamm, K. E., and Stull, J. T. (1985). The function of myosin and myosin light chain kinase phosphorylation in smooth muscle. *Ann. Rev. Pharmacol. Toxicol.* 25: 593–620.

Kamm, K. E., and Stull, J. T. (1986). Activation of smooth muscle contraction: Relation between myosin phosphorylation and stiffness. *Science* 232: 80–2.

Kamm, K. E., and Stull, J. T. (1989). Regulation of smooth muscle contractile elements by second messengers. *Ann. Rev. Physiol.* 51: 299–313.

Kargacin, G. J., Cooke, P. H., Abramson, S. B., and Fay, F. S. (1989). Periodic organization of the contractile apparatus in smooth muscle revealed by the motion of dense bodies in single cells. *J. Cell Biol.* 108: 1465–75.

Kargacin, G. J., and Fay, F. S. (1987). Physiological and structural properties of saponin-skinned single smooth muscle cells. *J. Gen. Physiol.* 90: 49–73.

Kelley, C. A., Takahashi, M., Yu, J. H., and Adelstein, R. S. (1993). An insert of seven amino acids confers functional differences between smooth muscle myosins from the intestines and vasculature. *J. Biol. Chem.* 268: 12848–54.

Kitazawa, T., Kobayashi, S., Horiuti, K., Somlyo, A. V., and Somlyo, A. P. (1989). Receptor-coupled, permeabilized smooth muscle. *J. Biol. Chem.* 264: 5339–42.

Klemt, P., Peiper, U., Speden, R. N., and Zilker, F. (1981). The kinetics of post-vibration tension recovery of the isolated rat portal vein. *J. Physiol. (Lond.)* 312: 281–96.

Kron, S. J., and Spudich, J. A. (1986). Fluorescent actin filaments move on myosin fixed to a glass surface. *Proc. Nat. Acad. Sci. USA* 83: 6272–6.

Kurumi, Y., and Chacko, S. (1995). Effect of unphosphorylated smooth muscle myosin on caldesmon-mediated regulation of actin filament velocity. *J. Muscle Res. Cell Motil.* 16: 11–19.

Mehta, D., Wu, M., and Gunst, S. J. (1996). Role of contractile protein activation in the length-dependent modulation of tracheal smooth muscle force. *Am. J. Physiol.* 270: C243–C252.

Meiss, R. A. (1971). Some mechanical properties of cat intestinal muscle. *Am. J. Physiol.* 220: 2000–7.

Meiss, R. A. (1975). Graded activation in rabbit mesotubarium smooth muscle. *Am. J. Physiol.* 229: 455–65.

Meiss, R. A. (1978). Dynamic stiffness of rabbit mesotubarium smooth muscle: Effect of isometric length. *Am. J. Physiol.* 234: C14–C26.

Meiss, R. A. (1982). Transient responses and continuous behavior of active smooth muscle during controlled stretches. *Am. J. Physiol.* 242: C146–C158.

Meiss, R. A. (1984a). Nonlinear force response of active smooth muscle subjected to small stretches. *Am. J. Physiol.* 246: C114–C124.

Meiss, R. A. (1984b). Solid-state optical scanning system for remote measurements in biomechanical systems. *Am. J. Physiol.* 247: C488–C494.

Meiss, R. A. (1987). Stiffness of active smooth muscle during forced elongation. *Am. J. Physiol.* 253: C484–C493.

Meiss, R. A. (1989). Mechanical properties of gastrointestinal smooth muscle. In: *Handbook of Physiology – The Gastrointestinal System I* (J. D. Wood, ed.). Bethesda, MD: American Physiological Society, pp. 273–329.

Meiss, R. A. (1990). The effect of tissue properties on smooth muscle mechanics. In: *Frontiers in Smooth Muscle Research* (N. Sperelakis and J. D. Wood, eds.). New York: Alan R. Liss, pp. 435–49.

Meiss, R. A. (1991). An analysis of length-dependent stiffness in smooth muscle strips. In: *Regulation of Smooth Muscle* (R. S. Moreland, ed.). New York: Plenum, pp. 425–34.

Meiss, R. A. (1992). Limits to shortening in smooth muscle tissues. *J. Muscle Res. Cell Motil.* 13: 190–8.

Meiss, R. A. (1993). Persistent mechanical effects of decreasing length during isometric contraction of ovarian ligament smooth muscle. *J. Muscle Res. Cell Motil.* 14: 205–18.

Meiss, R. A. (1994). Transient length-related mechanical states in smooth muscle. *Can. J. Physiol. Pharmacol.* 72: 1325–33.

Mulvany, M. J., and Warshaw, D. M. (1981). The anatomical location of the series elastic component in rat vascular smooth muscle. *J. Physiol (Lond.)* 314: 321–30.

Murphy, R. A. (1980). Mechanics of vascular smooth muscle. In: *Handbook of Physiology – The Cardiovascular System II* (D. F. Bohr, A. P. Somlyo, and H. V. Sparks, eds.). Bethesda, MD: American Physiological Society, pp. 325–51.

Murphy, R. A. (1994). What is special about smooth muscle? The significance of covalent crossbridge regulation. *FASEB J.* 8: 311–18.

Murphy, R. A., Herlihy, J. T., and Megerman, J. (1974). Force-generating capacity and contractile protein content of arterial smooth muscle. *J. Gen. Physiol.* 64: 691–705.

Nishimura, J., Kolber, M., and van Bremen, C. (1988). Norepinephrine and GTP-gamma-S increase myofilament Ca^{2+} sensitivity in alpha-toxin permeabilized arterial smooth muscle. *Biochem. Biophys. Res. Commun.* 157: 677–83.

Nishiye, E., Somlyo, A. V., Torok, K., and Somlyo, A. P. (1993). The effects of MgADP on cross-bridge kinetics: A laser flash photolysis study of guinea-pig smooth muscle. *J. Physiol. (Lond.)* 460: 247–71.

Oiwa, K., Chaen, S., Kamitsubo, E., Shimmen, T., and Sugi, H. (1990). Steady-state force–velocity relation in the ATP-dependent sliding movement of myosin-coated beads on actin cables *in vitro* studied with a centrifuge microscope. *Proc. Nat. Acad. Sci. USA* 87: 7893–7.

Packer, C. S., and Stephens, N. L. (1985). Mechanics of caudal artery relaxation in control and hypertensive rats. *Can. J. Physiol. Pharmacol.* 63: 209–13.

Paul, R. J. (1990). Smooth muscle energetics and theories of cross-bridge regulation. *Am. J. Physiol.* 258: C369–C375.

Pavalko, F. M., Adam, L. P., Wu, M.-F., Walker, T., and Gunst, S. J. (1995). Phosphorylation of dense-plaque proteins talin and paxillin during tracheal smooth muscle contraction. *Am. J. Physiol.* 268: C563–C571.

Peiper, U., Klemt, P., and Schleupner, R. (1978). The temperature dependence of parallel and series elastic elements in the vascular smooth muscle of the rat portal vein. *Pflügers Arch.* 378: 25–30.

Pratusevich, V. R., Seow, C., and Ford, L. E. (1995). Plasticity in canine airway smooth muscle. *J. Gen. Physiol.* 105: 73–94.

Prosser, C. L. (1967). Problems in the comparative physiology of non-striated muscles. In: *Invertebrate Nervous Systems* (C. A. G. Wiersma, ed.). Chicago: University of Chicago Press, pp. 133–49.

Prosser, C. L. (1973). Muscle. In: *Comparative Animal Physiology*, 3rd ed. (C. L. Prosser, ed.). Philadelphia: W. B. Saunders, pp. 719–88.

Prosser, C. L. (1982). Diversity of narrow-fibered and wide-fibered muscles. In: *Basic Biology of Muscles: A Comparative Approach* (B. M. Twarog, R. J. C. Levine, and M. M. Dewey, eds.). New York: Raven Press, pp. 381–97.

Ruegg, J. C. (1971). Smooth muscle tone. *Physiol. Rev.* 51: 201–48.

Siegman, M. J., Butler, T. M., and Mooers, S. U. (1984). Energetic, mechanical, and ultrastructural correlates of the length–tension relationship in smooth muscle. In: *Smooth Muscle Contraction* (N. L. Stephens, ed.). New York: Marcel Dekker, pp. 189–98.

Simmons, R. M., Finer, J. T., Warrick, H. M., Kralik, B., Chu, S., and Spudich, J. A. (1993). Force on single actin filaments in a motility assay measured with an optical trap. In: *Mechanism of Myofilament Sliding in Muscle Contraction* (H. Sugi and G. H. Pollack, eds.). New York: Plenum, pp. 331–7.

Singer, H. A., Kamm, K. E., and Murphy, R. A. (1986). Estimates of activation in arterial smooth muscle. *Am. J. Physiol.* 251: C465–C473.

Small, J. V., Fuerst, D. O., and De Mey, J. (1986). Localization of filamin in smooth muscle. *J. Cell Biol.* 102: 210–20.

Somlyo, A. P. (1993). Myosin isoforms in smooth muscle: How may they affect function and structure? *J. Muscle Res. Cell Motil.* 14: 557–63.

Somlyo, A. P., and Himpens, B. (1989). Cell calcium and its regulation in smooth muscle. *FASEB J.* 3: 2266–76.

Somlyo, A. P. and Somlyo, A. V. (1968). Vascular smooth muscle. I. Normal structure, pathology, biochemistry, and biophysics. *Physiol. Rev.* 20: 197–272.

Somlyo, A. P., and Somlyo, A. V. (1990). Flash photolysis studies of excitation–contraction coupling, regulation, and contraction in smooth muscle. *Annu. Rev. Physiol.* 52: 857–74.

Somlyo, A. P., and Somlyo, A. V. (1994). Smooth muscle: excitation–contraction coupling, contractile regulation, and the cross-bridge cycle. *Alcoholism: Clin. Exp. Res.* 18: 138–43.

Somlyo, A. V., and Franzini-Armstrong, C. (1985). New views of smooth muscle structure using freezing, deep-etching and rotary shadowing. *Experientia* 41: 841–56.

Somlyo, A. V., Goldman, Y. E., Fujimori, T., Bond, M., Trentham, D. R., and Somlyo, A. P. (1988). Cross-bridge kinteics, cooperativity, and negatively strained cross-bridges in vertebrate smooth muscle: A laser-flash photolysis study. *J. Gen. Physiol.* 91: 165–92.

Squire, J. M. (1994). The actomyosin interaction – Shedding light on structural events: 'Plus ca change, plus c'est la meme chose'. *J. Muscle Res. Cell Motil.* 15: 227–31.

Stephens, N. L., and Brutsaert, D. L. (1982). Maximal force potential of tetanized mammalian smooth muscle. *Am. J. Physiol.* 242: C283–C287.

Stephens, N. L., Claes, V. A., and Brutsaert, D. L. (1981). Relaxation of tetanized canine tracheal smooth muscle. *Pflügers Arch.* 390: 175–8.

Stephens, N. L., Kagan, M. L., and Packer, C. S. (1986). Time dependence of shortening velocity in tracheal smooth muscle. *Am. J. Physiol.* 251: C435–C442.

Stephens, N. L., Kroeger, E., and Mehta, J. A. (1969). Force–velocity characteristics of respiratory airway smooth muscle. *J. Appl. Physiol.* 26: 685–92.

Stull, J. T., Gallagher, P. J., Herring, B. P., and Kamm, K. E. (1991). Vascular smooth muscle contractile elements. Cellular regulation. *Hypertension* 17: 723–32.

Tanner, J. A., Haeberle, J. R., and Meiss, R. A. (1988). Regulation of glycerinated smooth muscle contraction and relaxation by myosin phosphorylation. *Am. J. Physiol.* 255: C34–C42.

Twarog B. M., Levine, R. J. C., and Dewey, M. M., eds. (1982). *Basic Biology of Muscles: A Comparative Approach.* New York: Raven.

Uvelius, B. (1976). Isometric and isotonic length–tension relations and variations in cell length in longitudinal smooth muscle from rabbit urinary bladder. *Acta Physiol. Scand.* 97: 1–12.

Uvelius, B., and Gabella, G. (1980). Relation between cell length and force production in urinary bladder smooth muscle. *Acta Physiol. Scand.* 110: 357–65.

Uvelius, B., and Hellstrand, P. (1980). Effects of phasic and tonic activation on contraction dynamics in smooth muscle. *Acta Physiol. Scand.* 109: 399–406.

van Bremen, C., and Saida, K. (1989). Cellular mechanisms regulating $Ca^{2+}{}_{I}$ in smooth muscle. *Ann. Rev. Physiol.* 51: 315–29.

VanBuren, P., Guilford, W., Kennedy, G., Wu, J., and Warshaw, D. M. (1995). Smooth muscle myosin: A high force-generating molecular motor. *Biophys. J.* 68: 256s–259s.

VanBuren, P., Work, S. S., and Warshaw, D. M. (1994). Enhanced force generation by smooth muscle myosin *in vitro. Proc. Nat. Acad. Sci. USA* 91: 202–5.

Wakabayashi, K., Sugimoto, Y., Tanaka, H., Ueno, Y., Takezawa, Y., and Amemiya, Y. (1994). X-ray diffraction evidence for the extensibility of actin and myosin filaments during muscle contraction. *Biophys. J.* 67: 2422–35.

Walsh, M. P. (1994). Calmodulin and the regulation of smooth muscle contraction. *Mol. Cell. Biochem.* 135: 21–41.

Warshaw, D. M. (1987). Force:velocity relationship in single isolated toad stomach smooth muscle cells. *J. Gen. Physiol.* 89: 771–89.

Warshaw, D. M., Desrosiers, J. M., Work, S. S., and Trybus, K. (1990). Smooth muscle myosin crossbridge interactions modulate actin filament sliding velocity *in vitro. J. Cell Biol.* 111: 453–63.

Warshaw, D. M., Desrosiers, J. M., Work, S. S., and Trybus, K. (1991). Effects of MgATP, MgADP, and P_i on actin movement by smooth muscle myosin. *J. Biol. Chem.* 266(36): 24339–43.

Warshaw, D. M., and Fay, F. S. (1983). Cross-bridge elasticity in single smooth muscle cells. *J. Gen. Physiol.* 82: 157–99.

Warshaw, D. M., McBride, W. J., and Work, S. S. (1987). Corkscrew-like shortening in single smooth muscle cells. *Science* 236: 1457–9.

Warshaw, D. M., Rees, D. D., and Fay, F. S. (1988). Characterization of cross-bridge elasticity and kinetics of cross-bridge cycling during force development in single smooth muscle cells. *J. Gen. Physiol.* 91: 761–79.

Warshaw, D. M., and Trybus, K. (1991). In vitro evidence for smooth muscle cross-bridge mechanical interactions. *Adv. Exp. Med. Biol.* 304: 53–9.

Wingard, C. J., Browne, A. K., and Murphy, R. A. (1995a). Dependence of force on length at constant cross-bridge phosphorylation in the swine carotid media. *J. Physiol. (Lond.)* 488(3): 729–39.

Wingard, C. J., Browne, A. K., Paul R. J., and Murphy R. A. (1995b). The ATP cost of covalent regulation in swine carotid medial rings. *Biophys. J.* 68(2): A164.

Yanagida, T., and Ishijima, A. (1995). Forces and steps generated by single myosin molecules. *Biophys. J.* 68: 312s–320s.

6

Regulation of Smooth Muscle Contraction by Myosin Phosphorylation

R. ANN WORD AND KRISTINE E. KAMM

1. Generalized scheme for smooth muscle contraction

The interaction between actin and myosin provides the molecular basis for muscle contraction. This interaction is regulated by Ca^{2+} in all muscle types, but mechanisms of regulation are fundamentally different for smooth muscle compared with skeletal or cardiac muscle. The discoveries that phosphorylation of smooth muscle myosin results in marked stimulation of actin-activated myosin MgATPase activity (Sobieszek 1977), and that myosin light chain kinase (MLCK), the enzyme responsible for myosin phosphorylation, requires Ca^{2+} and calmodulin (CaM) for activity (Dabrowska et al. 1978) led to the development of a model for regulation of smooth muscle contraction whereby the Ca^{2+}-dependent phosphorylation of myosin regulatory light chain (RLC) initiates smooth muscle contraction. In this scheme (illustrated in Figure 1), activators of smooth muscle contraction lead to increases in $[Ca^{2+}]_i$ and the formation of Ca^{2+}–CaM complexes that bind to and activate the enzyme MLCK. Activated MLCK catalyzes the phosphorylation of myosin RLC, which results in dramatic increases in actin-activated MgATPase activity of smooth muscle myosin and thereby initiates crossbridge cycling and mechanical output. Increases in Ca^{2+} may occur by influx of Ca^{2+} through Ca^{2+} channels in the plasma membrane or by release of Ca^{2+} from the sarcoplasmic reticulum (SR). Decreases in $[Ca^{2+}]_i$, brought about by Ca^{2+} extrusion or uptake into the SR, result in inactivation of MLCK, RLC dephosphorylation by myosin phosphatase (MLCP), and muscle relaxation. According to this simplified scheme, smooth muscle contraction is regulated by the relative activities of MLCK and myosin phosphatase leading to a particular value of RLC phosphorylation (Figure 2). Although Ca^{2+} homeostasis exerts predominant control over the MLCK:MLCP ratio, additional regulatory mechanisms can modulate the ratio, resulting in alterations in the Ca^{2+} sensitivity of RLC phosphorylation. In addition, the dependence of contractile force on RLC phosphorylation is also modulated under restricted conditions.

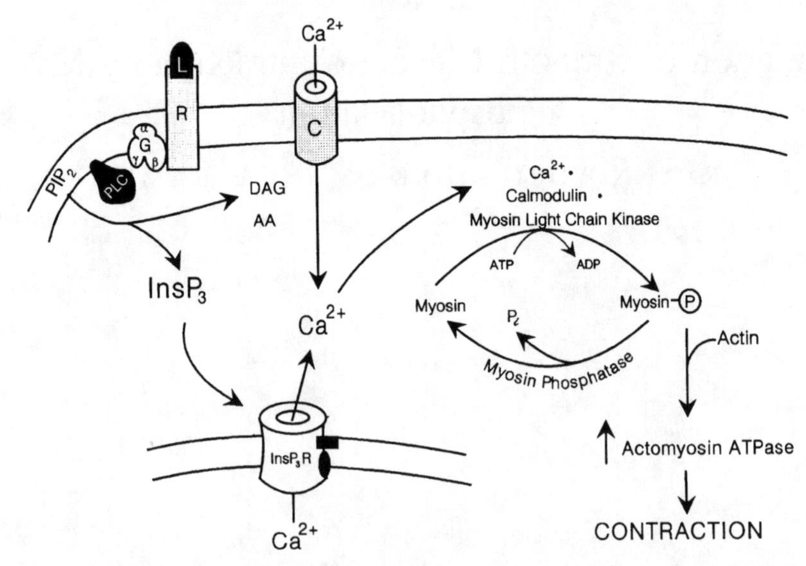

Figure 1. Scheme for initiation of smooth muscle contraction via myosin light chain phosphorylation. Increases in intracellular [Ca^{2+}] occur by ligand-receptor (LR)-activated phosphatidylinositol hydrolysis, $InsP_3$ formation, and release of Ca^{2+} from $InsP_3$-sensitive stores and-or by influx of extracellular Ca^{2+} *via* plasma membrane Ca^{2+} channels (C). This increase in cytoplasmic [Ca^{2+}] is the primary event that initiates the activation of smooth muscle contractile elements by Ca^{2+}-CaM-MLCK-dependent phosphorylation of myosin RLC (myosin–Ⓟ). Phosphorylated myosin undergoes ATP-dependent cyclic interactions with actin to produce contraction. Phosphorylation is reversed by myosin phosphatase. Additional regulation may occur by regulatory proteins associated with the $InsP_3$ receptor ($InsP_3R$), G proteins (G) associated with ligand receptors, or phospholipase C (PLC), and by regulation of MLCK or phosphatase activities.

The biochemical processes that alter the RLC phosphorylation–force relationship, however, are less well understood. It is the purpose of this chapter to review mechanisms by which smooth muscle contraction is regulated. The review is divided into two components. First, we describe proteins of the contractile system and particular studies that have elucidated biochemical elements of regulation. Second, we discuss regulation in the context of smooth muscle cells and tissues. Physiological studies have revealed that the regulatory apparatus responds to input from several second-messenger pathways. In general, cellular elements of regulation are consistent with biochemical properties of the contractile and regulatory proteins. However, there remain a number of unresolved aspects of regulation involving both the pathways and the effectors of regulation. These current issues are discussed throughout and summarized in the final section.

2. Proteins of the contractile system

The organization of contractile proteins and filaments in smooth muscle is much less evident than that of striated muscles. There are, however, three distinct types of filaments identified in smooth muscle cells: (1) thin filaments (6–8 nm in diameter) composed of polymerized double-helical strands of actin molecules with associated proteins tropomyosin and caldesmon and/or calponin arranged along the length of the double helix; (2) intermediate filaments (10 nm in diameter) composed of desmin and vimentin; and (3) thick filaments (12–18 nm in diameter) composed of polymerized myosin molecules. The composition and structure of thin and thick filaments are schematically illustrated in Figure 3. Immunocytochemical studies suggest that contractile and cytoskeletal filaments may be organized into distinct domains in smooth muscle cells (illustrated in Figure 4) (North et al. 1994). Intermediate filaments are involved in the formation of cytoskeletal networks and in the distribution of dense bodies to which contractile elements are anchored. Also found in this domain are thin filaments composed of a nonmuscle isoform of actin (β-actin). It is likely that this filamentous domain acts to distribute tension throughout the cell (Bárány et al. 1992), maintaining the shape and structural integrity of smooth muscle cells in the manner of a scaffold. Thick and thin filaments comprise the contractile elements of the contractile domain. They generate force and/or muscle shortening by a sliding-filament mechanism of contraction, in which the cyclic binding and release of actin by myosin heads (crossbridges) powered by ATP hydrolysis results in actin movement. The applicability of the sliding-filament mechanism to smooth muscle is supported by both structural and mechanical evidence (reviewed in Murphy 1979). Actin filaments in smooth muscle insert into the dense bodies much as they do in the Z bands of skeletal muscle, and smooth muscles exhibit length–tension and force–velocity relationships qualitatively similar to those in skeletal muscle (Walmsley and Murphy 1987; Wingard et al. 1995).

2.1. Thick filament system

2.1.1. Myosin – overall structure

Conventional myosin (myosin II) is the primary protein of smooth muscle thick filaments and is composed of two heavy chain subunits (approximately 200 kDa each) and two each of two types of light chain subunits (two 20-kDa RLCs and two 17-kDa essential light chains) (Hartshorne 1987) (Figure 3). The native polymer of hexameric myosin is configured as α-helical coiled tail regions of myosin (rods) forming the thick filament with the globular head regions protruding at regular intervals to form crossbridges. The head region, which can be isolated by proteolysis as subfragment 1 (S1), contains distinct sites for actin binding, ATP hydrolysis, and association of light chain subunits. The high-resolution structures of myosin S1 from skeletal myosin (Rayment et al. 1993) and the myosin regulatory domain from scallop myosin, containing the 10-kDa light chain–binding portion of the heavy chain with associated light chains (Xie et al. 1994), reveal striking structural features of the myosin molecule. The amino terminus of the S1 heavy chain constitutes the bulky end of the myosin head and contains both the actin and nucleotide binding sites (Rayment et al. 1993). This is often referred to as the *motor domain*, and has been shown to be sufficient for motility (Greene et al. 1983; Toyoshima et al.

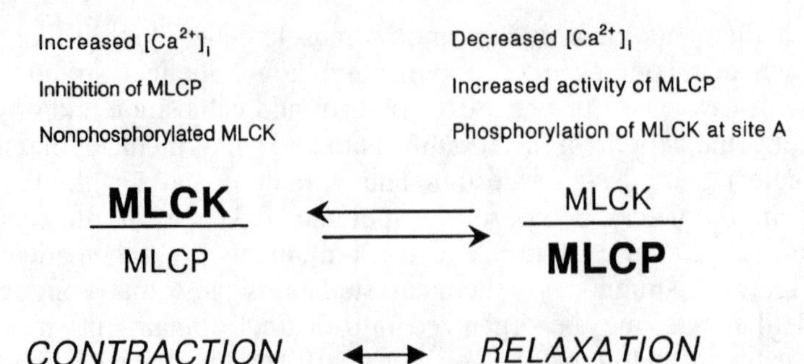

Figure 2. The ratio of activities of myosin light chain kinase and myosin phosphatase effects contraction and relaxation in smooth muscle. *Left:* Increased intracellular Ca^{2+} concentration ($[Ca^{2+}]_i$) activates CaM-MLCK, resulting in increased RLC phosphorylation and contraction. Inhibition of myosin phosphatase (MLCP) linked to receptor activation sensitizes myosin light chain phosphorylation to $[Ca^{2+}]_i$. *Right:* Decreased $[Ca^{2+}]_i$ inactivates CaM-MLCK, thereby decreasing phosphorylation and leading to relaxation. Phosphorylation of MLCK by a Ca^{2+}-CaM-dependent protein kinase (CaMK II) desensitizes MLCK to activation by Ca^{2+}-CaM.

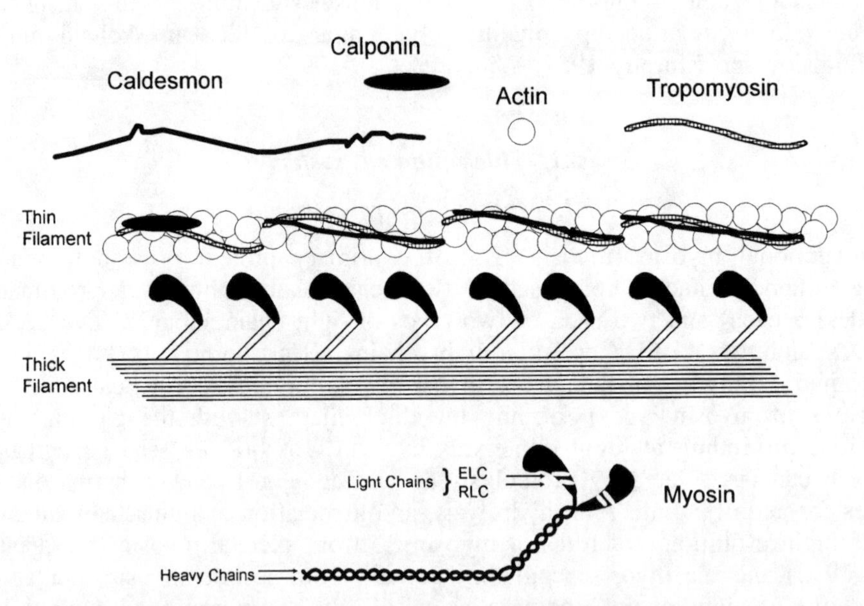

Figure 3. Contractile proteins in smooth muscle. Contractile proteins in smooth muscle are organized into thick and thin filaments as schematically illustrated. Myosin, composed of two heavy chains and two pairs of light chains—essential (ELC) and regulatory (RLC)—polymerizes into thick filaments. Thin filaments are composed of tropomyosin, caldesmon, and-or calponin bound to two intertwined strands of polymerized actin.

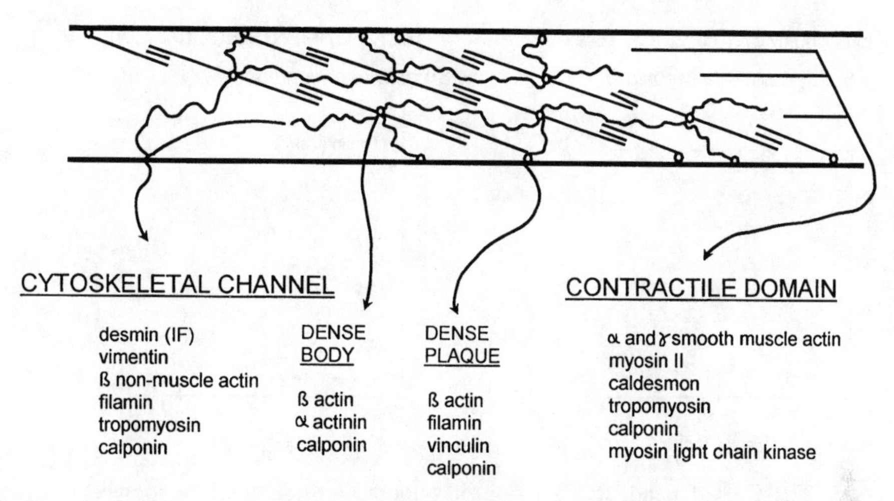

Figure 4. Localization of contractile and cytoskeletal components in smooth muscle cells. The contractile domain is composed of thin and thick filaments arranged in a sarcomeric array, and contains the proteins listed. The cytoskeletal channel is composed of intermediate filaments (IF) as well as actin-containing thin filaments and proteins listed. Thin filaments insert on dense bodies in the cytoplasm and on dense plaques on the sarcolemma. [After North et al. (1994).]

1987; Matsu-Ura and Ikebe 1995). Exiting the motor domain, the heavy chain forms an 85-Å α-helix (Rayment et al. 1993) that is stabilized by the binding of the two light chains. This complex is referred to as the *neck region* or *regulatory domain*. It has a gradual bend between the light chains and one sharp turn at the carboxyl terminus, where the two heads join at the head–rod junction. Like CaM and troponin-C, both the essential and regulatory light chains belong to the superfamily of EF-hand proteins (Kretsinger 1980). Each of these four proteins is folded into a dumbbell-like shape with two helix–loop–helix structural motifs (EF hands) in each globular end of the molecule connected by an extended central helix. Although each of the four EF hand motifs in CaM binds Ca^{2+}, three of these motifs are incompetent for divalent metal binding in myosin RLC (Nakayama et al. 1992). The essential light chains (ELCs) and RLCs bind in series to the myosin heavy chain in an antiparallel fashion with the ELC at the N-terminal portion and the RLC at the C-terminal end of the heavy chain α-helix. The carboxyl terminal half of the ELC is close to the motor domain, while the amino terminus of the RLC is located near the myosin head–rod junction, at least 10 nm from the active site. The ELC binds to the heavy chain such that it abuts a helical stretch in the motor domain formed from a region of highly conserved amino acids that undergo large rearrangements upon nucleotide binding (reviewed in Vibert and Cohen 1988). This arrangement may provide a linkage whereby phosphorylation of RLC effects ATP hydrolysis over a distance equal to half the length of the myosin head (Okamoto et al. 1986; Waller et al. 1995).

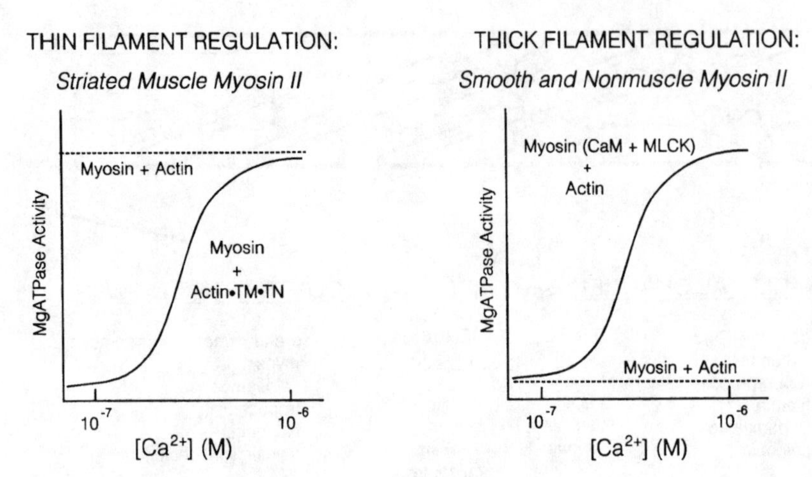

Figure 5. The Ca^{2+}-dependent regulation of smooth and nonmuscle actomyosin is fundamentally different from that of striated muscle. *Left:* striated muscle myosin has high level of actin-activated MgATPase (dashed line) at Ca^{2+} concentrations shown. The addition of thin-filament–binding proteins tropomyosin (TM) and troponin (TN) inhibits the ability of actin to activate myosin at low $[Ca^{2+}]_i$·Inhibition is reversed following Ca^{2+} binding to TN at elevated $[Ca^{2+}]_i$ (solid line). *Right:* Smooth and nonmuscle myosins have low levels of actin-activated MgATPase at Ca^{2+} concentrations shown (dashed line). Addition of calmodulin (CaM) and MLCK result in Ca^{2+}-CaM-dependent phosphorylation of myosin, thus allowing actin activation of MgATPase (solid line).

2.1.2. Regulation of smooth muscle actomyosin MgATPase by RLC

Myosin purified from smooth muscle has low MgATPase activity in the presence of filamentous actin (Hartshorne 1987). This is in contrast to the high levels of actin-activated MgATPase observed with purified skeletal muscle myosin. The MgATP-dependent crossbridge cycle is regulated by Ca^{2+} in both smooth and skeletal muscles, but the mechanisms of regulation are fundamentally different (illustrated in Figure 5). Ca^{2+} regulation in skeletal muscle resides in the troponin–tropomyosin system (bound to thin filaments) that inhibits actomyosin at low $[Ca^{2+}]_i$. In smooth muscle, Ca^{2+} regulation is mediated by the action of CaM–MLCK on thick filaments. Phosphorylation of RLC serine-19 by Ca^{2+}–CaM-dependent MLCK disinhibits smooth muscle myosin, thereby increasing actin-activated MgATPase activity (Hartshorne 1987). Phosphorylation of the 20-kDa myosin RLC is the key event necessary to initiate rapid force development in smooth muscle.

In vitro, RLC phosphorylation at serine-19 has profound effects on two aspects of smooth muscle myosin function. First, RLC phosphorylation regulates the ability of myosin to assemble into filaments; second, RLC phosphorylation results in a 500-fold increase in actin-activated myosin MgATPase activity.

Effects of RLC phosphorylation – filament assembly. In vitro, synthetic dephosphorylated myosin filaments are unstable in the presence of MgATP under physiological ionic strengths. The filament depolymerizes to yield monomeric myosin in which the α-helical tail is folded into thirds and the heads are bent toward the rod. These dephosphorylated monomers are enzymatically inactive and release the products of ATP hydrolysis slowly; the rate of phosphate release from single turnover experiments is about 0.0005 s^{-1} (Trybus 1991). *In vitro,* dephosphorylated myosin does not depolymerize in the presence of antirod monoclonal antibodies (Trybus and Henry 1989). Under these conditions, dephosphorylated myosin remains in filamentous form and the actin-activated MgATPase activity remains low (0.002 s^{-1}), but still five times that of the dephosphorylated monomer (0.0005 s^{-1}) (Trybus 1991). Phosphorylation of synthetic dephosphorylated monomers results in conformational changes such that the rod extends and assembles into filaments. Moreover, phosphorylation results in marked increases in actin-activated MgATPase activity to 0.3 s^{-1}, a value 500 times greater than dephosphorylated monomers and 100 times that of dephosphorylated filamentous myosin. These results clearly establish that RLC phosphorylation regulates myosin MgATPase activity independently of large changes in myosin conformation.

The role of filament depolymerization *in vivo* is not clear. A soluble monomeric myosin would present the advantage of diffusing throughout the cell to reassemble in a different location after phosphorylation. Such regulation may be more important in nonmuscle cells. Results from optical and electron microscopy experiments with smooth muscle suggest that assembly and disassembly can occur to a small extent during contraction and relaxation (Gillis et al. 1988; Godfraind-DeBecker and Gillis 1988); however, relaxed smooth muscle cells contain few dephosphorylated monomers (Somlyo et al. 1981; Vyas et al. 1992). Using monoclonal antibodies specific for the folded monomeric conformation, Fay and colleagues (Horowitz et al. 1994) found that only trace amounts of folded monomeric myosin are detected in either relaxed or contracted states of avian gizzard smooth muscle. The amount of monomer did not increase when α-toxin permeabilized gizzard was equilibrated in a solvent that disassembles filaments *in vitro*. Thus, assembly and disassembly may not play a major physiologic role in the regulation of contraction and relaxation in smooth muscle cells. Telokin, a myosin-binding protein, appears to stabilize thick filaments at physiological ionic strength (Shirinsky et al. 1993); however, telokin is not expressed in all smooth muscles (Gallagher and Herring 1991). Mechanisms responsible for the maintenance of thick-filament stability in smooth muscle cells remain the subject of investigation.

Effects of RLC phosphorylation – regulation of myosin MgATPase activity. In addition to its effects on the conformation of individual myosin molecules, RLC phosphorylation has profound effects on smooth muscle myosin MgATPase activity. Dephosphorylated RLC completely suppresses enzymatic activity of smooth muscle myosin. In contrast, the presence of RLC does not inhibit the high actin-activated activity of skeletal myosin MgATPase. Moreover, the light chains are not essential for enzymatic activity, but light chain–heavy chain interactions play an important role in the conversion of chemical energy into movement (Sellers 1985; Lowey et al. 1993). In smooth muscle myosin, specific interactions involving phosphoserine-19 and portions of the carboxyl terminal half of the RLC relieve the RLC-induced inhibition of myosin MgATPase and, in the presence of actin, ensure that high MgATPase activity is coupled to motility upon RLC phosphorylation (Sellers and Adelstein 1987; Sellers 1991; Trybus 1991; Trybus and Chatman 1993; Yang and Sweeney 1995).

Substrate determinants of RLC phosphorylation Synthetic peptides have been useful in defining the consensus sequence important for phosphorylation of RLC by MLCK (Kemp and Pearson 1990; Kennelly and Krebs 1991). Two groups of basic amino acid residues located amino-terminal to the phosphorylatable serine in either the smooth or skeletal muscle light chains confer substrate specificity for MLCK (Kemp and Pearson 1985; Michnoff et al. 1986). Basic residues N-terminal of the phosphoacceptor serine and the spatial arrangement of these basic residues (arginine 3 residues N-terminal to phosphoserine-19 [P-3] and three basic residues at P-6, P-7, P-8) are important for low K_m values (Figure 6). Hydrophobic C-terminal residues of phosphoserine are also important for peptide phosphorylation. Recent studies using intact RLC reveal that the substrate determinants are actually even more complex than those defined by synthetic peptides (Zhi et al. 1994). The V_{max} value of phosphotransferase reactions using synthetic peptides is 10% that of intact RLC. Additional substrate recognition motifs contained within subdomains I and II of RLC are important for RLC substrate recognition by smooth muscle MLCK and for high V_{max} values (Zhi et al. 1994). Although mutations of basic residues at P-6 to P-8 resulted in dramatic changes in K_m values for synthetic peptides, relatively modest changes in K_m values were found with the same mutations in the intact RLC. Further experiments confirmed the importance of arginine at P-3 and hydrophobic residues at P+1 to P+3. Thus, the serine phosphorylation of myosin RLC by MLCK requires a complex arrangement of substrate determinants (**R**XXⓈ**NVF** + subdomains I and II), resulting in a high degree of substrate specificity. Substrate determinants for RLC phosphorylation by MLCK are somewhat different from those determined with synthetic peptides (**KKR**XX**R**XXⓈ**NVF**).

A second phosphate can be incorporated into threonine-18 by MLCK in high concentrations. Whereas purified myosin containing such diphosphorylated RLCs undergoes further stimulation of MgATPase, diphosphorylation has no effect on contractile force, shortening velocity or *in vitro* motility of actomyosin (reviewed in Sellers 1991). Moreover, only trace amounts of phosphothreonine-18 are observed in intact smooth muscle (Colburn et al. 1988; Singer et al. 1989).

The isolated 20-kDa RLC is an excellent substrate for a number of other kinases, though many of these kinases (e. g., PKA and tyrosine kinases) cannot phosphorylate the light chain bound to MHC. However, purified myosin is phosphorylated *in vitro* by PKC and the multifunctional CaMK II (Nishikawa et al. 1984; Bengur et al. 1987; Ikebe et al. 1987; Edelman et al. 1990). Myosin RLC is phosphorylated by PKC at three sites distinct from the MLCK site. Sites of RLC phosphorylation are clustered at the amino terminus, as illustrated in Figure 6. Phosphorylation of myosin light chain by PKC has no direct effect on the actin-activated MgATPase activity of dephosphorylated myosin. However, PKC phosphorylation of myosin previously phosphorylated and activated by MLCK results in decreased affinity for actin and decreased actin-activated MgATPase activity (Nishikawa et al. 1984). PKC phosphorylation of RLC decreases the

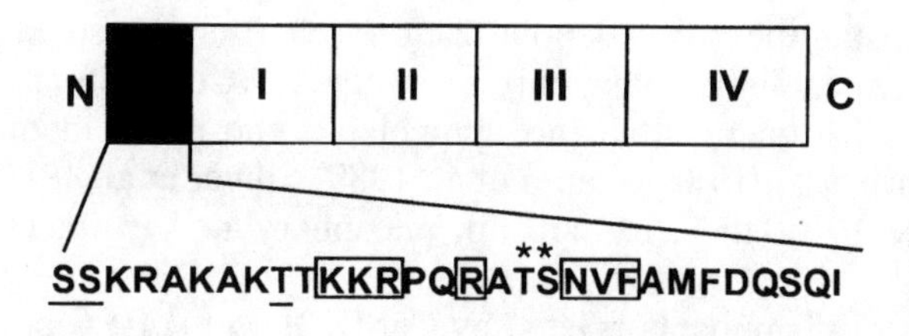

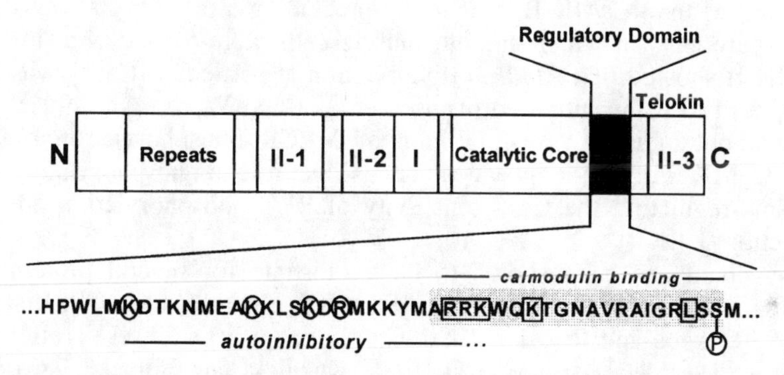

Figure 6. Domain organization of myosin regulatory light chain and myosin light chain kinase. The smooth muscle RLC (upper panel) contains four helix-loop-helix subdomains (I–IV). Sequence of the amino terminus of human smooth muscle RLC is shown (Kumar et al. 1989). Underlined residues indicate RLC phosphorylation sites demonstrated *in vitro* for PKC. Boxed residues indicate substrate determinants (as defined with synthetic peptides), and asterisks represent MLCK phosphorylation sites (threonine-18 and serine-19). The substrate determinants for serine-19 phosphorylation of intact myosin RLC by MLCK, however, are more complex than those of synthetic peptides requiring the consensus sequence (**R**XXⓈNVF) plus subdomains I and II. The smooth muscle MLCK (lower panel) contains several subdomains including the catalytic core and regulatory domain. Other structural motifs are described in the text. Sequence of the regulatory domain of rabbit smooth muscle MLCK is shown (Gallagher et al. 1991). Circled residues identify basic residues that bind to the catalytic core. Boxed residues are basic residues noted to be similar in sequential arrangement to the basic residues in the RLC that form part of the peptide consensus phosphorylation sequence. The shaded area indicates the CaM-binding domain, and the autoinhibitory region is noted to extend into the CaM-binding domain of MLCK. The phosphorylated serine residue –Ⓟ of MLCK denotes site A, the regulatory site of MLCK phosphorylation.

affinity of heavy meromyosin for MLCK, thereby inhibiting MLCK-induced phosphorylation. The significance of these *in vitro* effects is not clear because, in intact smooth muscle, contractions elicited by agonists (carbachol, 5-hydroxytryptamine, histamine) and depolarization (KCl) are associated with phosphorylation of myosin light chain only at serine-19. Contractions induced by agonists or depolarization result in no appreciable phosphate incorporation into PKC sites (Colburn et al. 1988; Kamm et al. 1989; Singer et al. 1989). Thus, sites other than serine-19 of the RLC are not phosphorylated in intact smooth muscle tissues.

Smooth muscle myosin is phosphorylated by CaMK II on the RLC at the same site phosphorylated by MLCK, with resultant enhancement of its actin-activated MgATPase activity (Edelman et al. 1990). The importance of this phosphorylation in regulating smooth muscle contraction has not been thoroughly investigated. The K_m of CaMK II for RLC is three to seven times that of MLCK, and the V_{max} 1/20 that of MLCK; moreover, MLCK has a hundredfold greater affinity for Ca^{2+}–CaM than CaMK II, potentially precluding a role for CaMK II in regulation. Experiments in which smooth muscle cells were treated with inhibitors of CaMK II showed that RLC phosphorylation at particular $[Ca^{2+}]_i$ was enhanced as opposed to being inhibited (Tansey et al. 1994; Word et al. 1994). This is brought about indirectly by inhibition of MLCK phosphorylation by CaMK II. Thus, CaMK II does not appear to play a direct role in phosphorylating RLC. Mechanisms regulating the Ca^{2+} sensitivity of RLC phosphorylation are discussed in Section 3.3.1.

Myosin heavy chain has been shown to be a substrate for several protein kinases, including PKC (Carroll and Wagner 1989; Kawamoto et al. 1989; Ludowyke et al. 1989; Ikebe and Reardon 1990; Conti et al. 1991), CK II (Trotter 1982; Murakami et al. 1984; Barylko et al. 1986; Kuznicki and Filipek 1988), and CaM kinase II (Keller and Mooseker 1982; Rieker et al. 1987). The phosphorylation sites on the heavy chain are located within the tail domain, and in most cases are in close proximity to the carboxyl terminus (Cross and Vandekerckhove 1986). In *Dictyostelium* and lower eukaryotic cells, heavy chain phosphorylation affects the interaction of myosin with actin as well as the assembly of myosin into filaments (reviewed in Tan et al. 1992). In rat basophilic leukemia cells and human platelets, activation of PKC results in increased levels of myosin heavy chain phosphorylation (Nachmias et al. 1987; Kawamoto et al. 1989), and it has been suggested that this phosphorylation may destabilize myosin filaments in these cells (Adelstein et al. 1990). Myosin heavy chains from macrophages, Ehrlich ascites tumor cells (Kuznicki and Filipek 1988), bovine brain (Murakami et al. 1990), and bovine aortic smooth muscle cells (Kelley and Adelstein 1990) have been shown to be phosphorylated by casein kinase II (CK II). Like PKC, the site of phosphorylation by CK II is located near the tip of the myosin tail but is distinct from the phosphorylation site by PKC (Conti et al. 1991). SM-1, but not SM-2 (smooth muscle myosin isoforms, see Section 2.1.3), can be phosphorylated by CK II (Kelley and Adelstein 1990). Phosphorylation does not affect filament formation, velocity of movement of actin filaments by myosin in *in vitro* motility assays, actin-activated MgATPase activity, or myosin conformation *in vitro* (Kelley and Adelstein 1990). In fact, smooth muscle myosins from chicken gizzard and rat aorta contain neither the PKC nor the CK II phosphorylation sites (Conti et al. 1991). Whether CK II phosphorylation affects some aspect of myosin function *in vivo* remains to be determined; however, there is no appreciable MHC phosphorylation in intact smooth muscle during relaxation or contraction (Colburn et al. 1988). In postconfluent smooth muscle cells in culture, though, significant levels of phosphorylated MHC are found (0.7 mol phosphate per 1 mol 204-kDa smooth muscle and 0.8 mol phosphate per 1 mol 196-kDa nonmuscle heavy chain) (Kawamoto and Adelstein 1988; Kamm et al. 1989). Taken together, under physiological conditions, phosphorylation of the RLC is the primary regulatory event in the initiation of smooth muscle contraction, and heavy chain phosphorylation is probably not involved in the regulation of contraction and relaxation in smooth muscle tissues. Phosphorylation of the RLC in sites other than serine-19 probably does not play an important role in the physiologic function of smooth muscle.

2.1.3. Myosin contents and isoforms

Conventional myosins may be broadly categorized into two subtypes: sarcomeric and nonsarcomeric. The biochemical properties of these are illustrated in Figure 5, with the former described as striated muscle myosin and the latter as smooth and nonmuscle myosin. Smooth muscles contain approximately 1/4 of the myosin in skeletal muscles, with cellular concentrations estimated at 30 μM and 120 μM, respectively (Murphy et al. 1974). Nevertheless, smooth muscle cells generate force per cross-sectional area comparable to skeletal muscle (Murphy et al. 1974). Greater stress per amount of myosin arises in part from greater numbers of crossbridges working in parallel because of longer myosin filaments (Ashton et al. 1975) and in part from enhanced force generation by smooth muscle myosin itself (VanBuren et al. 1994).

Isoforms of myosin arise from the specific subunit composition of heavy and light chains. Each of the four major nonsarcomeric MHC isoforms has been found to exist in vertebrate smooth muscle: SM-1 and SM-2, the smooth muscle isoforms found in abundance (Rovner et al. 1986b; Nagai et al. 1989), and MHC-A and MHC-B, the nonmuscle isoforms expressed primarily during development and in cultured cells (Rovner et al. 1986a; Gaylinn et al. 1989; Kuro-o et al. 1991; Frid et al. 1993). Whereas MHC-A and MHC-B are encoded by two distinct genes (Simons et al. 1991), SM-1 and SM-2 are alternatively spliced products of a single gene (Babij and Periasamy 1989). Alternative RNA splicing in the 3' region results in a 43–amino acid extension of the carboxyl terminus of SM-1. SM-1 and SM-2 differ in apparent molecular weight on SDS-PAGE gels with SM-2 (~200 kDa) migrating faster than SM-1 (~204 kDa). Although the heavy chain isoforms can form either homodimers (SM-1–SM-1 or SM-2–SM-2) or heterodimers (SM-1–SM-2) (Kelley et al. 1992; Tsao and Eddinger 1993), the heavy chain isoforms appear to exist as homodimers in smooth muscle (Kelley et al. 1992). Although there are many intriguing correlations between stress development and shortening velocities with various patterns of SM-1 and SM-2 expression patterns (Rovner et al., 1986a; Kawamoto and Adelstein 1987; Mohammad and Sparrow 1988; Sparrow et al. 1988; Schildmeyer and Seidel 1989; Eddinger and Wolf 1993; Hewitt et al. 1993), the functional significance of the two smooth muscle isoforms is uncertain.

Recently, additional smooth muscle MHC isoforms, produced by alternative splicing in the 5' region, have been characterized (Hamada et al. 1990; Babij 1993; Kelley et al. 1993; White et al. 1993). This alternative splicing produces an isoform that contains seven amino acids inserted in the myosin head (QGPSLAY, rabbit; QGPSFAY, rat; QGPSFSY, avian) in close proximity to the nucleotide binding site. This isoform has been termed a *visceral* isoform because it is found in smooth muscle of the gut but not in tonic vascular smooth muscle. The seven–amino acid insert confers higher MgATPase activity and faster movement of actin filaments in *in vitro* motility assays (Kelley et al. 1993). The faster *in vitro* motility of the inserted myosin, however, contained only the acidic isoform of the 17-kDa essential light chain that has also been associated with faster actomyosin kinetic properties, and the insert could be found in either SM-1 or SM-2 (Kelley et al. 1993). Exchange of the 17-kDa acidic isoform into avian aortic heavy chain that does not contain the head insert, however, does not increase actomyosin motility, implying that the faster kinetic properties of the visceral isoform are due to the insert and not the associated 17-kDa ELC isoform. Other studies using porcine aortic myosin revealed that exchange of the acidic 17-kDa ELC does indeed result in increased MgATPase activity (Hasegawa and Morita 1992). Thus, the relative contribution of heavy or light chain isoforms to faster MgATPase rates and physiologic rates of contraction and relaxation in various smooth muscle is not entirely clear.

It has been suggested that the seven–amino acid insert near the nucleotide binding site reduces the affinity for – and accelerates the release of – ADP from crossbridges (Somlyo 1993; Khromov et al. 1995). In support of this hypothesis is the finding that myosin of the phasic rabbit bladder smooth muscle contains the seven–amino acid insert, and myosin from this smooth muscle also has a lower affinity for MgADP than myosin from the rabbit femoral artery. Like many other phasic smooth muscles, however, the bladder myosin contains increased amounts of the acidic 17-kDa ECL which is sparse in the artery (Khromov et al. 1995). Thus, isoforms of the 17-kDa ELC, isoforms of MHC, or both may play a role in rates of contraction and relaxation in various smooth muscles.

2.2. Calmodulin

Calmodulin (CaM) is a 17-kDa Ca^{2+}-binding regulatory protein of 148 residues involved in a variety of intracellular Ca^{2+}-dependent signaling pathways. The binding of four Ca^{2+} ions results in conformational changes that allow CaM to modulate the activities and functions of a large number of enzymes and proteins, including MLCK. Ca^{2+}–CaM binds to many target proteins with high affinity. For example, the dissociation constant for MLCK is approximately 1 nM (Kamm and Stull 1985b). CaM exhibits broad specificity to basic amphiphilic α-helical peptides (O'Neil and DeGrado 1990). Smooth muscle cells contain large amounts of CaM (10–40 μM), and CaM is present in excess of MLCK (1–3 μM) (Rüegg et al. 1984; Tansey et al. 1994). The concentrations of CaM and kinase are 10,000 and 1,000 times greater, respectively, than the dissociation constant of Ca^{2+}–CaM for MLCK. Furthermore, the dissociation constants of CaM for Ca^{2+} are decreased by a factor of 10 to 50 when CaM binds to target proteins such as MLCK (Olwin et al. 1984). These cellular and biochemical properties result in positive cooperative activation of MLCK at Ca^{2+} concentrations less than 1 μM, which is the range associated with myosin light chain phosphorylation in smooth muscle cells. Recently, it has been found that most of CaM is bound in smooth muscle cells and that the free CaM concentration may be ≤1 μM (Tansey et al. 1994; Luby-Phelps et al. 1995; Zimmermann et al. 1995). Low values of free CaM suggest that Ca^{2+}–CaM may limit the extent of MLCK activation in smooth muscle cells. This finding is physiologically important because, in the presence of high concentrations of Ca^{2+}–CaM, the diminished Ca^{2+}–CaM affinity brought about by phosphorylation of MLCK would not affect the rate or extent of RLC phosphorylation in smooth muscle (Kamm and Stull, 1985b; see Section 3.3.1). Data from smooth muscle tissues and cells in culture, however, indicate that phosphorylation of MLCK at its regulatory site results in a decreased sensitivity to activating Ca^{2+} for RLC phosphorylation (Tansey et al. 1992; Word et al. 1994).

2.3. Myosin light chain kinase

MLCKs are Ca^{2+}–CaM-regulated serine or threonine protein kinases that catalyze the phosphorylation of myosin RLC. Two isoforms (skeletal and smooth/nonmuscle) have been well described (Stull et al. 1990a, 1996). In skeletal muscle, phosphorylation of RLC by MLCK has a modulatory role that accounts for contraction-induced potentiation of isometric twitch tension (Sweeney et al. 1993). In smooth muscle, phosphorylation of the 20-kDa RLC is important for the initiation of contraction (reviewed in Kamm and Stull 1985b). Inactive MLCK exists as a monomer with an apparent molecular mass ranging from 130,000 to 150,000 when isolated from smooth muscle tissues of various animal species. This isoform is expressed in both smooth and nonmuscle cells (Gallagher et al. 1991). The relative mobilities determined by SDS-PAGE are slightly larger than the masses predicted by the deduced primary sequences (Gallagher et al. 1991 1995; Kobayashi et al. 1992). For example, the rabbit smooth muscle MLCK cDNA encodes a protein with a predicted molecular mass of 126 kDa, whereas the recombinant and tissue forms migrate at 152 kDa on SDS-PAGE. In addition, recent studies indicate by immunoreactive properties that a 210-kDa protein, termed *embryonic* MLCK, is a unique form of MLCK expressed in mammalian embryonic tissues, stem cells, and proliferating cultured cells (Gallagher et al. 1995). In avian tissues, a high–molecular mass isoform of MLCK appears to arise as an alternative splice variant derived from the smooth/nonmuscle MLCK gene (Fisher and Ikebe 1995; Watterson et al. 1995).

MLCK is activated by the binding of 1 mol Ca^{2+}–CaM to 1 mol kinase. However, recent molecular studies suggest that the CaM lobes make distinct contributions to binding and activation of smooth muscle MLCK (Persechini et al. 1994). Both hydrophobic and electrostatic interactions are essential for CaM binding and activation in the smooth muscle kinase.

A comparison of the cDNAs of vertebrate smooth and nonmuscle MLCKs reveals a high degree of sequence homology. The domain organization of MLCK is schematically illustrated in Figure 6. The first 80 residues contain an actin-binding domain that is highly conserved in smooth muscle kinases cloned from three species (Kanoh et al. 1993). In fact, the first six residues predicted by the 5 coding region for the chicken, rabbit, and bovine cDNAs are identical, suggesting that all of the smooth muscle MLCKs have the same amino terminus. After the 80 N-terminal residues, mammalian MLCKs contain an insert not present in the avian smooth/nonmuscle MLCKs (Gallagher et al. 1991; Kobayashi et al. 1992; Potier et al. 1995). This insert consists of 5 to 20 tandem copies of a 12-residue repeat of unknown function. After the tandem repeats, four class-I and class-II structural motifs are present in smooth/nonmuscle MLCKs, but not in the skeletal muscle enzymes. These motifs are repeat structures of approximately 100 amino acids that are similar to those in the related family of giant muscle proteins, twitchin and titin (Olson et al. 1990; Labeit et al. 1992). Motif I is related to the type-III module of fibronectin (Campbell and Spitzfaden 1994); motif II belongs to the C-II set of the immunoglobulin superfamily. The third type-II motif is located at the C terminus of MLCK and, in some smooth muscle tissues, is expressed as an independent protein, telokin (Ito et al. 1989; Gallagher and Herring 1991; Collinge et al. 1992). Telokin is a 24-kDa acidic myosin-binding protein that has been shown to stabilize unphosphorylated smooth muscle myosin filaments (Shirinsky et al. 1993). Full-length MLCK is localized to microfilament bundles in cultured cells, whereas a truncation mutant containing only the catalytic and regulatory domains is diffusely distributed in the cytoplasm (Lin et al. 1996). More recent experiments have shown that the N terminus (actin-binding domain), but not the C terminus (telokin domain), is sufficient to result in localization to stress fibers. Thus, the function of the telokin domain in MLCK is not yet defined.

The MLCK gene has been mapped to a single somatic locus in the human genome (Potier et al. 1995). It is likely to be a large gene, exceeding 50 kb of genomic DNA, including several introns (Potier et al. 1995). The telokin gene is included in the MLCK gene with an independent promoter within the intronic sequence of the MLCK gene (Gallagher and Herring 1991; Collinge et al. 1992; Yoshikai and Ikebe 1992). Recent reports indicate that in select avian tissues a third transcript is derived from this gene resulting in a higher-mass form of MLCK with an additional N-terminal sequence of unknown function (Fisher and Ikebe 1995; Watterson et al. 1995).

MLCK contains a catalytic core that is highly homologous to other protein kinases. The catalytic cores of these kinases contain two lobes with the smaller lobe binding MgATP, leaving the γ phosphate positioned for catalytic transfer to the RLC. Phosphorylation occurs at a cleft between the two lobes, with the larger lobe providing binding sites for protein substrates. Limited tryptic cleavage of smooth muscle MLCK results in the generation of an active 61-kDa Ca^{2+}–CaM-*independent* MLCK (Pearson et al. 1988; Ikebe et al. 1987, 1989). A larger 64-kDa fragment is still regulated by Ca^{2+}–CaM. These studies indicated that MLCK contains an inhibitory region and a CaM-binding domain (Ikebe et al. 1987) that are removed during partial proteolysis. Synthetic peptides modeled after residues 794–800, including amino acid sequence Tyr-Met-Ala-Arg-Arg-Lys-Trp, inhibit the kinase activity of the constitutively active 61-kDa MLCK fragment (Ikebe et al. 1990). One of these peptides (783–804) is a competitive inhibitor with myosin RLC as substrate, probably because of similarity in structure related to three basic residues. Thus, it has been postulated that the inhibitory region mimics the light chain substrate, and is a pseudosubstrate prototype that binds to the active site of the kinase in the absence of CaM (Kemp et al. 1987). Pseudosubstrates have been proposed as a mechanism of regulating the activity of other allosterically regulated enzymes (Pearson et al. 1988). This mechanism of regulation, however, is not directly proven for MLCK (Shoemaker et al. 1990; Ito et al. 1991; Fitzsimons et al. 1992; Yano et al. 1993). Recent experiments indicate that a more complex model is required (Ikebe 1990; Gallagher et al. 1993; Krueger et al. 1995). Mutation analyses and the recently reported crystal structure of a related kinase, twitchin (Hu et al. 1994), have provided insights into the structural basis of intrasteric regulation of MLCK (for review, see Stull et al. 1996). Seven acidic residues following a distinct path across the surface of the catalytic core are implicated in binding to or near the inhibitory sequence, but not to basic residues in the RLC (Gallagher et al. 1993; Krueger et al. 1995). It appears that autoinhibition involves the folding of a C-terminal extension of the kinase back onto the catalytic core. Additional details of the autoinhibition will be resolved upon atomic resolution and structural analyses of smooth muscle MLCK in the presence and absence of Ca^{2+}–CaM.

2.4. Myosin phosphatase

The activity of myosin phosphatase is required to reverse the activation of smooth muscle myosin by RLC phosphorylation (Figure 1). As $[Ca^{2+}]_i$ falls Ca^{2+}-dependent MLCK is inactivated, shifting the balance of kinase to phosphatase activities resulting in dephosphorylation of myosin light chain and smooth muscle relaxation (Figure 2). In general, descriptions of structure, function, and mechanisms of regulation of kinases are more advanced than those of phosphatases. The identification of myosin phosphatase and mechanisms regulating its activity, as well as their physiological significance, have only been recently recognized.

Serine and threonine protein phosphatases (PP) are classified into four general categories (types 1, 2A, 2B, and 2C) according to specific activators or inhibitors (Cohen 1989). The activities of these protein phosphatases are regulated by binding to inhibitory proteins and regulatory subunits (Cohen 1989). In addition, targeting subunits interact with catalytic subunits to modulate phosphatase activity and determine its subcellular localization. PP1 and PP2A dephosphorylate a large number of phosphoproteins, whereas Ca^{2+}–CaM- or Mg^{2+}- dependent PP2B and PP2C, respectively, are much more specific. Type-1 and type-2A enzymes are classified into two groups, based on whether the enzyme catalyzes the dephosphorylation of the β subunit of phosphorylase kinase (specific for type 1) and on whether it is inhibited by two small heat- and acid-stable cytosolic proteins, inhibitor-1 and inhibitor-2 (also type 1). PP type 2A is characteristically insensitive to inhibitors 1 and 2 and preferentially dephosphorylates the subunit of phosphorylase kinase.

It has recently become clear that a type-1 protein phosphatase (PP1) is responsible for myosin dephosphorylation in smooth muscle. In intact smooth muscle, myosin dephosphorylation is better inhibited by calyculin A (more specific for type 1, Ishihara et al. 1989) than okadaic acid (more specific for type 2A, Takai et al. 1989), and PP type 1 (SMP-IV) dephosphorylates myosin light chain when added to skinned muscle fibers (Hoar et al. 1985). The importance of type-1 phosphatase in this context is further suggested by the finding that purified smooth muscle myosin contains associated PP1-like activity (Nomura et al. 1992; Mitsui et al. 1992).

A number of phosphatases have been purified from smooth muscle. First, four protein phosphatases were purified from gizzard smooth muscle that could dephosphorylate myosin light chain (termed SMP-I, -II, -III, and -IV, Pato and Adelstein 1983a,b). Only SMP-III and-IV dephosphorylated intact myosin. According to the Cohen classification, SMP-I and -II were classified as PP2A and PP2C, respectively (Pato et al. 1983; Pato and Adelstein 1983b). SMP-IV was reported to be a unique smooth muscle phosphatase, composed of subunits of 58 kDa and 40 kDa, that has high activity toward intact myosin (Pato and Kerc 1985) and high affinity to thiophosphorylated smooth muscle myosin (Sellers and Pato 1984). Originally, SMP-IV could not be categorized according to the Cohen criteria because it was reported that SMP-IV dephosphorylated the β subunit of phosphorylase kinase but was not sensitive to inhibitor-2 (Pato and Kerc 1985). SMP-III was classified as type-2 because it was resistant to inhibitor-2 (Tulloch and Pato 1991). Subsequently, it was found that SMP-IV is inhibited by inhibitor-1 with half-maximal inhibition at 10 nM and that SMP-IV can also be inhibited by inhibitor-2, although a higher concentration is required (Mitsui et al. 1992). Thus, SMP-IV was classified as a type-I phosphatase. The 58-kDa fragment was subsequently found to represent the N-terminal fragment of a 130-kDa myosin-binding subunit of smooth muscle myosin phosphatase (Okubo et al. 1994) that had been recently cloned and characterized (Shimizu et al. 1994; Shirazi et al. 1994).

A smooth muscle myosin–associated phosphatase (PP1) holoenzyme has been purified and cloned from avian and mammalian smooth muscles (Alessi et al. 1992; Shirazi et al. 1994). This phosphatase consists of subunits of 130, 37, and 20 kilodaltons. The 37-kDa subunit is the β isoform of the PP1 catalytic subunit, while the 130-and 20-kDa subunits form a complex that enhances dephosphorylation of myosin. The relationships between SMP-IV, the myosin-bound PP1-like enzyme, and the myosin-bound form of PP1 ($PP1_M$) are not entirely clear. Their similar properties suggest that they may be different versions of the same enzyme generated by partial proteolysis and dissociation–reassociation of subunits during the different purification protocols (Mumby and Walter, 1993). Nevertheless, experimental evidence is supportive of the view that PP1 (SMP-IV) protein phosphatase is the dominant PP that catalyzes the dephosphorylation of myosin light chains in smooth muscle. The possibility exists, however, that other protein phosphatases may be involved (Takai et al. 1989).

Type-1 protein phosphatases consist of multimeric structures of a catalytic subunit complexed to a number of accessory subunits. The free catalytic subunit has not been detected in cell or tissue extracts, suggesting that, *in vivo,* few (if any) uncomplexed catalytic subunits exist (Mumby and Walter 1993). Formation of complexes plays an important role in the regulation of catalytic subunit activity (Table 1). The first native form of PP1 to be characterized was isolated as a MgATP-dependent PP composed of a complex between the PP1 catalytic sub-

Table 1. *Type-1 protein phosphatase (PP1) enzymes*

Enzyme	Subunits
$PP1_I$/MgATP-dependent	I2 (23 kDa)·C (37 kDa)
$PP1_G$	G (121 kDa)·C (37 kDa)
$PP1_M$	M1 (130 kDa)·M2 (20 kDa)·Cβ (37 kDa)
$PP1_N$	N (16–18 kDa)·C (37 kDa)
$PP1_C$	C (37 kDa)

Key: $PP1_I$, catalytic subunit (C) and inhibitor 2 (I2); $PP1_G$, catalytic subunit and 121-kDa glycogen subunit; $PP1_M$, catalytic subunit and myosin binding subunits (M1 and M2); $PP1_N$, catalytic subunit and 16–18-kDa nuclear chromatin-associated inhibitors (N).

unit and inhibitor-2 (Ballou et al. 1983; Jurgensen et al. 1984; Tung and Cohen 1984). This form of PP1 has low intrinsic activity, and is activated by phosphorylation of inhibitor-2 by glycogen synthase kinase-3. This form of PP1 has also been referred to as $PP1_I$. $PP1_G$ consists of a stoichiometric complex between the PP1 catalytic subunit and a glycogen binding protein termed the *G subunit* (Stralfors et al. 1985). The G subunit targets the catalytic subunit to membranes of the sarcoplasmic reticulum by a hydrophobic domain near the C terminus (Hubbard et al. 1990). PP1 activity associated with the sarcoplasmic reticulum in skeletal and cardiac muscle is similar, if not identical, to $PP1_G$.

The PP1 catalytic subunit is also associated with the actomyosin contractile apparatus in skeletal and cardiac muscles by a myosin-binding protein termed the *M subunit*. Association of PP1 with actomyosin enhances activity toward the phosphorylated light chain subunit of myosin (Chisholm and Cohen 1988). In smooth muscle, $PP1_M$ is composed of $PP1_C$ complexed to 130- and 20-kDa proteins; this is known as the *M complex*. Both skeletal and smooth muscle M subunits cause a severalfold increase in $PP1_C$ catalytic activity toward smooth muscle myosin, suppression of activity toward phosphorylase, and decreased sensitivity to the cytosolic inhibitor proteins. The significance of these biochemical measurements was recently confirmed in permeabilized smooth muscle tissues, where rates of relaxation were significantly enhanced by perfusion with $PP1_M$–$PP1_C$ complex compared with $PP1_C$ subunit alone (Shirazi et al. 1994; see also Figure 7). Myosin-targeting subunits differ in skeletal and smooth muscles. Skeletal M subunit enhances the dephosphorylation of smooth and skeletal muscle myosin by factors of 3 and over 20, respectively. Smooth muscle M complex does not enhance the dephosphorylation of skeletal muscle myosin (Dent et al. 1992).

Myosin light chain dephosphorylation and relaxation are enhanced by the complex of $PP1_C$ with M subunits (Figure 7). Dissociation of catalytic subunit from M subunits (by phosphorylation or other second messengers) leads to inactivation of myosin phosphatase and enhanced RLC phosphorylation. Inactivation of $PP1_C$ occurs by formation of complexes with cytosolic inhibitor proteins. These complexes may serve as a reservoir for $PP1_C$ that may be liberated after phosphorylation of the inhibitor by other protein kinases. The identification of the kinases

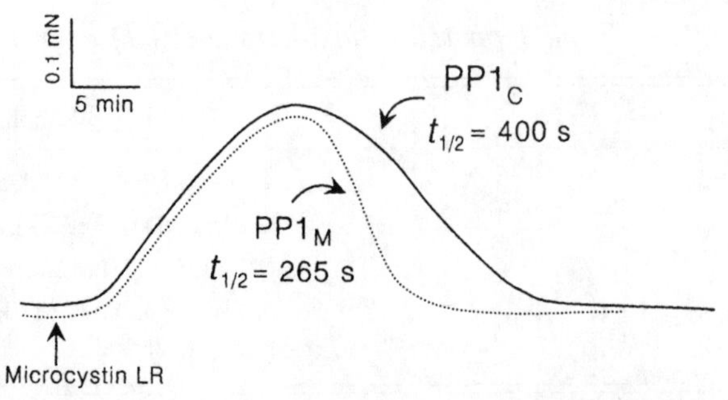

Figure 7. Rates of relaxation of microcystin LR-induced contractions in Triton X–permeabilized smooth muscle. Fibers were treated with the phosphatase inhibitor microcystin-LR (1 μM) to induce Ca^{2+}-independent increases in force. Permeabilized tissues were then treated with the catalytic subunit of PP1 (37 kDa) to effect relaxation (solid line). Rates of relaxation were increased with addition of the catalytic subunit complexed to the 20- and 130-kDa subunits of $PP1_M$ (dotted line). [Redrawn from Shirazi et al. (1994).]

and phosphatases involved in the regulation of these protein–protein interactions will provide much new information on the regulation of smooth muscle contraction and relaxation.

2.5. Thin filament system

Native thin filaments isolated from smooth muscle tissues activate myosin MgATPase activity and are regulated in a Ca^{2+}-dependent manner (Marston and Lehman 1985). Native thin filaments appear to be organized into two distinct classes, both of which contain actin and tropomyosin, but which are enriched in either caldesmon or calponin (Lehman 1991; Makuch et al. 1991). Immunocytochemical studies have shown that these two thin filament populations are localized in different cellular domains, as illustrated in Figure 4 (North et al. 1994). The presence of Ca^{2+}-dependent thin filament modulatory proteins in all smooth muscles suggests that thin-filament–linked regulation plays a role in smooth muscle contraction, although the physiological significance of thin-filament regulation remains to be established.

2.5.1. Actin

All cells contain actin, which supports a variety of motile and cytoskeletal functions. Actin is a highly conserved protein of 42,000 kDa that polymerizes in the presence of physiological salt concentrations to form a two-stranded F-actin helix, which is the backbone of the thin filament (Pollard 1990). Six different actin isoforms have been identified as products of distinct genes in vertebrate species (Korn 1982). Slight differences have been demonstrated by isoelectric focusing among various actin isoforms; the most acidic form (based on isoelectric points) is α, followed by β and γ (Whalen et al. 1976; Rubenstein and Spudich 1977). These variants are further categorized as follows: α-skeletal, α-cardiac, and α-smooth; β-cytoplasmic; and γ-enteric and γ-cytoplasmic. The actin isoforms exhibit 95% amino acid–sequence homology and, in general, the properties of these actins are similar. Most smooth muscles contain a mixture of actin isoforms including β-cytoplasmic actin (Whalen et al. 1976; Rubenstein and Spudich 1977; Hartshorne and Gorecka 1980). Alpha actin is the predominant actin isoform in vascular smooth muscle, whereas the γ isoform is more abundant in enteric and uterine smooth muscle (Fatigati and Murphy 1984). The β-cytoplasmic isoform appears to be localized to the cytoskeletal domain, whereas the α-smooth muscle actin is enriched in the contractile domain (Figure 4).

There have been numerous reports on proteins which bind to actin that either modify polymerization or induce severing of actin filaments (Craig and Pollard 1982; Korn 1982; Weeds 1988), thus influencing the assembly and disassembly of thin filaments. Their role in the modification of specific thin filaments in smooth muscle, and thus their effects on contractile function, are not known.

Smooth muscles contain twice the amount of actin as skeletal muscle, with cellular concentrations estimated at 1 mM and 500 μM, respectively (Murphy et al. 1974). The mole ratios of actin to tropomyosin are similar for smooth and skeletal muscles, indicating that F actin is saturated with tropomyosin (Murphy et al. 1974). Taking into account the relatively low myosin contents of smooth muscle, its 15:1 ratio of thin to thick filaments differs markedlyfrom the 6:1 ratio found in striated muscle.

2.5.2. Tropomyosin

Although proteins equivalent to the Ca^{2+} regulatory complex troponin (found in cardiac and skeletal muscle) are not present in smooth muscle, tropomyosin is associated with the thin filaments of both skeletal and smooth muscle at a stoichiometry of approximately one tropomyosin molecule for each seven actin molecules. The structure of tropomyosin is conserved among smooth and skeletal types, and – despite the absence of troponin – tropomyosin retains its troponin- and actin-binding sites in smooth muscle (Parry and Squire 1973; Hartshorne et al. 1977). Tropomyosin is a dimer (M_r = 66,000) of elongated, α-helical polypeptides coiled around each other. The protein is located in the grooves between two strands of the actin double helix. It has been shown that smooth muscle tropomyosin modifies actomyosin MgATPase activity (in general, a tendency for activation) (Hartshorne and Gorecka 1980). The position of tropomyosin on smooth muscle thin filaments (as studied by X-ray diffraction) is altered upon activation of smooth muscle (Vibert et al. 1972; Parry and Squire 1973). Thus, it is postulated that tropomyosin may play a role in the regulation of the formation of actomyosin crossbridges; however, the biochemical mechanism in smooth muscle (in the absence of troponin) is not completely understood.

2.5.3. Caldesmon and calponin

Detailed information regarding the biochemical and molecular properties of the thin-filament proteins caldesmon and calponin, as well as their roles in the regulation of smooth muscle function, are provided elsewhere in this volume (see Chapter 7). There are two primary problems in the understanding of the physiological role of caldesmon (and calponin) in the regulation of force development. First, the affinity of caldesmon or calponin for Ca^{2+}–CaM is two to three orders of magnitude lower than that of MLCK for Ca^{2+}–CaM. Thus, it is unlikely that CaM will bind to these proteins at $[Ca^{2+}]_i$ sufficient for full or partial activation of MLCK. Although the low affinity of caldesmon and calponin for Ca^{2+}–CaM brings into doubt the importance of CaM in thin filament regulation, it is possible that other Ca^{2+}-binding proteins may act to sensitize smooth muscle thin filaments (Pritchard and Marston 1993). Second, at resting values of $[Ca^{2+}]_i$, addition of a partially proteolyzed, CaM-independent MLCK to skinned fibers results in maximal force of contraction (Walsh et al. 1982). The modulatory role of caldesmon may be difficult to discern in these experiments, however, because caldesmon undergoes significant proteolysis in skinned muscle preparations. Nevertheless, microinjection of CaM-independent MLCK into single smooth muscle cells also results in cell shortening (Itoh et al. 1989). The binding of calponin or caldesmon to the thin filaments (in the presence of low Ca^{2+} concentrations) should act to inhibit contractions induced by phosphorylated myosin; yet, the muscle contracts. Thus, more investigation is needed to establish a physiologic role for the thin-filament regulatory proteins in smooth muscle contraction.

3. Regulation of contraction in smooth muscle cells and tissues

In all muscle cells, $[Ca^{2+}]_i$ exerts primary control over the initiation, time course, and extent of force or shortening. In smooth muscles, the excitatory stimulus leads to elevations in $[Ca^{2+}]_i$ by influx of Ca^{2+} through voltage-dependent Ca^{2+} channels in the sarcolemma or by release of Ca^{2+} through $InsP_3$-dependent Ca^{2+} channels in the sarcoplasmic reticulum. Particular pathways of excitation–contraction coupling differentially affect Ca^{2+} homeostasis (Figure 8). Specific mechanisms of Ca^{2+} homeostasis in smooth muscle are discussed elsewhere (see Chapter 2 in this volume). Elevation of intracellular $[Ca^{2+}]_i$ sets in motion activation of the contractile elements, primarily or exclusively by the Ca^{2+}–CaM-dependent phosphorylation of myosin, and there is for all smooth muscles a positive relation between $[Ca^{2+}]_i$ and force. Alterations in the Ca^{2+} sensitivity of force may arise from effects on RLC phosphorylation or other contractile proteins (Figure 8).

3.1. Activation of smooth muscle contraction

Smooth muscle tissues that utilize different sources of activating Ca^{2+} have been used to study the precise temporal relationships between increases in $[Ca^{2+}]_i$, RLC phosphorylation, and force development in intact smooth muscle (Word et al. 1994). In bovine trachealis, electrical stimulation results in synchronous release of neurotransmitters from the distal terminal neurons, hydrolysis of glycerophosphoinositides, release of Ca^{2+} from internal stores, and the generation of contractile force (Miller-Hance et al. 1988). In myometrial tissues, electrical depolarization results in rapid and synchronous depolarization of smooth muscle cells, influx of extracellular Ca^{2+}, and force development. The homogeneity of activation in these two preparations facilitated investigations of precise temporal

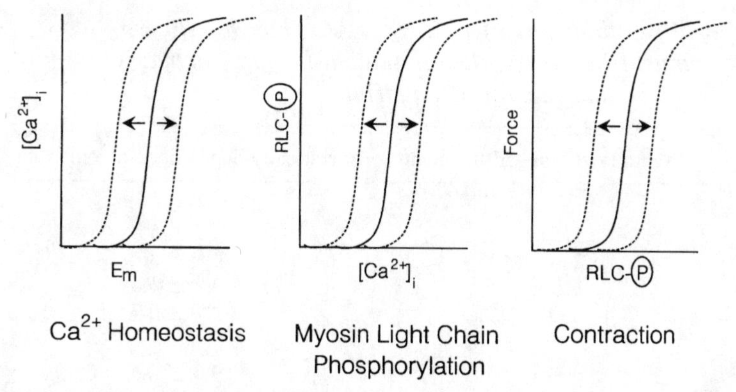

Figure 8. Regulation of smooth muscle contraction occurs at several steps. *Left:* Stimulation is generally associated with alterations in Ca^{2+} homeostasis. Intracellular Ca^{2+} concentration ($[Ca^{2+}]_i$) increases as a function of membrane potential (E_m). Agonists may increase $[Ca^{2+}]_i$ by altering E_m or by changing the relationship between E_m and $[Ca^{2+}]_i$. *Center:* $[Ca^{2+}]_i$ determines the fraction of phosphorylated myosin RLC (RLC-P). The sensitivity of myosin RLC-P to $[Ca^{2+}]_i$ can be modified by the action of second messengers. For example, inhibition of myosin phosphatase shifts the relationship between $[Ca^{2+}]_i$ and RLC-P to the left. Inhibition of MLCK activity shifts the curve to the right. *Right:* Steady-state contractile force is generally dependent on myosin RLC-P. Second-messenger pathways can also modify the sensitivity of force to myosin RLC-P, perhaps by altering the action of thin-filament proteins.

relationships among $[Ca^{2+}]_i$, protein phosphorylation, and mechanical output. Increases in $[Ca^{2+}]_i$ preceded light chain phosphorylation and force development in both phasic and tonic smooth muscles. In both preparations, electrical stimulation resulted in immediate increases in $[Ca^{2+}]_i$, reaching maximal values in 9 s (Table 2). After a period of latency (200–350 ms), significant increases in RLC phosphorylation and force occurred. It was hypothesized that the rate of activation of MLCK by Ca^{2+}–CaM may be an important factor in determining latency (Kamm and Stull 1986; Miller-Hance et al. 1988).

The kinetics of RLC phosphorylation during initiation of contraction in intact trachealis are described by a pseudo–first-order rate of 1 s^{-1}, showing no evidence of ordered or cooperative phosphorylation of the RLCs on myosin (Kamm and Stull 1986; Persechini et al. 1986). The rate of RLC phosphorylation, however, is more rapid than either force development or $[Ca^{2+}]_i$ increase in both trachealis and myometrium (Table 2). During force development, the extent of light chain phosphorylation reaches steady-state values while $[Ca^{2+}]_i$ continues to increase, indicating diminished sensitivity of MLCK to activation by Ca^{2+}–CaM. Desensitization arises following phosphorylation of MLCK by CaMK II, as discussed in Section 3.3.1. These results indicate that MLCK is sensitive to small increases in intracellular Ca^{2+} during the initiation of contraction in both tonic and phasic smooth muscle. After initiation of contraction, the enzyme becomes desensitized to Ca^{2+}–CaM and limits the extent of light chain phospho-

Table 2. *Temporal relationships in [Ca^{2+}]$_i$, RLC phosphorylation, and force development in electrically stimulated bovine trachealis and human myometrium*

	Neurally stimulated trachealis	Electrically depolarized myometrium
[Ca^{2+}]$_i$		
resting (nM)	115 ± 2	121 ± 5
maximal (nM)	254 ± 4	495 ± 36
t_{max} (s)	8.9 ± 1.5	9.2 ± 1.6
$t_{1/2}$ (s)	2.0 ± 0.3	2.7 ± 0.7
Force		
$t_{1/2}$ (s)	4.9 ± 1.1	9.6 ± 0.6
RLC-P		
resting (%)	4.9 ± 0.4	1.2 ± 0.4
maximal (%)	67 ± 2.3	53.6 ± 3.9
t_{max} (s)	2.0	2.0
Latency		
[Ca^{2+} to force (ms)]	343 ± 80	241 ± 61

rylation. Upon cessation of electrical stimulation RLC phosphorylation declines more rapidly than force with half-times on the order of 2.5 s and 5 s, respectively (Kamm and Stull 1985a).

Experiments in permeabilized smooth muscles using photolytically released Ca^{2+} and ATP to produce step changes in concentration have yielded additional insights regarding kinetics of activation (Somlyo et al. 1988; Somlyo and Somlyo 1994; Zimmermann et al. 1995). In agreement with the foregoing, latency to force development was ~200 ms in permeabilized rabbit portal vein following photolytic release of Ca^{2+} in the presence of 40 μm CaM; in addition, photolysis of caged ATP following pre-equilibration with Ca^{2+}–CaM shortened the latent period to 40 ms, supporting the notion that activation and isomerization of MLCK contribute to latency (Zimmermann et al. 1995). Exogenous CaM accelerated the rate of RLC phosphorylation in response to photolytic release of Ca^{2+} at both 0 μM and 40 μM added CaM, indicating that the slower rate observed in intact cells may result in part from recruitment of CaM from a slowly diffusable component (Tansey et al. 1994; Luby-Phelps et al. 1995; Zimmermann et al. 1995).

RLC phosphorylation results in crossbridge attachment, cycling, and force development and/or shortening (Kamm and Stull 1986; Murphy 1994). Rates of contraction initiated by photolytic release of ATP in the presence of Ca^{2+} are accelerated tenfold by prior thiophosphorylation of RLC, suggesting that (1) initial rates of contraction are dominated by the myosin phosphorylation reaction, and (2) the dynamic reactions of the contractile machinery immediately follow phosphorylation (Somlyo et al. 1988). Similar studies showed that phasic muscles develop force more rapidly than tonic muscles (Horiuti et al. 1989), and that this difference in crossbridge cycling rates may be accounted for by a higher ADP affinity (therefore slower off-rate) of force-generating crossbridges from tonic than phasic smooth muscles (Khromov et al. 1995).

3.2. Regulation of force maintenance by phosphorylated and nonphosphorylated crossbridges

Direct dependence of force on the amount of phosphorylated myosin does not adequately describe all of the physiological and cellular processes involved in smooth muscle contraction (Kamm and Stull 1985b; Hartshorne 1987; Hai and Murphy 1989). Although Ca^{2+}-dependent myosin phosphorylation is the primary effector of smooth muscle contraction, it is clear from a variety of experiments that nonphosphorylated myosin crossbridges participate in the contractile process. This was first hypothesized by Murphy and co-workers (reviewed in Murphy 1994) following the discovery that myosin RLC phosphorylation in arterial smooth muscle increased only transiently in response to stimulation, leading to high sustained isometric force with only small amounts of phosphorylated "active" myosin. The decline in myosin light chain phosphorylation has also been correlated with a decrease in maximal velocity of shortening, an estimate of crossbridge cycling rates. Although initially the importance of RLC phosphorylation for force maintenance was disputed, subsequent studies showed that in the steady state there is a steep dependence of force on phosphorylation between 0.07 and 0.2 moles of phosphate per mole of RLC, with maximal force achieved between 0.2 and 0.3 mol/mol (Di Blasi et al. 1992; reviewed in Murphy 1994). Assuming that maximal force reflects the activity of most of the crossbridges, this result indicates that nonphosphorylated cross bridges contribute to force maintenance. It is now generally accepted that nonphosphorylated crossbridges participate in contraction; however, the fundamental properties and regulation of these crossbridges remain the subjects of active investigation.

Theories for regulation of the crossbridge cycle in smooth muscle are illustrated in Figure 9. The simplest scheme describing the cyclic interaction of myosin with actin is a two-state model in which actin (A) and myosin (M) are bound (A·M) or not bound (A+M). This simple two-state cycle, illustrated in panel C of Figure 9, is applicable to striated muscle myosins that are constitutively active. In striated muscle, Ca^{2+}-dependent regulation is mediated by the troponin–tropomyosin complex on the thin filaments. Because smooth and nonmuscle myosins are regulated by phosphorylation, the crossbridge cycle must include four states – as first formalized by Murphy and co-workers (reviewed in Murphy 1994) and illustrated in panel A of Figure 9. Both attached and detached myosins are substrates for MLCK and myosin phosphatase. Myosin phosphorylation activates crossbridge attachment and rapid cycling. In this model, dephosphorylated attached crossbridges (latch bridges) detach slowly and thus, along with cycling phosphorylated crossbridges, contribute to force maintenance. Slow detachment rates following dephosphorylation may result from relatively slow release of ADP from force-generating crossbridges (Khromov et al. 1995). The latch-bridge hypothesis requires only one regulatory mechanism, phosphorylation of myosin RLC; nonphosphorylated attached crossbridges arise only from previously phosphorylated attached bridges, and not via direct attachment. Experiments with permeabilized fibers have led to a second model, illustrated in panel B of Figure 9. Here, cycling nonphosphorylated crossbridges are regulated by a cooperative mechanism in which a few phosphorylated crossbridges pro-

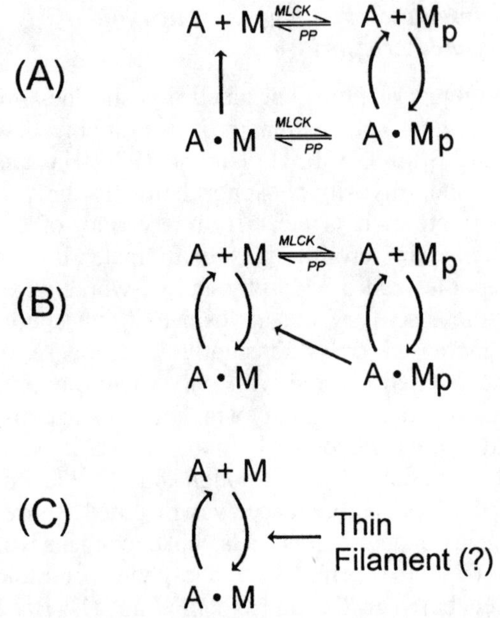

(A)

A + M ⇌ A + Mp (MLCK / PP)

A • M ⇌ A • Mp (MLCK / PP)

(B)

A + M ⇌ A + Mp (MLCK / PP)

A • M A • Mp

(C)

A + M

A • M

Thin Filament (?)

Figure 9. Models for regulation of the crossbridge cycle in smooth muscle. The cyclic interaction of myosin (M) or phosphorylated myosin (Mp) crossbridges with actin (A) in thin filaments is illustrated. A and M alternate between bound (A•M or A•Mp) force-bearing states and unbound (A + M or A + Mp) states. M and Mp are substrates for MLCK and myosin phosphatase (PP), respectively. (A) In the latch model, myosin RLC phosphorylation is required for crossbridge attachment and force generation. (B) In the cooperative model, nonphosphorylated crossbridges are activated to attach and cycle by a small population of phosphorylated crossbridges. (C) In the phosphorylation-independent model, nonphosphorylated crossbridges attach and cycle.

mote attachment and cycling of many nonphosphorylated crossbridges (Somlyo et al. 1988; Kenney et al. 1990; Vyas et al., 1992, 1994). In protocols with permeabilized fibers, myosin RLC is thiophosphorylated, rendering RLC resistant to the action of myosin phosphatase. The full range of forces and ATP consumption are achieved with RLC phosphorylation values similar to those for the steady state in intact arterial muscle (DiBlasi et al. 1992). Under these conditions, phosphorylated crossbridges are prevented from becoming latch bridges, suggesting that nonphosphorylated bridges must attach and cycle (Kenney et al. 1990; Vyas et al. 1992). Cooperative activation of nonphosphorylated bridges by attached phosphorylated crossbridges, in analogy with cooperative effects of crossbridge activation in striated muscle, has been proposed (Somlyo et al. 1988; Vyas et al. 1992). Finally, experiments with skinned fibers have demonstrated that nonphosphorylated rigor crossbridges detached by ATP release and then reattach and cycle (Somlyo et al. 1988). This simple scheme is illustrated in panel C of Figure 9. Rembold and Murphy (1993) have mathematically modeled combinations of these regulatory schemes and conclude that isometric force produced by nonphosphorylated crossbridges could be explained by either the latch model or the cooperative model in which only phosphorylated crossbridges effect activation, leaving Ca^{2+}-dependent RLC phosphorylation as the primary regulatory mechanism. The role of thin-filament regulatory proteins in modulating any or all of these crossbridge cycles remains to be determined, as well as the relative contributions of the phosphorylated, nonphosphorylated, and latch-bridge cycles to force maintenance *in vivo.*

3.3. Regulation of the Ca^{2+} sensitivity of smooth muscle contractile elements

The precise relation between $[Ca^{2+}]_i$ and force (the Ca^{2+} sensitivity of force) may vary according to a particular stimulus and its second-messenger pathways. For example, whereas agonists increase both $[Ca^{2+}]_i$ and force, activators of PKC are widely reported to increase force with little or no change in $[Ca^{2+}]_i$ (Kamm and Grange 1996). Processes involved in force regulation are illustrated in two relations: (1) the Ca^{2+} dependence of RLC phosphorylation; and (2) the RLC phosphorylation dependence of force (Figure 8). Alterations in the former indicate that second-messenger pathways act to modify MLCK or MLCP activities, and thus RLC phosphorylation, at any given $[Ca^{2+}]_i$. Alterations in the latter suggest that regulatory elements in addition to RLC phosphorylation are involved. These may or may not be Ca^{2+}-dependent themselves.

3.3.1. Ca^{2+} sensitivity of myosin light chain phosphorylation

The activities of both smooth muscle MLCK and MLCP are subject to modulation, as is demonstrated in biochemical as well as physiological experiments. The sensitivity of MLCK to activation by Ca^{2+}–CaM is diminished upon phosphorylation at a regulatory site A (Conti and Adelstein 1981). As predicted, phosphorylation of MLCK has been shown to desensitize RLC phosphorylation to $[Ca^{2+}]_i$ in smooth muscle cells (Tansey et al. 1994; Word et al., 1994). MLCP is regulated by the association of the catalytic subunit with its targeting subunit *in vitro* (Alessi et al. 1992), and experiments in cells and tissues have shown that agonists activate second-messenger pathways that inhibit MLCP activity, thus sensitizing the Ca^{2+} dependence of RLC phosphorylation (Somlyo and Somlyo 1994). These mechanisms will influence the ratio of MLCK to MLCP activities (Figure 2) and thereby modulate the Ca^{2+} sensitivity of RLC phosphorylation.

Ca^{2+} dependence of myosin light chain kinase activity – site-specific phosphorylation of MLCK increases the [Ca^{2+}–CaM] required for activation. MLCK is a Ca^{2+}–CaM-regulated phosphotransferase that contains ATP-, RLC-, and CaM-binding domains. Unlike other protein kinases that have many protein substrates, MLCK is highly specific for myosin RLC. The domain organization of MLCK is illustrated in Figure 6. Activation of MLCK upon binding Ca^{2+}–CaM results from removal of autoinhibition conferred by the regulatory region of the enzyme (Stull et al. 1995). The concentration of Ca^{2+}–CaM required for half-maximal activation (K_{CaM}) is 1 nM. The Ca^{2+} sensitivity of RLC phosphorylation

measured in cells and tissues (100–400 nM) results in part from the Ca^{2+} affinities of CaM and the Ca^{2+}–CaM affinity of MLCK. Diminished Ca^{2+} sensitivity of RLC phosphorylation may arise by an increase in K_{CaM} (Tansey et al. 1994).

Purified MLCK can be phosphorylated at two sites C-terminal to the CaM-binding domain (site A and site B; Figure 6). These sites are phosphorylated by PKA, PKC, and CaMK II (Conti and Adelstein 1981; Nishikawa et al. 1984; Hashimoto and Soderling 1990; Ikebe and Reardon 1990a). MLCK is desensitized to activation by Ca^{2+}-CaM after phosphorylation of site A (tenfold increases in K_{CaM}), but phosphorylation at site B alone is insufficient to alter the Ca^{2+}-CaM activation properties (reviewed in Kotlikoff and Kamm 1996). MLCK is also phosphorylated by cGMP-dependent protein kinase (not at site A); however, its activation properties are not affected by phosphorylation by cGMP-dependent protein kinase (Nishikawa et al. 1984). Site A is located at the immediate C-terminal end of the CaM-binding domain, and is thus situated to affect the affinity of MLCK for Ca^{2+}–CaM upon the addition of negative charge with phosphorylation (Figure 6). Site B, on the other hand, is located 13 additional residues C-terminal to the site-A phosphoserine (Lukas et al. 1986; Payne et al. 1986). Purified MLCK is dephosphorylated by PP types 1 and 2A (Pato and Kerc 1985; Nomura et al. 1992). Experiments with intact tissues suggest that MLCK is dephosphorylated by the same phosphatase that dephosphorylates myosin RLC (Tang et al. 1992, 1993).

cAMP does not cause relaxation via phosphorylation of MLCK in smooth muscle. Initial investigations on the phosphorylation-dependent desensitization of MLCK focused on this as a potential mechanism whereby increases in cAMP and activation of PKA would lead to relaxation of smooth muscle (Conti and Adelstein 1981). However, agents that elevate cAMP (or activate PKC) in smooth muscle tissues have negligible effects on either phosphorylation of site-A or Ca^{2+} activation properties of MLCK (Miller et al. 1983; Stull et al. 1990b; Tang et al. 1992; Van Riper et al. 1995). The use of Ca^{2+} indicators in smooth muscle has shown that elevated cAMP generally brings about relaxation by lowering $[Ca^{2+}]_i$, particularly in agonist-stimulated muscle (reviewed in Kotlikoff and Kamm 1996). Nevertheless, several studies with permeabilized smooth muscle fibers have demonstrated desensitization of force to $[Ca^{2+}]_i$ in response to cAMP (Nishimura and van Breemen 1989) or addition of the catalytic subunit of cAMP-dependent protein kinase (Rüegg and Paul 1982; Meisheri and Rüegg 1983). Moreover, in KCl-depolarized intact smooth muscles, isoproterenol or forskolin inhibit contraction and RLC phosphorylation without changing $[Ca^{2+}]_i$ (Tang et al. 1992; Van Riper et al. 1995). Measurement of the Ca^{2+} activation properties of MLCK isolated from these tissues showed that desensitization of RLC phosphorylation to $[Ca^{2+}]_i$ was not, however, associated with changes in phosphorylation of MLCK. These results suggest that under certain conditions PKA acts to desensitize contractile elements – possibly by stimulating myosin phosphatase activity, although biochemical mechanisms responsible for such activation remain to be identified.

MLCK is phosphorylated at site A by CaM kinase II in contracting smooth muscle. Investigations of MLCK phosphorylation in tracheal smooth muscle tissues involved measurements of site-specific incorporation of ^{32}P by phosphopeptide mapping of enzyme isolated from labeled tissues that were treated with agents activating different protein kinase pathways (Stull et al., 1990b). In addition, the Ca^{2+}–CaM activation properties of MLCK isolated from treated muscles were assessed by an activity ratio assay (Stull et al. 1990b). In contrast to the effects of activators of PKA or PKC, agents that elevated $[Ca^{2+}]_i$ resulted in stoichiometric phosphorylation of MLCK at site A that was associated with diminished enzymatic activation by Ca^{2+}–CaM (Stull et al. 1990b). Similar results were subsequently reported for the carotid artery (Van Riper et al. 1995). Alterations in K_{CaM} were found to be Ca^{2+}-dependent in both intact and permeabilized tracheal smooth muscle strips and cells, indicating that site A is phosphorylated by a Ca^{2+}-dependent kinase (Tang et al. 1992; Tansey et al. 1992, 1994). The concentration of Ca^{2+} required for half-maximal phosphorylation was greater for MLCK (500 nM) than for RLC (250 nM). Cells treated with inhibitors of the Ca^{2+}–CaM-dependent protein kinase II showed no phosphorylation of MLCK in response to elevated $[Ca^{2+}]_i$. Consistent with inhibition of desensitization of MLCK by these inhibitors, the half-maximal $[Ca^{2+}]_i$ for RLC phosphorylation was decreased to 170 nM (Tansey et al. 1992, 1994; Word et al. 1994). It is concluded that MLCK is phosphorylated by CaMK II when $[Ca^{2+}]_i$ is elevated to high values, resulting in down-regulation of the Ca^{2+} effector, MLCK.

The activation properties of MLCK vary during cycles of contraction and relaxation of both tonic (tracheal) and phasic (myometrial) smooth muscles (Word et al., 1994). During the initial rapid phase of RLC phosphorylation (between 0.5 and 2 seconds in electrically stimulated tissues), there are no changes in MLCK activation properties. Thereafter, the K_{CaM} increases some threefold while the Ca^{2+} sensitivity of RLC phosphorylation is diminishing (Word et al. 1994). Times to recover to half-resting values of K_{CaM} are similar to those for RLC dephosphorylation (approximately 10 s) during relaxation. Thus, Ca^{2+}-dependent phosphorylation of MLCK acts to inhibit high levels of RLC phosphorylation on a contraction-to-contraction basis.

Differences in Ca^{2+} sensitivity of RLC phosphorylation between agonists and depolarization have been shown to arise from regulation of myosin phosphatase activity. In intact tissues, MLCK site-A phosphorylation has greater sensitivity to $[Ca^{2+}]_i$ with stimulation by agonists than with depolarization (Tang et al. 1992). In permeabilized strips, both GTPγS (nonhydrolyzable form of GTP) and carbachol increase the Ca^{2+} sensitivity of MLCK phosphorylation, suggesting that both myosin RLC and MLCK are dephosphorylated by the same protein phosphatase (Tang et al. 1993).

Regulation of myosin phosphatase activity. The Ca^{2+} sensitivity of myosin RLC phosphorylation is affected by myosin phosphatase activity (Figure 8). Recent studies have shown that inhibitory regulation of myosin phosphatase activity – through a guanine nucleotide-binding protein-dependent pathway – leads to increased sensitivity of RLC phosphorylation to $[Ca^{2+}]_i$, thus potentiating contraction (Figure 2) (Somlyo and Somlyo 1994; Kamm and Grange 1996).

Protein serine or threonine phosphatases are regulated by protein–protein interactions, with formation of oligomeric complexes that direct phosphatase catalytic subunits toward specific substrates by association with regulatory proteins. Moreover, phosphatase subunits and their inhibitor proteins are also regulated by phosphorylation. For example, the phosphorylation of the glycogen-binding subunit of PP1 by PKA results in dissociation of the catalytic subunit from glycogen and inactivation of the enzyme. Phosphorylation of inhibitor-1 by PKA results in activation of the inhibitor and inactivation of PP1. Recent experiments involving expressed cDNAs for three isoforms of $PP1_C$ (α, β, and γ) revealed that all three isoforms of the catalytic subunit were capable of forming complexes with each targeting subunit, but expressed PP1α and PP1γ interact with M complexes only if they are first converted to PP1, complexed to inhibitor-2, and then reactivated by incubation with glycogen synthase kinase-3 and MgATP (Alessi et al. 1993). These experiments indicate that inhibitor-2 functions as a molecular chaperone to ensure that $PP1_C$ adopts its correct conformation prior to its delivery to a specific targeting subunit.

A guanine nucleotide-binding protein (G protein) may be involved in the regulation of myosin phosphatase activity. Treatment of permeabilized smooth muscle with GTPγS increases the sensitivity of the contractile response to activation by Ca^{2+}, and results in a leftward shift of the Ca^{2+}–myosin light chain phosphorylation relationship (Nishimura et al., 1988; Fujiwara et al. 1989; Kitazawa et al. 1989); these effects are inhibited by GDPβS (Kitazawa et al. 1991). Treatment of smooth muscle from the portal vein (also permeabilized with α-toxin) with GTPγS results in increased myosin light chain phosphorylation (Kitazawa et al. 1991). These results are supportive of the hypothesis that G protein Ca^{2+}-sensitizing effects are secondary to increases in myosin light chain phosphorylation. In skinned bovine tracheal smooth muscle, GTPγS likewise induces myosin light chain phosphorylation at only the MLCK sites, indicating that PKC or other kinases are not responsible for the increase in myosin light chain phosphorylation (Kubota et al. 1992). Moreover, GTPγS does not enhance okadaic acid-induced contractions. Taken together, these results are consistent with GTPγS-mediated inhibition of myosin phosphatase with the resultant increase in myosin light chain phosphorylation and force of contraction. Currently, the GTP-binding protein is not known, although experimental evidence suggests that rho p21 or small guanine nucleotide-binding proteins of the ras family may be involved (Hirata et al. 1992; Satoh et al. 1993; Itagaki et al. 1995; Takai et al. 1995; Gong et al. 1996). In permeabilized smooth muscle fibers, arachidonic acid (50–100 μM) stimulates RLC phosphorylation without an obligatory increase in cytoplasmic Ca^{2+} and slows the rate of RLC dephosphorylation and muscle relaxation (Gong et al. 1992). *In vitro,* arachidonic acid also inhibits myosin dephosphorylation by releasing $PP1_C$ from the M subunit (Gong et al. 1992). These findings suggest that Ca^{2+}-sensitizing agents could inhibit dephosphorylation of RLC *in vivo* by dissociating phosphatase activity from their targeting subunits. However, the physiologic role of arachidonic acid at subcellular concentrations is not clear.

Dissociation of $PP1_C$ from the M complex may also occur by phosphorylation of M subunits. This type of regulation in smooth muscle is supported by the recent findings of Trinkle-Mulcahy et al. (1995), where pretreatment of α-toxin permeabilized rabbit portal vein with ATPγS increased the subsequent Ca^{2+} sensitivity for RLC phosphorylation and force generation. This change in Ca^{2+} sensitivity was not associated with thiophosphorylation of myosin light chain ($<1\%$), but did result in a fivefold decrease in myosin phosphatase activity and thiophosphorylation of the 130-kDa subunit of myosin phosphatase. These results suggest that phosphorylation of the myosin-binding subunit leads to a decrease in the activity of the phosphatase, and that phosphorylation and dephosphorylation of the subunit may play a role in regulation of myosin phosphatase activity (Figure 10). There are several potential phosphorylation sites on the M130 subunit for PKA, PKC, $p34^{cdc2}$ kinase, and glycogen synthase kinase-3 (Shimizu et al. 1994), but the effects (if any) of phosphorylation by any of these kinases are not known.

Finally, β-adrenergic agents under conditions of depolarization have been shown to desensitize RLC phosphorylation to $[Ca^{2+}]_i$. Because this desensitization does not result from phosphorylation of MLCK (Tang et al. 1992; Van Riper et al. 1995), it may be associated with activation of myosin phosphatase (Kotlikoff and Kamm 1996). Specific pathways involved in regulating myosin phosphatase activity will remain a subject of ongoing investigation.

3.3.2. Relation between myosin light chain phosphorylation and force

The dependence of steady-state force on RLC phosphorylation under most conditions supports the thesis that myosin phosphorylation is the fundamental mechanism for regulating smooth muscle contraction. Nevertheless, a number of experimental conditions have been shown to alter the RLC phosphorylation–force relation (Figure 8), pointing to the involvement of collateral regulation, presumably mediated by thin-filament–associated proteins.

Sensitization of force to RLC phosphorylation has been most frequently reported when smooth muscles are stimulated to contract with activators of PKC. In ferret portal vein, 12-deoxyphorbol 13-isobutyrate 20-acetate acts to cause a slow contraction of similar magnitude to that of K^+, but light chain phosphorylation is not increased (Jiang and Morgan 1987). Similar results have been shown for other smooth muscle preparations (Singer and Baker 1987; Adam et al. 1989; Sato et al. 1992; Fulginiti et al. 1993), although PKC activators may also stimulate contractions with concomitant increases in $[Ca^{2+}]_i$ and RLC phosphorylation (Rembold and Murphy 1988; Ozaki et al. 1990; Singer 1990; Itoh et al. 1993). Moreover, PKC activators do not always elicit contractions (Kamm et al. 1989b). It has been reported that treatment of the ferret aorta with $PGF_{2\alpha}$ results in increased levels of light chain phosphorylation relative to $[Ca^{2+}]_i$ as well as in more force development for a given level of light chain phosphorylation than that obtained with KCl (Suematsu et al. 1991). In this unique example, both the $[Ca^{2+}]_i$ light chain phosphorylation and its relation to force were shifted to the left of KCl values by agonist treatment.

Desensitization of force to RLC phosphorylation has also been described. In canine tracheal smooth muscle, myosin light chain phosphorylation is completely dissociated from force development when muscles are stimulated with carbachol in Ca^{2+}-free physiological salt solution and then contracted by restoring $CaCl_2$ (Gerthoffer 1987). In the absence of extracellular Ca^{2+}, carbachol induces an increase in myosin light chain phosphorylation, yet the muscle remains relaxed. The addition of Ca^{2+} to the bathing medium results in an increase in force, and the light chain remains phosphorylated. The addition of okadaic acid (a protein phosphatase inhibitor isolated from the marine sponge *Halichondria*) to bovine tracheal smooth muscle contracted with carbachol results in a decrease in $[Ca^{2+}]_i$ and relaxation, in spite of maintenance of high levels of myosin light chain

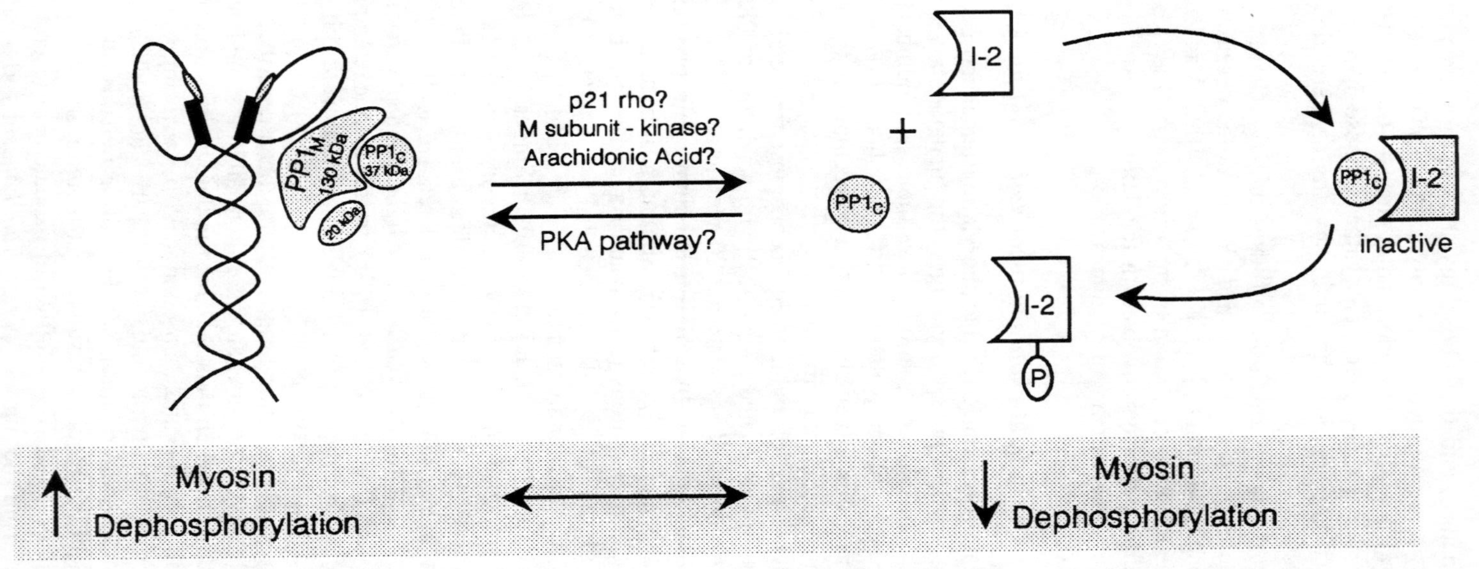

Figure 10. Proposed regulation of $PP1_M$ in smooth muscle. Myosin phosphatase is localized to the myofibrils by way of targeting subunits (130- and 20-kDa $PP1_M$). Phosphorylation of the targeting subunits by kinases or exposure to increased concentrations of arachidonic acid lead to release of the free catalytic subunit ($PP1_C$) into the cytoplasm and to formation of complexes with cytosolic inhibitor proteins such as inhibitor-2 (I-2). Although I-2/$PP1_C$ complexes are inactive, phosphorylation of I-2 by glycogen synthase kinase-3 leads to release of the catalytic subunit, which is then free to reassociate with the targeting subunit and actomyosin. Association of $PP1_C$ with $PP1_M$ increases the affinity for myosin light chain and decreases affinity for phosphorylase. Dissociation of $PP1_C$ from $PP1_M$ leads to decreased affinity for myosin light chain and increased levels of light chain phosphorylation, thereby favoring contraction.

phosphorylation (Tansey et al. 1990). Both arterial and uterine muscles are relaxed upon addition of high external Mg^{2+} without reduction in RLC phosphorylation (D'Angelo et al. 1992; Bárány and Bárány, 1993), and sodium nitroprusside has been shown to relax arterial smooth muscle without lowering RLC phosphorylation (McDaniel et al. 1992). Redistribution of phosphate into RLC sites that do not activate myosin was not associated with these relaxations as ^{32}P was incorporated into serine-19, the site normally phosphorylated by MLCK. It has been proposed that complex interactions of the contractile proteins with thin filament regulatory proteins may alter the MLC phosphorylation–force relationship.

4. Summary and perspectives

There have been major advances in our understanding of the regulation and physiological functions of contractile proteins in smooth muscles in recent years. Phosphorylation of the RLC of myosin by Ca^{2+}–CaM-dependent MLCK plays a pivotal role in activation. The simple view – that contractile force in smooth muscles is proportionate to $[Ca^{2+}]_i$ and myosin light chain phosphorylation – has been modified as additional experiments have provided insights into mechanisms of regulation of the contractile elements. It is apparent that, although $[Ca^{2+}]_i$ is undoubtedly the master mediator of regulation, additional pathways act to modulate the regulatory components. The sensitivity of RLC phosphorylation to $[Ca^{2+}]_i$ can be shifted by second messengers modulating activities of MLCK and myosin phosphatase. MLCK is phosphorylated, which desensitizes its activation by Ca^{2+}–CaM; protein phosphatase activity toward myosin is also regulated, perhaps by phosphorylation and dephosphorylation of regulatory subunits or cytosolic inhibitors. The dependence of force on RLC phosphorylation is established in the steady state of tonic contractions; however, specific interventions can shift the relation, indicating that the number of crossbridges can be altered independent of RLC phosphorylation. Thin-filament regulation is a good candidate for such a modulatory effector. The physiologic importance and relative contributions of these regulatory components for smooth muscle function *in vivo* are not entirely clear. Moreover, the role of these mechanisms in pathophysiologic states remains to be defined. The diversity of different types of smooth muscle is apparent by differences in ligand receptor expression, electrical properties, and Ca^{2+} homeostatic mechanisms. In addition, there is growing appreciation for differences in contractile protein contents and differing patterns of expressed contractile protein isoforms between tonic and phasic smooth muscles. These differences may well extend to variations in the relative contributions of regulatory components to contractile responses of these tissues. A wealth of knowledge awaits us.

Acknowledgments

This work was supported in part by grants from the National Institutes of Health HD11149 and HD30497 (R.A.W.) and HL54891 (K.E.K.).

References

Adam, L. P., Haeberle, J. R., and Hathaway, D. R. (1989). Phosphorylation of caldesmon in arterial smooth muscle. *J. Biol. Chem.* 264: 7698–7703.

Adelstein, R. S., Peleg, I., Ludowyke, R., Kawamoto, S., and Conti, M. A. (1990). Phosphorylation of vertebrate nonmuscle myosin heavy chains by protein kinase C. *Adv. Second Messenger Phosphoprotein Res.* 24: 405–10.

Alessi, D., Macdougall, L. K., Sola, M. M., Ikebe, M., and Cohen, P. (1992). The control of protein phosphatase-1 by targetting subunits. *Eur. J. Biochem.* 210: 1023–35.

Alessi, D. R., Street, A. J., Cohen, P., and Cohen, P. T. (1993). Inhibitor-2 functions like a chaperone to fold three expressed isoforms of mammalian protein phosphatase-1 into a conformation with the specificity and regulatory properties of the native enzyme. *Eur. J. Biochem.* 213: 1055–66.

Ashton, F. T., Somlyo, A. V., and Somlyo, A. P. (1975). The contractile apparatus of vascular smooth muscle: intermediate high voltage stereo electron microscopy. *J. Mol. Biol.* 98: 17–29.

Babij, P. (1993). Tissue-specific and developmentally regulated alternative splicing of a visceral isoform of smooth muscle myosin heavy chain. *Nucleic Acids Res.* 21: 1467–71.

Babij, P., and Periasamy, M. (1989). Myosin heavy chain isoform diversity in smooth muscle is produced by differential RNA processing. *J. Mol. Biol.* 210: 673–9.

Ballou, L. M., Brautigan, D. L., and Fischer, E. H. (1983). Subunit structure and activation of inactive phosphorylase phosphatase. *Biochemistry* 22: 3393–9.

Bárány, M., and Bárány, K. (1993). Dissociation of relaxation and myosin light chain dephosphorylation in porcine uterine muscle. *Arch. Biochem. Biophys.* 305: 202–4.

Bárány, M., Polyak, E., and Bárány, K. (1992). Protein phosphorylation during the contraction-relaxation-contraction cycle of arterial smooth muscle. *Arch. Biochem. Biophys.* 294: 571–8.

Barylko, B., Tooth, P., and Kendrick-Jones, J. (1986). Proteolytic fragmentation of brain myosin and localisation of the heavy-chain phosphorylation site. *Eur. J. Biochem.* 158: 271–82.

Bengur, A. R., Robinson, E. A., Appella, E., and Sellers, J. R. (1987). Sequence of the sites phosphorylated by protein kinase C in the smooth muscle myosin light chain. *J. Biol. Chem.* 262: 7613–17.

Campbell, I. D., and Spitzfaden, C. (1994). Building proteins with fibronectin type III modules. *Structure* 2: 333–7.

Carroll, A. G., and Wagner, P. D. (1989). Protein kinase C phosphorylation of thymus myosin. *J. Muscle Res. Cell Motil.* 10: 379–84.

Chisholm, A. A. K., and Cohen, P. (1988). The myosin-bound form of protein phosphatase 1 (PP-1_M) is the enzyme that dephosphorylates native myosin in skeletal and cardiac muscles. *Biochim. Biophys. Acta* 971: 163–9.

Cohen, P. (1989). The structure and regulation of protein phosphatases. *Annu. Rev. Biochem.* 58: 453–508.

Colburn, J. C., Michnoff, C. H., Hsu, L., Slaughter, C. A., Kamm, K. E., and Stull, J. T. (1988). Sites phosphorylated in myosin light chain in contracting smooth muscle. *J. Biol. Chem.* 263: 19166–73.

Collinge, M., Matrisian, P. E., Zimmer, W. E., Shattuck, R. L., Lukas, T. J., Van Eldik, L. J., and Watterson, D. M. (1992). Structure and expression of a calcium-binding protein gene contained within a calmodulin-regulated protein kinase gene. *Mol. Cell. Biol.* 12: 2359–71.

Conti, M. A., and Adelstein, R. S. (1981). The relationship between calmodulin binding and phosphorylation of smooth muscle myosin kinase by the catalytic subunit of 3': 5' cAMP-dependent protein kinase. *J. Biol. Chem.* 256: 3178–81.

Conti, M. A., Sellers, J. R., Adelstein, R. S., and Elzinga, M. (1991). Identification of the serine residue phosphorylated by protein kinase C in vertebrate nonmuscle myosin heavy chains. *Biochemistry* 30: 966–70.

Craig, S. W., and Pollard, T. D. (1982). Actin-binding proteins. *TIBS* 7: 88–92.

Cross, R. A., and Vandekerckhove, J. (1986). Solubility-determining domain of smooth muscle myosin rod. *FEBS Letts.* 200: 355–60.

Dabrowska, R., Sherry, J. M. F., Aromatorio, D. K., and Hartshorne, D. J. (1978). Modulator protein as a component of the myosin light chain kinase from chicken gizzard. *Biochemistry* 17: 253–8.

D'Angelo, D. K. G., Singer, H. A., and Rembold, C. M. (1992). Magnesium relaxes arterial smooth muscle by decreasing intracellular Ca^{2+} without changing intracellular Mg^{2+}. *J. Clin. Invest.* 89: 1988–94.

Dent, P., Macdougall, L. K., Mackintosh, C., Campbell, D. G., and Cohen, P. (1992). A myofibrillar protein phosphatase from rabbit skeletal muscle contains the β isoform of protein phosphatase-1 complexed to a regulatory subunit which greatly enhances the dephosphorylation of myosin. *Eur. J. Biochem.* 210: 1037–44.

Di Blasi, P., Van Riper, D., Kaiser, R., Rembold, C. M., and Murphy, R. A. (1992). Steady-state dependence of stress on cross-bridge phosphorylation in the swine carotid media. *Am. J. Physiol.* 262: C1388–C1391.

Eddinger, T. J., and Wolf, J. A. (1993). Expression of four myosin heavy chain isoforms with development in mouse uterus. *Cell Motil. Cytoskeleton* 25: 358–68.

Edelman, A. M., Lin, W.-H., Osterhout, D. J., Bennett, M. K., Kennedy, M. B., and Krebs, E. G. (1990). Phosphorylation of smooth muscle myosin by type II Ca^{2+}/calmodulin-dependent protein kinase. *Mol. Cell. Biochem.* 97: 87–98.

Fatigati, V., and Murphy, R. A. (1984). Actin and tropomyosin variants in smooth muscles. Dependence on tissue type. *J. Biol. Chem.* 259: 14383–8.

Fisher, S. A., and Ikebe, M. (1995). Developmental and tissue distribution of expression of nonmuscle and smooth muscle isoforms of myosin light chain kinase. *Biochem. Biophys. Res. Commun.* 217: 696–703.

Fitzsimons, D. P., Herring, B. P., Stull, J. T., and Gallagher, P. J. (1992). Identification of basic residues involved in activation and calmodulin binding of rabbit smooth muscle myosin light chain kinase. *J. Biol. Chem.* 267: 23903–9.

Frid, M. G., Printesva, O. Y., Chiavegato, A., Faggin, E., Seatena, M., Koteliansky, V. E., Pauletto, P., Glukhova, M. A., and Sartore, S. (1993). Myosin heavy-chain isoform composition and distribution in developing and adult human aortic smooth muscle. *J. Vasc. Res.* 30: 279–92.

Fujiwara, T., Itoh, T., Kubota, Y., and Kuriyama, H. (1989). Effects of guanosine nucleotides on skinned smooth muscle tissue of the rabbit mesenteric artery. *J. Physiol. (Lond.)* 408: 535–47.

Fulginiti, J., III, Singer, H. A., and Moreland, R. S. (1993). Phorbol ester-induced contractions of swine carotid artery are supported by slowly cycling crossbridges which are not dependent on calcium or myosin light chain phosphorylation. *J. Vasc. Res.* 30: 315–22.

Gallagher, P. J., Garcia, J. G. N., and Herring, B. P. (1995). Expression of a novel myosin light chain kinase in embryonic tissues and cultured cells. *J. Biol. Chem.* 270: 29090–5.

Gallagher, P. J., and Herring, B. P. (1991). The C-terminus of the smooth muscle myosin light chain kinase is expressed as an independent protein, telokin. *J. Biol. Chem.* 266: 23945–52.

Gallagher, P. J., Herring, B. P., Griffin, S. A., and Stull, J. T. (1991). Molecular characterization of a mammalian smooth muscle myosin light chain kinase. *J. Biol. Chem.* 266: 23936–44.

Gallagher, P. J., Herring, B. P., Trafny, A., Sowadski, J., and Stull, J. T. (1993). A molecular mechanism for autoinhibition of myosin light chain kinases. *J. Biol. Chem.* 268: 26578–82.

Gaylinn, B. D., Eddinger, T. J., Martino, P. A., Monical, P. L., Hunt, D. F., and Murphy, R. A. (1989). Expression of nonmuscle myosin heavy and light chains in smooth muscle. *Am. J. Physiol.* 257: C997–C1004.

Gerthoffer, W. T. (1987). Dissociation of myosin phosphorylation and active tension during muscarinic stimulation of tracheal smooth muscle. *J. Pharmacol. Exp. Ther.* 240: 8–15.

Gillis, J. M., Cao, M. L., and Godfraind-De Becker, A. (1988). Density of myosin filaments in the rat anacoccygeus muscle, at rest and in contraction. II. *J. Muscle Res. Cell Motil.* 9: 18–29.

Godfraind-De Becker, A., and Gillis, J. M. (1988). Analysis of the birefringence of the smooth muscle anococcygeus of the rat, at rest and in contraction. I. *J. Muscle Res. Cell Motil.* 9: 9–17.

Gong, M. C., Fuglsang, A., Alessi, D., Kobayashi, S., Cohen, P., Somlyo, A. V., and Somlyo, A. P. (1992). Arachidonic acid inhibits myosin light chain phosphatase and sensitizes smooth muscle to calcium. *J. Biol. Chem.* 267: 21492–8.

Gong, M. C., Iizuka, K., Nixon, G., Browne, J. P., Hall, A., Eccleston, J. F., Sugai, M., Kobayashi, S., Somlyo, A. V., and Somlyo, A. P. (1996). Role of guanine nucleotide-binding proteins—ras-family or trimeric proteins or both—in Ca^{2+} sensitization of smooth muscle. *Proc. Nat. Acad. Sci. USA* 93: 1340–5.

Greene, L. E., Sellers, J. R., Eisenberg, E., and Adelstein, R. S. (1983). Binding of gizzard smooth muscle myosin subfragment 1 to actin in the presence and absence of adenosine 5'-triphosphate. *Biochemistry* 22: 530–5.

Hai, C.-M., and Murphy, R. A. (1989). Ca^{2+}, crossbridge phosphorylation, and contraction. *Annu. Rev. Physiol.* 51: 285–98.

Hamada, Y., Yanagisawa, M., Katsuragawa, Y., Coleman, J. R., Nagata, S., Matsuda, G., and Masaki, T. (1990). Distinct vascular and intestinal smooth muscle myosin heavy chain mRNAs are encoded by a single-copy gene in the chicken. *Biochem. Biophys. Res. Commun.* 170: 53–8.

Hartshorne, D. J. (1987). Biochemistry of the contractile process in smooth muscle. In: *Physiology of the Gastrointestinal Tract*. 2nd ed. (L. R. Johnson, ed.). New York: Raven, pp. 423–82.

Hartshorne, D. J., and Gorecka, A. (1980). The biochemistry of the contractile proteins of smooth muscle. In: *Handbook of Physiology – The Cardiovascular System* (D. F. Bohr, A. P. Somlyo, and H. V. Sparks, eds.). Bethesda, MD: American Physiological Society, pp. 93–120.

Hartshorne, D. J., Gorecka, A., and Aksoy, M. O. (1977). Aspects of the regulatory mechanism in smooth muscle. In: *Excitation–Contraction Coupling in Smooth Muscle*. (R. Casteels, ed.). Amsterdam: Elsevier, pp. 377–84.

Hasegawa, Y., and Morita, F. (1992). Role of 17-kDa essential light chain isoforms of aorta smooth muscle myosin. *J. Biochem* 111: 804–9.

Hashimoto, Y., and Soderling, T. R. (1990). Phosphorylation of smooth muscle myosin light chain kinase by Ca^{2+}/calmodulin-dependent protein kinase-II. Comparative study of the phosphorylation sites. *Arch. Biochem. Biophys.* 278: 41–5.

Hewett, T. E., Martin, A. F., and Paul, R. J. (1993). Correlations between myosin heavy chain isoforms and mechanical parameters in rat myometrium. *J. Physiol. (Lond.)* 460: 351–64.

Hirata, K., Kikuchi, A., Sasaki, T., Kuraoda, S., Kaibuchi, K., Matsura, Y., Seki, H., Saida, K., and Takai, Y. (1992). Involvement of rho p21 in the GTP-enhanced calcium ion sensitivity of smooth muscle contraction. *J. Biol. Chem.* 267: 8719–22.

Hoar, P. E., Pato, M. D., and Kerrick, W. G. L. (1985). Myosin light chain phosphatase. Effect on the activation and relaxation of gizzard smooth muscle skinned fibers. *J. Biol. Chem.* 260: 8760–4.

Horiuti, K., Somlyo, A. V., Goldman, Y. E., and Somlyo, A. P. (1989). Kinetics of contraction initiated by flash photolysis of caged adenosine triphosphate in tonic and phasic smooth muscles. *J. Gen. Physiol.* 94: 769–81.

Horowitz, A., Trybus, K. M., Bowman, D. S., and Fay, F. S. (1994). Antibodies probe for folded monomeric myosin in relaxed and contracted smooth muscle. *J. Cell Biol.* 126: 1195–200.

Hu, S.-H., Parker, M. W., Lel, J. Y., Wilce, M. C. J., Benian, G. M., and Kemp, B. E. (1994). Insights into autoregulation from the crystal structure of twitchin kinase. *Nature* 369: 581–4.

Hubbard, M. J., Dent, P., Smythe, C., and Cohen, P. (1990). Targetting of protein phosphatase 1 to the sarcoplasmic reticulum of rabbit skeletal muscle by a protein that is very similar or identical to the G subunit that directs the enzyme to glycogen. *Eur. J. Biochem.* 189: 243–9.

Ikebe, M. (1990). Mode of inhibition of smooth muscle myosin light chain kinase by synthetic peptide analogs of the regulatory site. *Biochem. Biophys. Res. Commun.* 168: 714–20.

Ikebe, M., Maruta, S., and Reardon, S. (1989). Location of the inhibitory region of smooth muscle myosin light chain kinase. *J. Biol. Chem.* 264: 6967–71.

Ikebe, M., and Reardon, S. (1990a). Phosphorylation of smooth myosin light chain kinase by smooth muscle Ca^{2+}/calmodulin-dependent multifunctional protein kinase. *J. Biol. Chem.* 265: 8975–8.

Ikebe, M., and Reardon, S. (1990b). Phosphorylation of bovine platelet myosin by protein kinase-C. *Biochemistry* 29: 2713–20.

Ikebe, M., Stepinska, M., Kemp, B. E., Means, A. R., and Hartshorne, D. J. (1987). Proteolysis of smooth muscle myosin light chain kinase. Formation of inactive and calmodulin-independent fragments. *J. Biol. Chem.* 262: 13828–34.

Ishihara, H., Ozaki, H., Sato, K., Hori, M., Karaki, H., Watabe, S., Kato, Y., Fusetani, N., Hashimoto, K., Uemura, D., and Hartshorne, D. J. (1989). Calcium-independent activation of contractile apparatus in smooth muscle by calyculin-A. *J. Pharmacol. Exp. Ther.* 250: 388–96.

Itagaki, M., Komori, S., Unno, T., Syuto, B., and Ohsashi, H. (1995). Possible involvement of a small G-protein sensitive to exoenzyme C3 of clostridium botulinum in the regulation of myofilament Ca^{2+} sensitivity in β-escin skinned smooth muscle of guinea pig ileum. *Jpn. J. Pharmacol.* 67: 1–7.

Ito, M., Dabrowska, R., Guerriero, V., Jr., and Hartshorne, D. J. (1989). Identification in turkey gizzard of an acidic protein related to the C-terminal portion of smooth muscle myosin light chain. *J. Biol. Chem.* 264: 13971–4.

Ito, M., Guerriero, V., Jr., Chen, X., and Hartshorne, D. J. (1991). Definition of the inhibitory domain of smooth muscle myosin light chain kinase by site-directed mutagenesis. *Biochemistry* 30: 3498–3503.

Itoh, H., Shimomura, A., Okubo, S., Ichikawa, K., Ito, M., Konishi, T., and Nakano, T. (1993). Inhibition of myosin light chain phosphatase during Ca^{2+}-independent vasocontraction. *Am. J. Physiol.* 265: C1319–C1324.

Itoh, T., Ikebe, M., Kargacin, G. J., Hartshorne, D. J., Kemp, B. E., and Fay, F. S. (1989). Effects of modulators of myosin light-chain kinase activity in single smooth muscle cells. *Nature* 338: 164–7.

Jiang, M. J., and Morgan, K. G. (1987). Intracellular calcium levels in phorbol ester-induced contractions of vascular muscle. *Am. J. Physiol.* 253: H1365–H1371.

Jurgensen, S., Shacter, E., Huang, C. Y., Chock, P. B., Yang, S., Vandenheede, J. R., and Merlevede, W. (1984). On the mechanism of activation of the ATP×Mg(II)-dependent phosphoprotein phosphatase by kinase F_A. *J. Biol. Chem.* 259: 5864–70.

Kamm, K. E., and Grange, R. W. (1996). Ca^{2+} sensitivity of contraction. In: *Biochemistry of Smooth Muscle Contraction* (M. Bárány, ed.). Orlando, FL: Academic Press, pp. 355–65.

Kamm, K. E., Hsu, L.-C., Kubota, Y., and Stull, J. T. (1989). Phosphorylation of smooth muscle myosin heavy and light chains: Effects of phorbol dibutyrate and agonists. *J. Biol. Chem.* 264: 21223–9.

Kamm, K. E., and Stull, J. T. (1985a). Myosin phosphorylation, force, and maximal shortening velocity in neurally stimulated tracheal smooth muscle. *Am. J. Physiol.* 249: C238–C247.

Kamm, K. E., and Stull, J. T. (1985b). The function of myosin and myosin light chain kinase phosphorylation in smooth muscle. *Annu. Rev. Pharmacol. Toxicol.* 25: 593–620.

Kamm, K. E., and Stull, J. T. (1986). Activation of smooth muscle contraction: Relation between myosin phosphorylation and stiffness. *Science* 232: 80–2.

Kanoh, S., Ito, M., Niwa, E., Kawano, Y., and Hartshorne, D. J. (1993). Actin-binding peptide from smooth muscle myosin light chain kinase. *Biochemistry* 32: 8902–7.

Kawamoto, S., and Adelstein, R. S. (1987). Characterization of myosin heavy chains in cultured aorta smooth muscle cells. A comparative study. *J. Biol. Chem.* 262: 7282–8.

Kawamoto, S., and Adelstein, R. S. (1988). The heavy chain of smooth muscle myosin is phosphorylated in aorta cells. *J. Biol. Chem.* 263: 1099–102.

Kawamoto, S., Bengur, A. R., Sellers, J. R., and Adelstein, R. S. (1989). *In situ* phosphorylation of human platelet myosin heavy and light chains by protein kinase C. *J. Biol. Chem.* 264: 2258–65.

Keller, T. C., III, and Mooseker, M. S. (1982). Ca^{++}-calmodulin-dependent phosphorylation of myosin, and its role in brush border contraction in vitro. *J. Cell Biol.* 95: 943–59.

Kelley, C. A., and Adelstein, R. S. (1990). The 204-kDa smooth muscle myosin heavy chain is phosphorylated in intact cells by casein kinase II on a serine near the carboxyl terminus. *J. Biol. Chem.* 265: 17876–82.

Kelley, C. A., Sellers, J. R., Goldsmith, P. K., and Adelstein, R. S. (1992). Smooth muscle myosin is composed of homodimeric heavy chains. *J. Biol. Chem.* 267: 2127–30.

Kelley, C. A., Takahashi, M., Yu, J. H., and Adelstein, R. S. (1993). An insert of seven amino acids confers enzymatic differences between smooth muscle myosins from the intestines and vasculature. *J. Biol. Chem.* 268: 12848–54.

Kemp, B. E., and Pearson, R. B. (1985). Spatial requirements for location of basic residues in peptide substrates for smooth muscle myosin light chain kinase. *J. Biol. Chem.* 260: 3355–9.

Kemp, B. E., and Pearson, R. B. (1990). Protein kinase recognition sequence motifs. *Trends Biochem. Sci.* 15: 342–6.

Kemp, B. E., Pearson, R. B., Guerriero, V., Jr., Bagchi, I. C., and Means, A. R. (1987). The calmodulin binding domain of chicken smooth muscle myosin light chain kinase contains a pseudosubstrate sequence. *J. Biol. Chem.* 262: 2542–8.

Kennelly, P. J., and Krebs, E. G. (1991). Consensus sequences as substrate specificity determinants for protein kinases and protein phosphatases. *J. Biol. Chem.* 266: 15555–8.

Kenney, R. E., Hoar, P. E., and Kerrick, W. G. L. (1990). The relationship between ATPase activity, isometric force, and myosin light-chain phosphorylation and thiophosphorylation in skinned smooth muscle fiber bundles from chicken gizzard. *J. Biol. Chem.* 265: 8642–9.

Khromov, A., Somlyo, A. V., Trentham, D. R., Zimmermann, B., and Somlyo, A. P. (1995). The role of MgADP in force maintenance by dephosphorylated cross-bridges in smooth muscle: A flash photolysis study. *Biophys. J.* 69: 2611–22.

Kitazawa, T., Gaylinn, B. D., Denney, G. H., and Somlyo, A. P. (1991). G-protein-mediated Ca^{2+} sensitization of smooth muscle contraction through myosin light chain phosphorylation. *J. Biol. Chem.* 266: 1708–15.

Kitazawa, T., Kobayashi, S., Horiuti, K., Somlyo, A. V., and Somlyo, A. P. (1989). Receptor-coupled, permeabilized smooth muscle. Role of the phosphatidylinositol cascade, G-proteins, and modulation of the contractile response to Ca^{2+}. *J. Biol. Chem.* 264: 5339–42.

Kobayashi, H., Inoue, A., Mikawa, T., Kuwayama, H., Hotta, Y., Masaki, T., and Ebashi, S. (1992). Isolation of cDNA for bovine stomach 155 kDa protein exhibiting myosin light chain kinase activity. *J. Biochem.* 112: 786–91.

Korn, E. D. (1982). Actin polymerization and its regulation by proteins from nonmuscle cells. *Physiol. Rev.* 62: 672–737.

Kotlikoff, M. I., and Kamm, K. E. (1996). Molecular mechanisms of β adrenergic relaxation of airway smooth muscle. *Annu. Rev. Physiol.* 58: 115–41.

Kretsinger, R. H. (1980). Structure and evolution of calcium-modulated proteins. *Crit. Rev. Biochem.* 8: 119–74.

Krueger, J. K., Padre, R. C., and Stull, J. T. (1995). Intrasteric regulation of myosin light chain kinase. *J. Biol. Chem.* 270: 16848–53.

Kubota, Y., Nomura, M., Kamm, K. E., Mumby, M. C., and Stull, J. T. (1992). GTPgammaS-dependent regulation of smooth muscle contractile elements. *Am. J. Physiol.* 262: C405–C410.

Kumar, C. C., Mohan, S. R., Zavodny, P. J., Narula, S. K., and Leibowitz, P. J. (1989). Characterization and differential expression of human vascular smooth muscle myosin light chain 2 isoform in nonmuscle cells. *Biochemistry* 28: 4027–35.

Kuro-o, M., Nagai, R., Nakahara, K., Katoh, H., Tsai, R., Tsuchimochi, H., Yazaki, Y., Ohkubo, A., and Takaku, F. (1991). cDNA cloning of a myosin heavy chain isoform in embryonic smooth muscle and its expression during vascular development and in arteriosclerosis. *J. Biol. Chem.* 266: 3768–73.

Kuznicki, J., and Filipek, A. (1988). Purification of myosin from Ehrlich ascites tumor cells (phosphorylation of its light chain and heavy chain). *Int. J. Biochem.* 20: 1203–1209.

Labeit, S., Gautel, M., Lakey, A., and Trinick, J. (1992). Towards a molecular understanding of titin. *EMBO J.* 11: 1711–16.

Lehman, W. (1991). Calponin and the composition of smooth muscle thin filaments. *J. Muscle Res. Cell Motil.* 12: 221–4.

Lin, P.-J., Luby-Phelps, K., and Stull, J. T. (1996). Binding of myosin light chain kinase to microfilament bundles in living smooth muscle cells. *Biophys. J.* 70: A52 (abstract).

Lowey, S., Waller, G. S., and Trybus, K. M. (1993). Function of skeletal muscle myosin heavy and light chain isoforms by an *in vitro* motility assay. *J. Biol. Chem.* 268: 20414–18.

Luby-Phelps, K., Hori, M., Phelps, J. M., and Won, D. (1995). Ca^{2+}-regulated dynamic compartmentalization of calmodulin in living smooth muscle cells. *J. Biol. Chem.* 270: 21532–8.

Ludowyke, R. I., Peleg, I., Beaven, M. A., and Adelstein, R. S. (1989). Antigen-induced secretion of histamine and the phosphorylation of myosin by protein kinase C in rat basophilic leukemia cells. *J. Biol. Chem.* 264: 12492–12501.

Lukas, T. J., Burgess, W. H., Prendergast, F. G., Lau, W., and Watterson, D. M. (1986). Calmodulin binding domains: Characterization of a phosphorylation and calmodulin binding site from myosin light chain kinase. *Biochemistry* 25: 1458–64.

Makuch, R., Birukov, K., Shirinsky, V., and Dabrowska, R. (1991). Functional interrelationship between calponin and caldesmon. *Biochem. J.* 280: 33–8.

Marston, S. B., and Lehman, W. (1985). Caldesmon is a Ca^{2+}-regulatory component of native smooth-muscle thin filaments. *Biochem. J.* 231: 517–22.

Matsu-Ura, M., and Ikebe, M. (1995). Requirement of the two-headed structure for the phosphorylation dependent regulation of smooth muscle myosin. *FEBS Letts.* 363: 246–50.

McDaniel, N. L., Chen, X.-L., Singer, H. A., Murphy, R. A., and Rembold, C. M. (1992). Nitrovasodilators relax arterial smooth muscle by decreasing $[Ca^{2+}]_i$ and uncoupling stress from myosin phosphorylation. *Am. J. Physiol.* 263: C461–C467.

Meisheri, K. D., and Rüegg, J. C. (1983). Dependence of cyclic-AMP induced relaxation on Ca^{2+} and calmodulin in skinned smooth muscle of guinea pig *Taenia coli*. *Pflügers Arch.* 399: 315–20.

Michnoff, C. H., Kemp, B. E., and Stull, J. T. (1986). Phosphorylation of synthetic peptides by skeletal muscle myosin light chain kinases. *J. Biol. Chem.* 261: 8320–6.

Miller, J. R., Silver, P. J., and Stull, J. T. (1983). The role of myosin light chain kinase phosphorylation in β-adrenergic relaxation of tracheal smooth muscle. *Mol. Pharmacol.* 24: 235–42.

Miller-Hance, W. C., Miller, J. R., Wells, J. N., Stull, J. T., and Kamm, K. E. (1988). Biochemical events associated with activation of smooth muscle contraction. *J. Biol. Chem.* 263: 13979–82.

Mitsui, T., Inagaki, M., and Ikebe, M. (1992). Purification and characterization of smooth muscle myosin-associated phosphatase from chicken gizzards. *J. Biol. Chem.* 267: 16727–35.

Mohammad, M. A., and Sparrow, M. P. (1988). Changes in myosin heavy chain stoichiometry in pig tracheal smooth muscle during development. *FEBS Letts.* 228: 109–12.

Mumby, M. C., and Walter, G. (1993). Protein serine/threonine phosphatases: Structure, regulation, and functions in cell growth. *Physiol. Rev.* 73: 673–99.

Murakami, N., Healy-Louie, G., and Elzinga, M. (1990). Amino acid sequence around the serine phosphorylated by casein kinase II in brain myosin heavy chain. *J. Biol. Chem.* 265: 1041–7.

Murakami, N., Matsumura, S., and Kumon, A. (1984). Purification and identification of myosin heavy chain kinase from bovine brain. *J. Biochem.* 95: 651–60.

Murphy, R. A. (1979). Filament organization and contractile function in vertebrate smooth muscle. *Annu. Rev. Physiol.* 41: 737–48.

Murphy, R. A. (1994). What is special about smooth muscle? The significance of covalent crossbridge regulation. *FASEB J.* 8: 311–18.

Murphy, R. A., Herlihy, J. T., and Megerman, J. (1974). Force-generating capacity and contractile protein content of arterial smooth muscle. *J. Gen. Physiol.* 64: 691–705.

Nachmias, V. T., Yoshida, K.-I., and Glennon, M. C. (1987). Lowering pH in blood platelets dissociates myosin phosphorylation from shape change and myosin association with the cytoskeleton. *J. Cell Biol.* 105: 1761–9.

Nagai, R., Kuro-o, M., Babij, P., and Periasamy, M. (1989). Identification of two types of smooth muscle myosin heavy chain isoforms by cDNA cloning and immunoblot analysis. *J. Biol. Chem.* 264: 9734–7.

Nakayama, S., Moncrief, N. D., and Kretsinger, R. H. (1992). Evolution of EF-hand calcium-modulated proteins. II. Domains of several subfamilies have diverse evolutionary histories. *J. Mol. Evol.* 34: 416–48.

Nishikawa, M., de Lanerolle, P., Lincoln, T. M., and Adelstein, R. S. (1984a). Phosphorylation of mammalian myosin light chain kinases by the catalytic subunit of cyclic AMP-dependent protein kinase and by cyclic GMP-dependent protein kinase. *J. Biol. Chem.* 259: 8429–36.

Nishikawa M., Sellers, J. R., Adelstein R. S., and Hidaka H. (1984b). Protein kinase C modulates *in vitro* phosphorylation of the smooth muscle heavy meromyosin by myosin light chain kinase. *J. Biol. Chem.* 259: 8808–14.

Nishimura, J., Kolber, M., and van Breemen, C. (1988). Norepinephrine and GTPgammaS increase myofilament Ca^{2+} sensitivity in α-toxin permeabilized arterial smooth muscle. *Biochem. Biophys. Res. Commun.* 157: 677–83.

Nishimura, J., and van Breemen, C. (1989). Direct regulation of smooth muscle contractile elements by second messengers. *Biochem. Biophys. Res. Commun.* 163: 929–35.

Nomura, M., Stull, J. T., Kamm, K. E., and Mumby, M. C. (1992). Site-specific dephosphorylation of smooth muscle myosin light chain kinase by protein phosphatases 1 and 2A. *Biochemistry* 31: 11915–20.

North, A. J., Gimona, M., Cross, R. A., and Small, J. V. (1994). Calponin is localised in both the contractile apparatus and the cytoskeleton of smooth muscle cells. *J. Cell Sci.* 107: 437–44.

Okamoto, Y., Sekine, T., Grammer, J., and Yount, R. G. (1986). The essential light chains constitute part of the active site of smooth muscle myosin. *Nature* 324: 78–80.

Okubo, S., Ito, M., Takashiba, Y., Ichikawa, K., Miyahara, M., Shimizu, H., Konishi, T., Shima, H., Nagao, M., and Hartshorne, D. J., et al. (1994). A regulatory subunit of smooth muscle myosin bound phosphatase. *Biochem. Biophys. Res. Commun.* 200: 429–34.

Olson, N. J., Pearson, R. B., Needleman, D. S., Hurwitz, M. Y., Kemp, B. E., and Means, A. R. (1990). Regulatory and structural motifs of chicken gizzard myosin light chain kinase. *Proc. Nat. Acad. Sci. USA* 87: 2284–8.

Olwin, B. B., Edelman, A. M., Krebs, E. G., and Storm, D. R. (1984). Quantitation of energy coupling between Ca^{2+}, calmodulin, skeletal muscle myosin light chain kinase, and kinase substrates. *J. Biol. Chem.* 259: 10949–55.

O'Neil, K. T., and DeGrado, W. F. (1990). How calmodulin binds its targets: Sequence independent recognition of amphiphilic α-helices. *Trends Biochem. Sci.* 15: 59–63.

Ozaki, H., Ohyama, T., Sato, K., and Karaki, H. (1990). Ca^{2+}-dependent and independent mechanisms of sustained contraction in vascular smooth muscle of rat aorta. *Jpn. J. Pharmacol.* 52: 509–12.

Parry, D. A. D., and Squire, J. M. (1973). Structural role of tropomyosin in muscle regulation: Analysis of the X-ray diffraction patterns from relaxed and contracting muscle. *J. Mol. Biol.* 75: 33–55.

Pato, M. D., and Adelstein, R. S. (1983a). Purification and characterization of a multisubunit phosphatase from turkey gizzard smooth muscle. The effect of calmodulin binding to myosin light chain kinase on dephosphorylation. *J. Biol. Chem.* 258: 7047–54.

Pato, M. D., and Adelstein, R. S. (1983b). Characterization of a Mg^{2+}-dependent phosphatase from turkey gizzard smooth muscle. *J. Biol. Chem.* 258: 7055–8.

Pato, M. D., Adelstein, R. S., Crouch, D., Safer, B., Ingebritsen, T. S., and Cohen, P. (1983). The protein phosphatases involved in cellular regulation. 4. Classification of two homogeneous myosin light chain phosphatases from smooth muscle as protein phosphatase-2A1 and 2C, and a homogeneous protein phosphatase from reticulocytes active on protein synthesis initiation factor eIF-2 as protein phosphatase-2A2. *Eur. J. Biochem.* 132: 283–7.

Pato, M. D., and Kerc, E. (1985). Purification and characterization of a smooth muscle myosin phosphatase from turkey gizzards. *J. Biol. Chem.* 260: 12359–66.

Payne, M. E., Elzinga, M., and Adelstein, R. S. (1986). Smooth muscle myosin light chain kinase. Amino acid sequence at the site phosphorylated by adenosine cyclic 3',5'-phosphate-dependent protein kinase whether or not calmodulin is bound. *J. Biol. Chem.* 261: 16346–50.

Pearson, R. B., Wettenhall, R. E. H., Means, A. R., Hartshorne, D. J., and Kemp, B. E. (1988). Autoregulation of enzymes by pseudosubstrate prototopes: Myosin light chain kinase. *Science* 241: 970–3.

Persechini, A., Kamm, K. E., and Stull, J. T. (1986). Different phosphorylated forms of myosin in contracting tracheal smooth muscle. *J. Biol. Chem.* 261: 6293–9.

Persechini, A., McMillan, K., and Leakey, P. (1994). Activation of myosin light chain kinase and nitric oxide synthase activities by calmodulin fragments. *J. Biol. Chem.* 269: 16148–54.

Pollard, T. D. (1990). Actin. *Curr. Opin. Cell Biol.* 2: 33–40.

Potier, M.-C., Chelot, E., Pekarsky, Y., Gardiner, K., Rossier, J., and Turnell, W. G. (1995). The human myosin light chain kinase (MLCK) from hippocampus: Cloning, sequencing, expression, and localization to 3qcen-q21. *Genomics* 29: 562–70.

Pritchard, K., and Marston, S. B. (1993). The Ca^{2+}-sensitizing component of smooth muscle thin filaments: Properties of regulatory factors that interact with caldesmon. *Biochem. Biophys. Res. Commun.* 190: 668–73.

Rayment, I., Rypniewski, W. R., Schmidt-Bäse, K., Smith, R., Tomchick, D. R., Benning, M. M., Winkelmann, D. A., Wesenberg, G., and Holden, H. M. (1993). Three-dimensional structure of myosin subfragment-1: A molecular motor. *Science* 261: 50–8.

Rembold, C. M., and Murphy, R. A. (1988). Myoplasmic [Ca^{2+}] determines myosin phosphorylation in agonist-stimulated swine arterial smooth muscle. *Circ. Res.* 63: 593–603.

Rembold, C. M., and Murphy, R. A. (1993). Models of the mechanism for crossbridge attachment in smooth muscle. *J. Muscle Res. Cell Motil.* 14: 325–33.

Rieker, J. P., Swanljung-Collins, H., Montibeller, J., and Collins, J. H. (1987). Brush border myosin heavy chain phosphorylation is regulated by calcium and calmodulin. *FEBS Letts.* 212: 154–8.

Rovner, A. S., Murphy, R. A., and Owens, G. K. (1986a). Expression of smooth muscle and nonmuscle myosin heavy chains in cultured vascular smooth muscle cells. *J. Biol. Chem.* 261: 14740–5.

Rovner, A. S., Thompson, M. M., and Murphy, R. A. (1986b). Two different heavy chains are found in smooth muscle myosin. *Am. J. Physiol.* 250: C861–C870.

Rubenstein, P. A., and Spudich, J. A. (1977). Actin microheterogeneity in chick embryo fibroblasts. *Proc. Nat. Acad. Sci. USA* 74: 120–3.

Rüegg, J. C., and Paul, R. J. (1982). Vascular smooth muscle. Calmodulin and cyclic AMP-dependent protein kinase alter calcium sensitivity in porcine carotid skinned fibers. *Circ. Res.* 50: 394–9.

Rüegg, J. C., Pfitzer, G., Zimmer, M., and Hofmann, F. (1984). The calmodulin fraction responsible for contraction in an intestinal smooth muscle. *FEBS Letts.* 170: 383–6.

Sato, K., Hori, M., Ozaki, H., Takano-Ohmuro, H., Tsuchiya, T., Sugi, H., and Karaki, H. (1992). Myosin phosphorylation-independent contraction induced by phorbol ester in vascular smooth muscle. *J. Pharmacol. Exp. Ther.* 261: 497–505.
Satoh, S., Rensland, H., and Pfitzer, G. (1993). Ras proteins increase Ca^{2+}-responsiveness of smooth muscle. *FEBS Letts.* 324: 211–5.
Schildmeyer, L. A., and Seidel, C. L. (1989). Quantitative and qualitative heterogeneity in smooth muscle myosin heavy chains. *Life Sci.* 45: 1617–25.
Sellers, J. R. (1985). Mechanism of the phosphorylation-dependent regulation of smooth muscle heavy meromyosin. *J. Biol. Chem.* 260: 15815–19.
Sellers, J. R. (1991). Regulation of cytoplasmic and smooth muscle myosin. *Curr. Opin. Cell Biol.* 3: 98–104.
Sellers, J. R., and Adelstein, R. S. (1987). Regulation of contractile activity. In: *The Enzymes* (P. D. Boyer and E. G. Krebs, eds.). Orlando, FL: Academic Press, pp. 381–418.
Sellers, J. R., and Pato, M. D. (1984). The binding of smooth muscle myosin light chain kinase and phosphatases to actin and myosin. *J. Biol. Chem.* 259: 7740–6.
Shimizu, H., Ito, M., Miyahara, M., Ichikawa, K., Okubo, S., Konishi, T., Naka, M., Tanaka, T., Hirano, K., Hartshorne, D. J., and Nakano, T. (1994). Characterization of the myosin-binding subunit of smooth muscle myosin phosphatase. *J. Biol. Chem.* 269: 30407–11.
Shirazi, A., Iizuka, K., Fadden, P., Mosse, C., Somlyo, A. P., Somlyo, A. V., and Haystead, T. A. J. (1994). Purification and characterization of the mammalian myosin light chain phosphatase holoenzyme. *J. Biol. Chem.* 269: 31598–1606.
Shirinsky, V. P., Vorotnikov, A. V., Birukov, K. G., Nanaev, A. K., Collinge, M., Lukas, T. J., Sellers, J. R., and Watterson, D. M. (1993). A kinase-related protein stabilizes unphosphorylated smooth muscle myosin minifilaments in the presence of ATP. *J. Biol. Chem.* 268: 16578–83.
Shoemaker, M. O., Lau, W., Shattuck, R. L., Kwiatkowski, A. P., Matrisian, P. E., Guerra-Santos, L., Wilson, E., Lukas, T. J., Van Eldik, L. J., and Watterson, D. M. (1990). Use of DNA sequence and mutant analyses and antisense oligodeoxynucleotides to examine the molecular basis of nonmuscle myosin light chain kinase autoinhibition, calmodulin recognition, and activity. *J. Cell Biol.* 111: 1107–25.
Simons, M., Want, M., McBride, W., Kawamoto, S., Yamakawa, K., Gdula, D., Adelstein, R. S., and Weir, L. (1991). Human nonmuscle myosin heavy chains are encoded by two genes located on different chromosomes. *Circ. Res.* 69: 530–9.
Singer, H. A. (1990). Protein kinase C activation and myosin light chain phosphorylation in ^{32}P-labeled arterial smooth muscle. *Am. J. Physiol.* 259: C631–C639.
Singer, H. A., and Baker, K. M. (1987). Calcium dependence of phorbol 12,13-dibutyrate-induced force and myosin light chain phosphorylation in arterial smooth muscle. *J. Pharmacol. Exp. Ther.* 243: 814–21.
Singer, H. A., Oren, J. W., and Benscoter, H. A. (1989). Myosin light chain phosphorylation in ^{32}P-labeled rabbit aorta stimulated by phorbol 12,13-dibutyrate and phenylephrine. *J. Biol. Chem.* 264: 21215–22.
Sobieszek, A. (1977). Ca-linked phosphorylation of a light chain of vertebrate smooth-muscle myosin. *Cell Motil.* 73: 477–83.
Somlyo, A. P. (1993). Myosin isoforms in smooth muscle: How may they affect function and structure? *J. Muscle Res. Cell Motil.* 14: 557–63
Somlyo, A. P., and Somlyo, A. V. (1994). Signal transduction and regulation in smooth muscle. *Nature* 372: 231–6.
Somlyo, A. V., Butler, T. M., Bond, M., and Somlyo, A. P. (1981). Myosin filaments have nonphosphorylated light chains in relaxed smooth muscle. *Nature* 294: 567–9.

Somlyo, A. V., Goldman, Y. E., Fujimori, T., Bond, M., Trentham, D. R., and Somlyo, A. P. (1988). Cross-bridge kinetics, cooperativity, and negatively strained cross-bridges in vertebrate smooth muscle. A laser-flash photolysis study. *J. Gen. Physiol.* 91: 165–92.

Sparrow, M. P., Mohammad, M. A., Arner, A., Hellstrand, P., and Rüegg, J. C. (1988). Myosin composition and functional properties of smooth muscle from the uterus of pregnant and non-pregnant rats. *Pflügers Arch.* 412: 624–33.

Stralfors, P., Hiraga, A., and Cohen, P. (1985). The protein phosphatases involved in cellular regulation. Purification and characterisation of the glycogen-bound form of protein phosphatase-1 from rabbit skeletal muscle. *Eur. J. Biochem.* 149: 295–303.

Stull, J. T., Bowman, B. F., Gallagher, P. J., Herring, B. P., Hsu, L. C., Kamm, K. E., Kubota, Y., Leachman, S. A., Sweeney, H. L., and Tansey, M. G. (1990a). Myosin phosphorylation in smooth and skeletal muscles: Regulation and function. *Prog. Clin. Biol. Res.* 327: 107–26.

Stull, J. T., Hsu, L.-C., Tansey, M. G., and Kamm, K. E. (1990b). Myosin light chain kinase phosphorylation in tracheal smooth muscle. *J. Biol. Chem.* 265: 16683–90.

Stull, J. T., Krueger, J. K., Kamm, K. E., Gao, Z.-H., Zhi, G., and Padre, R. (1996). Myosin light chain kinase. In: *Biochemistry of Smooth Muscle Contraction* (M. Bárány, ed.). Orlando, FL: Academic Press, 119–30.

Stull, J. T., Krueger, J. K., Zhi, G., and Gao, Z.-H. (1995). Molecular properties of myosin light chain kinases. In: *Calcium as Cell Signal* (K. Maruyama, Y. Nonomura, and K. Kohama, eds.). Tokyo: Igaku-Shoin, pp. 175–84.

Suematsu, E., Resnick, M., and Morgan, K. G. (1991). Change of Ca^{2+} requirement for myosin phosphorylation by prostaglandin $F_{2\alpha}$. *Am. J. Physiol.* 261: C253–C258.

Sweeney, H. L., Bowman, B. F., and Stull, J. T. (1993). Myosin light chain phosphorylation in vertebrate striated muscle: Regulation and function. *Am. J. Physiol.* 264: C1085–C1095.

Takai, A., Troschka, M., Mieskes, G., and Somlyo, A. V. (1989). Protein phosphatase composition in the smooth muscle of guinea-pig ileum studied with okadaic acid and inhibitor 2. *Biochem. J.* 262: 617–23.

Takai, Y., Sasaki, T., Tanaka, K., and Nakanishi, H. (1995). Rho as a regulator of the cytoskeleton. *Trends Biochem. Sci.* 20: 227–31.

Tan, J. L., Ravid, S., and Spudich, J. A. (1992). Control of nonmuscle myosins by phosphorylation. *Annu. Rev. Biochem.* 61: 721–59.

Tang, D.-C., Kubota, Y., Kamm, K. E., and Stull, J. T. (1993). GTPgammaS-induced phosphorylation of myosin light chain kinase in smooth muscle. *FEBS Letts.* 331: 272–5.

Tang, D.-C., Stull, J. T., Kubota, Y., and Kamm, K. E. (1992). Regulation of the Ca^{2+} dependence of smooth muscle contraction. *J. Biol. Chem.* 267: 11839–45.

Tansey, M. G., Hori, M., Karaki, H., Kamm, K. E., and Stull, J. T. (1990). Okadaic acid uncouples myosin light chain phosphorylation and tension in smooth muscle. *FEBS Letts.* 270: 219–21.

Tansey, M. G., Luby-Phelps, K., Kamm, K. E., and Stull, J. T. (1994). Ca^{2+}-dependent phosphorylation of myosin light chain kinase decreases the Ca^{2+} sensitivity of light chain phosphorylation within smooth muscle cells. *J. Biol. Chem.* 269: 9912–20.

Tansey, M. G., Word, R. A., Hidaka, H., Singer, H. A., Schworer, C. M., Kamm, K. E., and Stull, J. T. (1992). Phosphorylation of myosin light chain kinase by the multifunctional calmodulin-dependent protein kinase II in smooth muscle cells. *J. Biol. Chem.* 267: 12511–16.

Toyoshima, Y. Y., Kron, S. J., McNally, E. M., Niebling, K. R., Toyoshima, C., and Spudich, J. A. (1987). Myosin subfragment-1 is sufficient to move actin filaments *in vitro*. *Nature* 328: 536–9.

Trinkle-Mulcahy, L., Ichikawa, K., Hartshorne, D. J., and Siegman, M. J. (1995). Thiophosphorylation of the 130-kDa subunit is associated with a decreased activity of myosin light chain phosphatase in alpha-toxin-permeabilized smooth muscle. *J. Biol. Chem.* 270: 18191–4.

Trotter, J. A. (1982). Living macrophages phosphorylate the 20,000 Dalton light chains and heavy chains of myosin. *Biochem. Biophys. Res. Commun.* 106: 1071–7.

Trybus, K. M. (1991). Regulation of smooth muscle myosin. *Cell Motil. Cytoskeleton* 18: 81–5.

Trybus, K. M., and Chatman, T. A. (1993). Chimeric regulatory light chains as probes of smooth muscle myosin function. *J. Biol. Chem.* 268: 4412–19.

Trybus, K. M., and Henry, L. (1989). Monoclonal antibodies detect and stabilize conformational states of smooth muscle myosin. *J. Cell Biol.* 109: 2879–86.

Tsao, A. E., and Eddinger, T. J. (1993). Smooth muscle myosin heavy chains combine to form three different native myosin isoforms. *Am. J. Physiol.* 264: H1653–H1662.

Tulloch, A. G., and Pato, M. D. (1991). Turkey gizzard smooth muscle myosin phosphatase-III is a novel protein phosphatase. *J. Biol. Chem.* 266: 20168–20174.

Tung, H. Y., and Cohen, P. (1984). The protein phosphatases involved in cellular regulation. Comparison of native and reconstituted Mg-ATP-dependent protein phosphatases from rabbit skeletal muscle. *Eur. J. Biochem.* 145: 57–64.

VanBuren, P., Work, S. S., and Warshaw, D. M. (1994). Enhanced force generation by smooth muscle myosin *in vitro*. *Proc. Nat. Acad. Sci. USA* 91: 202–5.

Van Riper, D. A., Weaver, B. A., Stull, J. T., and Rembold, C. M. (1995). Myosin light chain kinase phosphorylation in swine carotid artery contraction and relaxation. *Am. J. Physiol.* 268: H1–H10.

Vibert, P., and Cohen, C. (1988). Domains, motions and regulation in the myosin head. *J. Muscle Res. Cell Motil.* 9: 296–305.

Vibert, P. J., Haselgrove, J. C., Lowry, J., and Poulsen, F. R. (1972). Structural changes in actin-containing filaments of muscle. *J. Mol. Biol.* 71: 757–67.

Vyas, T. B., Mooers, S. U., Narayan, S. R., Siegman, M. J., and Butler, T. M. (1994). Cross-bridge cycling at rest and during activation. *J. Biol. Chem.* 269: 7316–22.

Vyas, T. B., Mooers, S. U., Narayan, S. R., Witherell, J. C., Siegman, M. J., and Butler, T. M. (1992). Cooperative activation of myosin by light chain phosphorylation in permeabilized smooth muscle. *Am. J. Physiol.* 263: C210–C219.

Waller, G. S., Ouyang, G., Swafford, J., Vibert, P., and Lowey, S. (1995). A minimal motor domain from chicken skeletal muscle myosin. *J. Biol. Chem.* 270: 15348–52.

Walmsley, J. G., and Murphy, R. A. (1987). Force-length dependence of arterial lamellar, smooth muscle, and myofilament orientations. *Am. J. Physiol.* 253: H1141–H1147.

Walsh, M. P., Bridenbaugh, R., Hartshorne, D. J., and Kerrick, W. G. L. (1982). Phosphorylation-dependent activated tension in skinned gizzard muscle fibers in the absence of Ca^{2+}. *J. Biol. Chem.* 257: 5987–90.

Watterson, D. M., Collinge, M., Lukas, T. J., Van Eldik, L. J., Birukov, K. G., Stepanova, O. V., and Shirinsky, V. P. (1995). Multiple gene products are produced from a novel protein kinase transcription region. *FEBS Letts.* 373: 217–20.

Weeds, A. (1988). Actin-binding proteins – Regulators of cell architecture and motility. *Nature* 2296: 811–16.

Whalen, R. G., Butler-Browne, G. S., and Gros, F. (1976). Protein synthesis and actin heterogeneity in calf muscle cells in culture. *Proc. Nat. Acad. Sci. USA* 73: 2018–22.

White, S., Martin, A. E., and Periasamy, M. (1993). Identification of a novel smooth muscle myosin heavy chain cDNA; Isoform diversity in the S1 head region. *Am. J. Physiol.* 264: C1252–C1258.

Wingard, C. J., Browne, A. K., and Murphy, R. A. (1995). Dependence of force on length at constant cross-bridge phosphorylation in the swine carotid media. *J. Physiol. (Lond.)* 488: 729–39.

Word, R. A., Tang, D.-C., and Kamm, K. E. (1994). Activation properties of myosin light chain kinase during contraction/relaxation cycles of tonic and phasic smooth muscles. *J. Biol. Chem.* 269: 21596–1602.

Xie, X., Harrison, D. H., Schlichting, I., Sweet, R. M., Kalabokis, V. N., Szent-Györgyi, A. G., and Cohen, C. (1994). Structure of the regulatory domain of scallop myosin at 2.8 Å resolution. *Nature* 368: 306–12.

Yang, Z., and Sweeney, H. L. (1995). Restoration of phosphorylation-dependent regulation to the skeletal muscle myosin regulatory light chain. *J. Biol. Chem.* 270: 24646–9.

Yano, K., Araki, Y., Hales, S. J., Tanaka, M., and Ikebe, M. (1993). Boundary of the autoinhibitory region of smooth muscle myosin light-chain kinase. *Biochemistry* 32: 12054–61.

Yoshikai, S.-I., and Ikebe, M. (1992). Molecular cloning of the chicken gizzard telokin gene and cDNA. *Arch. Biochem. Biophys.* 299: 242–7.

Zhi, G., Herring, B. P., and Stull, J. T. (1994). Structural requirements for phosphorylation of myosin regulatory light chain from smooth muscle. *J. Biol. Chem.* 269: 24723–7.

Zimmermann, B., Somlyo, A. V., Ellis-Davies, C. R., Kaplan, J. H., and Somlyo, A. P. (1995). Kinetics of prephosphorylation reactions and myosin light chain phosphorylation in smooth muscle. *J. Biol. Chem.* 270: 23966–74.

7

Structure and Function of the Thin Filament Proteins of Smooth Muscle

JOSEPH M. CHALOVICH AND GABRIELE PFITZER

The goal of this chapter is to explore possible models by which caldesmon and calponin may alter the force-producing interaction between myosin and actin and how the inhibitory activity of caldesmon and calponin may be controlled. Contraction may be potentially regulated by altering the properties of either myosin or actin. In both cases, one or more of many possible transitions in the cycle of ATP hydrolysis by actomyosin may be affected, including (but not limited to) the binding of myosin to actin, the binding of ATP, Pi, and ADP to actomyosin, and cooperative transitions between the active and inactive forms of the actin–tropomyosin filament. A detailed account of these transitions is not possible here but may be found elsewhere (Chalovich 1992).

1. Evidence for actin filament–mediated regulation

Phosphorylation of the 20-kDa myosin light chain (MLC_{20}) by the Ca^{2+}- and calmodulin-dependent myosin light chain kinase (MLCK) is generally thought to be the primary event initiating smooth muscle contraction (for reviews see Kamm and Stull 1985; Somlyo and Somlyo 1994; and Chapter 6 of this volume). In fact, there is evidence that MLC_{20} phosphorylation is sufficient to trigger smooth muscle contraction (Itoh et al. 1989). On the other hand, numerous physiological studies have shown that there is no fixed relationship between isometric force and MLC_{20} phosphorylation. During prolonged contraction MLC_{20} phosphorylation, crossbridge cycling rates, and intracellular Ca^{2+} decrease while force is fully maintained by the so-called latch state (Dillon et al. 1981; see also Kamm and Stull 1985 for a review). Together with the observation that isolated native thin filaments from smooth muscle are Ca^{2+}-regulated (see Marston and Smith 1985), this dissociation of force and phosphorylation led to the proposal that smooth muscle contraction is dually regulated by MLC_{20} phosphorylation and a second Ca^{2+}-dependent mechanism (Gerthoffer and Murphy 1983). However, more recently it was proposed that the latch state (or, more generally, any high force–low phosphorylation state) does not require a second Ca^{2+}-dependent regulatory mechanism. High force output at low levels of MLC_{20} phosphorylation could occur if a small fraction of phosphorylated myosin led to the cooperative attachment of a maximum number of myosin molecules or, alternatively, if force-bearing dephosphorylated crossbridges were generated by the dephosphorylation of attached phosphorylated crossbridges (see Remboldt and Murphy 1993; Somlyo and Somlyo 1994). Nevertheless, there are a number of observations that are more readily explained in terms of a second regulatory mechanism. First, several laboratories demonstrated the existence of low force–high phosphorylation states (Gerthoffer 1987; Rüegg et al. 1989; Tansey et al. 1990; Steusloff et al. 1995). Second, relaxation may occur without apparent changes in MLC_{20} phosphorylation levels (Fischer and Pfitzer 1989). Third, the coupling between force and phosphorylation depends on the type of agonist (Suematsu et al. 1991). Fourth, under certain conditions, vascular smooth muscle cells contract at constant Ca^{2+} levels which are at or below resting levels (Collins et al. 1992). These observations indicate that there must be one or more mechanisms regulating the attachment of both phosphorylated and unphosphorylated crossbridges. These regulatory functions could be mediated by the thin filament–associated proteins, caldesmon and calponin.

2. Evidence that caldesmon affects contractility

Experiments in solution suggest that caldesmon has the potential of regulating smooth muscle contraction. This is supported by the following experimental approaches: (i) studies using skinned fibers; (ii) *in vitro* motility assays, and (iii) correlation of caldesmon phosphorylation with contraction in intact smooth muscle.

The synthetic peptide corresponding to residues Gly-651–Ser-667 (GS17C) of caldesmon binds to both Ca^{2+}–calmodulin and F actin but does not inhibit actomyosin ATPase activity. Incubation of saponin-permeabilized ferret aorta smooth muscle cells with GS17C resulted in activation even at constant low concentrations of Ca^{2+} (i.e., pCa 9; Katsuyama et al. 1992). The authors suggested that GS17C displaced endogenous caldesmon, thus antagonizing its normal inhibitory action. This implies that endogenous caldesmon inhibits force production in resting smooth muscle. Furthermore, incubation of Triton-skinned gizzard fibers with caldesmon produced relaxation (Szpacenko et al. 1985) or inhibition of tension development at constant MLC_{20} phosphorylation (Pfitzer et al. 1993). Tension development was also inhibited by the 20-kDa actin-binding fragment but not by a myosin-binding fragment from caldesmon. The relation between force and phosphorylation was shifted toward higher levels of phosphorylation in the presence of the 20-kDa actin-binding fragment (Pfitzer et al. 1993). Both approaches suggest that caldesmon is not involved in maintenance of the latch state. Rather, caldesmon appears to induce relaxation or inhibit force.

Although these studies suggest a role for caldesmon in the regulation of smooth muscle, they do not establish it. The study of Katsuyama et al. (1992)

did not directly show that the GS17C peptide actually displaces endogenous caldesmon. The study of Pfitzer et al. (1993) demonstrated an effect of exogenously added caldesmon that was probably at an unphysiologically high concentration of caldesmon. It should be noted, however, that loading skinned fibers with an excess of a protein is similar to overexpression of a cellular protein without the disadvantage of changes in the cell phenotype. The latter type of study has been widely used to study protein function.

Caldesmon has also been shown to inhibit the movement of actin filaments over myosin coated surfaces in the *in vitro* motility assay (Okagaki et al. 1991; Haeberle et al. 1992; Shirinsky et al. 1992). In addition, caldesmon inhibits force production when added to skeletal muscle fibers that normally lack caldesmon (Brenner et al. 1991). These studies are discussed in more detail in later sections.

If caldesmon does inhibit contraction, there must be a mechanism for disinhibiting caldesmon. Several possible mechanisms do in fact exist and will be discussed in more detail later. Phosphorylation of caldesmon is particularly interesting, because caldesmon is phosphorylated in intact tracheal and vascular smooth muscle in response to a number of agonists including KCl, carbachol, phorbol esters, endothelin, and norepinephrine (Park and Rasmussen 1986; Adam et al. 1989, 1990; Abe et al. 1991; Bárány et al. 1992a). In resting smooth muscle, caldesmon phosphorylation was 0.5 mol Pi per 1 mol caldesmon, increasing to 1.0–1.6 mol Pi per mole in stimulated tissue (Adam et al. 1989; Bárány et al. 1992a). Phosphorylation of caldesmon occurs at a slower rate than phosphorylation of MLC_{20} and approximately parallels force development (Adam et al. 1989). Although it is clear that caldesmon is specifically phosphorylated during activation of smooth muscle, the relationship is complex: (i) there is no correlation between the stoichiometry of caldesmon phosphorylation and tension developed (Adam et al. 1989); (ii) relaxation precedes dephosphorylation of caldesmon (Adam et al. 1989) and may even occur without dephosphorylation (Abe et al. 1991; Bárány et al. 1992a); and (iii) in the presence of the stimulus, phosphorylation of caldesmon is high irrespective of whether the contraction is tonic, as in the case of stimulation with KCl, or phasic, as in the case of stimulation with angiotensin II (Adam et al. 1990).

Of the several protein kinases that phosphorylate caldesmon *in vitro*, only MAP kinase appears to phosphorylate caldesmon *in vivo* (Adam et al. 1989; Adam and Hathaway, 1993). Mitogen-activated protein (MAP) kinase and other members of the MAP kinase signaling pathway are present in smooth muscle (Childs et al. 1992; Adam and Hathaway 1993). These findings led to a model in which activation of smooth muscle eventually leads to activation of MAP kinase, which then phosphorylates caldesmon (Adam and Hathaway 1993). However, to date it has not been shown that there is a causal relation between phosphorylation of caldesmon by MAP kinase and modulation of smooth muscle contraction.

Other important points relevant to the function of caldesmon are its concentration and distribution. Caldesmon is found in both smooth muscle and nonmuscle cells (see Sobue and Sellers 1991). Contractile smooth muscle expresses a high–molecular mass isoform (h-caldesmon), whereas de-differentiated smooth muscle cells and nonmuscle cells express a low–molecular mass isoform (l-caldesmon). There is some controversy about the caldesmon content of smooth muscle cells, as well as conflicting views on the fraction of cellular actin that contains bound caldesmon (Haeberle, Hathaway, and Smith 1992; Lehman et al. 1992). Haeberle, Hathaway, and Smith reported that the ratio of caldesmon to actin is considerably higher in visceral muscle (phasic) than in vascular smooth muscle (tonic). The other view is that the ratio of caldesmon is constant (0.1) in all smooth muscles. Haeberle, Hathaway, and Smith argue that both groups agree on the caldesmon content but that they have extracted a greater fraction of cellular actin, giving lower ratios of caldesmon to actin (Haeberle, Hathaway, and Smith 1992).

There is agreement that, in smooth muscle cells, caldesmon is associated with those actin filaments which contain tropomyosin and myosin and which are devoid of intermediate filament proteins (Furst et al. 1986; North et al. 1994; Mabuchi et al. 1996). Caldesmon has the proper location for either a regulatory protein or a structural protein involved in the organization of the contractile apparatus.

3. Caldesmon binding to actin

The task of obtaining suitable binding data for caldesmon and calponin is complicated because both proteins bind to multiple actin monomers and both can crosslink actin filaments. The problem with crosslinking can be dealt with experimentally. The problem of binding to a site consisting of n actin monomers ($n > 1$) can be handled by fitting an appropriate model to the data. Such a model is available (McGhee and vonHippel, 1974).

Caldesmon was first identified as a protein that binds to actin and calmodulin (Sobue et al. 1981). The first detailed study of the binding of caldesmon to actin reported two classes of binding sites (Smith et al. 1987). The high-affinity binding site ($K \approx 10^7$ M^{-1}) had a stoichiometry of 1 caldesmon to 22 actin monomers (corrected to 93,000 M_r for caldesmon), whereas the low-affinity site ($K \approx 10^5$ M^{-1}) had a stoichiometry of one caldesmon per four or five actin monomers. The high-affinity binding was initially thought to be present in both the absence and presence of tropomyosin (Smith et al. 1987), although subsequent reports refer to the high-affinity binding only in the presence of tropomyosin. In a subsequent study, only a single population of binding sites was observed, with a stoichiometry of about 1:7 for both actin and actin-tropomyosin (Velaz et al. 1989). Analysis of caldesmon binding by the model of McGhee and vonHippel gave values of $\omega = 6$ and $K = 6 \times 10^5$ M^{-1} for pure actin, and $\omega = 5$ and $K = 1.5 \times 10^6$ M^{-1} for actin–tropomyosin (Velaz et al. 1989). In this model, ω is the interaction parameter for adjacent caldesmon monomers and K is the association constant to an isolated site of n actin monomers. The product ωK (4×10^6 and 8×10^6 M^{-1} in the absence and presence of tropomyosin, respectively) should be compared to the earlier values obtained without this analysis. Figure 1 compares some of the actin binding data from these and other laboratories. The data have been corrected to the current values of the molecular mass of caldesmon: 93 kDa for chicken gizzard (Graceffa et al. 1988) and 90 kDa for turkey gizzard caldesmon (Stafford et al. 1994). There is little difference among these corrected data. The ar-

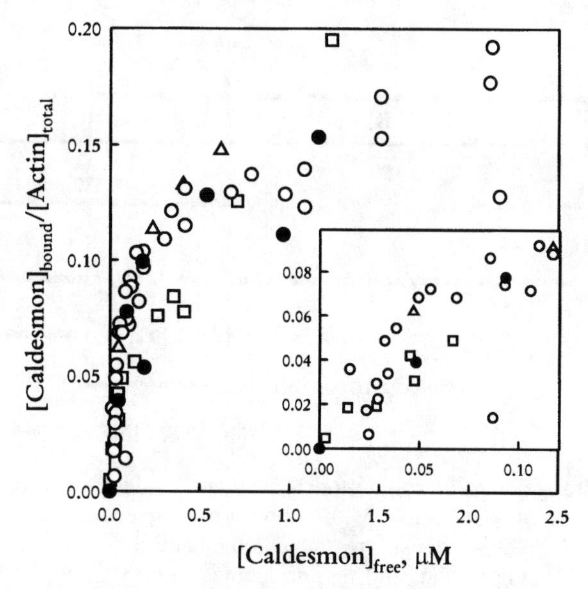

Figure 1. Binding of caldesmon to actin-tropomyosin as measured in several laboratories and corrected to the true molecular weight of caldesmon; open circles (Velaz et al. 1989), solid circles (Wang et al. 1994), squares (Smith et al. 1987) and triangles (Ngai and Walsh, 1987). The inset is an expanded view of the data.

gument for two populations of sites comes from the slight deviation of the squares from a smooth hyperbola. A greater deviation from a single binding population was observed with a recombinant COOH-terminal actin-binding fragment of caldesmon (658C), which bound to actin–tropomyosin in a biphasic manner with a strong site saturating at 1 per 37 actin monomers (Redwood and Marston 1993). Fragment 658C is similar to the 10-kDa CNBr fragment shown in Figure 2. In contrast, the chymotryptic 20-kDa fragment bound actin-tropomyosin with a single phase, giving saturation at one fragment per two to three actin monomers (Velaz et al. 1993). No evidence of multiple sites was observed in a study of the kinetics of binding of caldesmon to actin–tropomyosin (Chalovich et al. 1995). In that study, caldesmon bound to actin and actin-tropomyosin with an association rate constant near 1×10^7 M^{-1} s^{-1}; the only effect of tropomyosin was to reduce the dissociation rate constant from 18 s^{-1} to about 6 s^{-1}.

The structure of the actin–caldesmon complex has not been completely described. Hydrodynamic studies (Graceffa et al. 1988) and electron-microscopic images (Mabuchi et al. 1993; Mabuchi and Wang 1991) of caldesmon are consistent with dimensions of ~740 × 19 Å. Electron microscopy of reconstituted or native thin filaments indicated that caldesmon is arranged along the filament in association with the actin–tropomyosin backbone (Furst et al. 1986; Moody et al. 1990; Mabuchi et al. 1993; Vibert et al. 1993). The periodicity of caldesmon on actin is similar to that of tropomyosin (Lehman et al. 1989). One laboratory reported electron-microscopic evidence that both ends of caldesmon appear to bind to actin (Mabuchi et al. 1993). However, the strong actin inter-

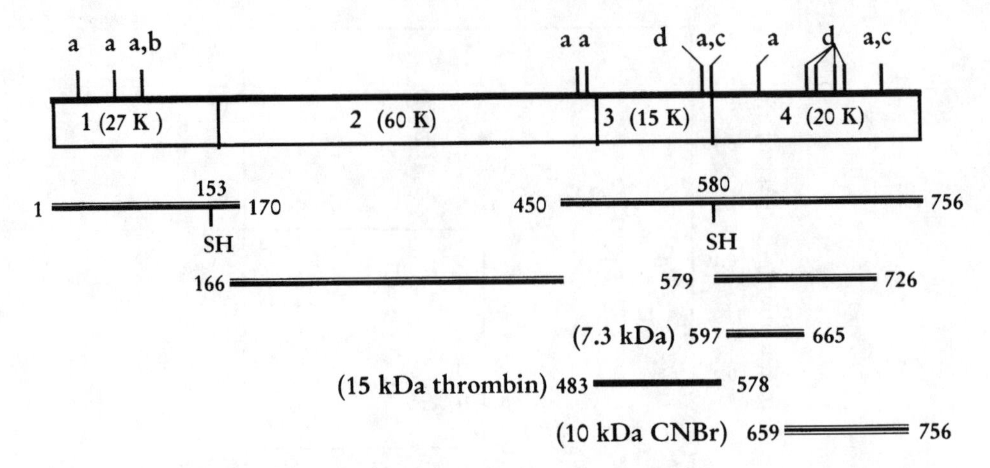

Figure 2. Structural regions of caldesmon. The intact caldesmon molecule can be described as consisting of four regions with approximate molecular weights shown in parentheses. Region 1 binds to myosin, region 2 is the central helical region, region 3 is a low-affinity actin-binding region, and region 4 binds more tightly to actin and inhibits ATPase activity. Several sites of phosphorylation are shown, as indicated by the letters a–d: "a" (Ca^{2+} calmodulin protein kinase II), "b" (casein kinase), "c" (protein kinase C) and "d" (CDC or MAP kinase). Several useful caldesmon fragments are also shown. Those fragments produced by digestion with thrombin and CNBr are labeled; the remaining five fragments are produced by digestion with chymotrypsin. The numbering of the caldesmon sequence is that of Bryan et al. (1989).

action is clearly at the COOH-terminal region of caldesmon. Several researchers have proposed that the NH_2-terminal region of caldesmon may project from the actin filament, although this has been very difficult to demonstrate. Marston showed that intact myosin and myosin rod filaments can coaggregate with thin filaments and actin–caldesmon (Marston et al. 1992), but the fine structure of these complexes could not be resolved. A more recent study (Katayama and Ikebe 1995) succeeded in demonstrating numerous whiskerlike projections from reconstituted and native thin filaments which could be labeled by antibodies against the N terminus of caldesmon and which bound heavy meromyosin. Together with the finding that caldesmon can induce filament formation of dephosphorylated myosin (Katayama et al. 1995), these recent results suggest that caldesmon has the potential to organize the actomyosin domain of smooth muscle. There are indications that caldesmon may be flexible enough to form a flexed conformation in solution whereby both the NH_2- and COOH-terminal ends could be near each other (Lynch et al. 1987; Horiuchi and Chacko 1988; Martin et al. 1991; Crosbie et al. 1995). However, this does not appear to be a state of high probability, particularly in the presence of actin.

Several regions of actin have been implicated in the binding to caldesmon, but the picture is far from complete. In particular, there is little information regarding (i) the number of actin monomers that have close interactions with caldesmon and (ii) whether the same groups on all actin monomers within a single caldesmon binding site form the same type of contacts. The seven NH_2-terminal residues of actin have been implicated in binding to caldesmon by antibody competition and chemical modification (Adams et al. 1990) as well as by chemical crosslinking (Bartegi et al. 1990b) and nuclear magnetic resonance (NMR) spectroscopy (Mornet et al. 1995). Caldesmon does bind to a mutated actin that has a charge change in the 1–7 region (Crosbie et al. 1994). Therefore, the 1–7 region must be only part of the binding site. NMR studies have further indicated that residues 1–7 and residues 20–41 of actin interact with caldesmon (Mornet et al. 1995). Other regions of actin appear to be involved in caldesmon binding, including the COOH-terminal region (Crosbie et al. 1991, 1992; Makuch et al. 1992; Graceffa et al. 1993) and Lys-373 (Kolakowski et al. 1992).

The region of caldesmon responsible for binding to actin is contained within the 35-kDa fragment shown in Figure 2 (residues 450–756). This fragment binds to actin and inhibits ATPase activity in a Ca^{2+}–calmodulin dependent manner (Szpacenko and Dabrowska 1986; Fujii et al. 1987). Region 3, the 15-kDa thrombin or submaxillaris arginase-C protease fragment containing residues 483–578 (Leszyk et al. 1989), has low actin affinity (Mornet et al. 1988; Hayashi et al. 1991) and does not inhibit ATP hydrolysis. Region 4 (residues ~579–756) is largely encompassed by the 20-kDa chymotryptic fragment, has high actin affinity, and inhibits ATP hydrolysis (Makuch et al. 1989; Riseman et al. 1989; Bartegi et al. 1990a; Hayashi et al. 1991; Velaz et al. 1993). Region 4 is not as potent an inhibitor as the intact region 3+4. Therefore, although region 3 does little by itself, it does substantially enhance the effectiveness of region 4.

The 7.3-kDa chymotryptic fragment, Leu-597–Phe-665 (Chalovich et al. 1992) and the 10-kDa CNBr fragment, Trp-659–Pro756 (Bartegi et al. 1990a) bind to calmodulin and actin and inhibit ATPase activity. The COOH–terminal 7 amino acid residues of the 7.3-kDa fragment overlap with the first seven residues of the 10-kDa fragment; together, these fragments comprise essentially all of region 4 of caldesmon. A synthetic peptide analog of the overlap region (Gly-651–Ser-667) binds to both actin and calmodulin but does not inhibit ATPase activity (Zhan et al. 1991). A synthetic polypeptide corresponding to the Leu-693–Trp-722 (LW30) region of caldesmon within the 10-kDa region binds to actin ($K = 4 \times 10^3\ M^{-1}$) and inhibits the acto-S1 ATPase activity by more than 90% in the presence of tropomyosin (Mezgueldi et al. 1994). It is unclear at present whether the 10-kDa and 7.3-kDa polypeptides bind to separate actin monomers or if they cooperate to produce a single actin binding site. NMR studies support the view that the COOH region is folded (Levine et al. 1990; Mornet et al. 1995). In particular, residues 1–7 of actin bind to residues in region 4 of caldesmon while residues 20–41 of actin bind to region 3 (Mornet et al. 1995).

Several laboratories have shown that caldesmon binds directly to tropomyosin and that this might contribute to the enhancement of caldesmon binding by tropomyosin. It is not clear to what extent caldesmon and tropomyosin interact with each other on the actin filament. Actin appears to eliminate binding of the NH_2-terminal region of caldesmon to residues 142–281 of tropomyosin, but enhances the interaction of the COOH-terminal region of caldesmon to tropomyosin (Tsuruda et al. 1995).

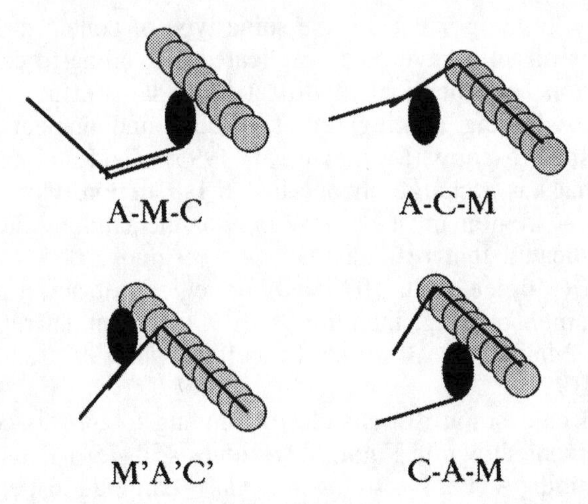

Figure 3. Possible ternary complexes of caldesmon (C), myosin (M), and actin (A). Complex A–M–C has normal activity. Complex M'A'C' is active and is a case where caldesmon facilitates the productive binding of myosin to actin. Complex A–C–M is inactive, since actin is not bound to myosin but is tethered. The degree to which complex C–A–M may form and the degree to which it is active is a matter of debate. If caldesmon inhibits without dissociating actin from myosin, then this complex forms but is inactive.

4. Caldesmon binding to myosin

The discovery of binding of caldesmon to myosin resulted from the different effects that caldesmon had on the apparent binding of skeletal S1 and smooth muscle HMM to actin in sedimentation assays. Thus, the inhibition of ATPase activity by caldesmon was correlated with a decrease in S1–ATP binding to actin (Chalovich et al. 1987) but an increase in smooth muscle myosin HMM–ATP binding to actin (Lash et al. 1986; Hemric and Chalovich 1988). Furthermore, whereas caldesmon enhanced HMM–ATP binding, the binding of the higher-affinity HMM–AMP–PNP to actin was inhibited (Hemric and Chalovich 1988). These data could be explained if caldesmon could bind simultaneously to both actin and myosin as shown in Figure 3. This creates ambiguities in binding studies, since in some methods the tethering of myosin to actin by caldesmon (A–C–M) is indistinguishable from a complex where both caldesmon and myosin are bound to the same actin monomer (C–A–M). The type of complex formed depends on the relative affinities of the three proteins.

Proof of the binding of caldesmon to myosin came from affinity chromatography (Hemric and Chalovich 1988; Ikebe and Reardon 1988) and direct binding studies of [^{14}C]-labeled caldesmon with filamentous smooth muscle myosin (Hemric and Chalovich 1990). The direct measurements indicated that two caldesmon molecules were bound per myosin molecule with an affinity of 5.6×10^4 M^{-1}. Quantitation of binding by electrophoresis gave an association constant of 2.5×10^5 M^{-1} at an ionic strength of about 40 mM (Marston and Redwood 1992). A much higher affinity, 4.5×10^7 M^{-1}, was determined by fluorescence measurements of the binding of acrylodan-labeled caldesmon to smooth muscle HMM (Mani et al. 1992).

ATP changes the stoichiometry of binding of gizzard caldesmon to one caldesmon molecule per myosin molecule, and increases the affinity about twentyfold to 1×10^6 M^{-1} (Hemric and Chalovich 1990; Lu and Chalovich 1995). The effect of ATP is apparently due to a specific change in smooth muscle myosin structure that is different from the 10S–6S change (Lu and Chalovich 1995). One group reported that ATP did not affect the binding of sheep aorta caldesmon to myosin (Marston et al. 1992).

The myosin binding region has been localized to the N-terminal region of caldesmon between residues 1 and 170 (Velaz et al. 1990). Another myosin binding site has been proposed in the COOH-terminal region of caldesmon (Yamakita et al. 1992) between residues 467 and 578 (Huber et al. 1993). This is probably a minor site, since a mutant caldesmon lacking the 159 COOH-terminal amino acids has only a slightly reduced affinity for myosin (Wang et al. 1994). The primary locus on myosin for caldesmon binding is the S2 region (Ikebe and Reardon 1988; Hemric and Chalovich 1988, 1990). However, S1 does compete with the binding of caldesmon to myosin (Hemric and Chalovich 1990), indicating that there is a weak association with the S1 region of myosin.

5. Inhibition of actin-activated ATPase activty

Many early observations regarding the inhibition of actin-activated myosin ATPase activity by caldesmon are reviewed elsewhere (Chalovich et al. 1990). In particular, it is now known that inhibition of ATPase activity is not due to the bundling of actin by caldesmon. It is also known that tropomyosin facilitates the inhibition by caldesmon and that, in the case of filamentous myosin, little effect of caldesmon occurs in the absence of tropomyosin. In the case of soluble myosin subfragments, tropomyosin is facilitative but unessential for inhibition by caldesmon. Tropomyosin increases the affinity of caldesmon for actin by a factor of 2 to 4 and decreases to 33% the amount of *bound* caldesmon required for half-maximal inhibition of ATPase activity (Smith et al. 1987; Velaz et al. 1989; Marston and Redwood 1993). There are two important questions: how does caldesmon inhibit ATPase activity, and how is this inhibition facilitated by tropomyosin?

One factor contributing to inhibition of ATPase activity by caldesmon is the inhibition of the weak binding of skeletal S1–ATP to actin and actin–tropomyosin at low (24 mM) and moderate (66 mM) ionic strengths (Chalovich et al. 1987). In many studies, there is a close relationship between the decrease in ATPase activity and the decrease in S1–ATP binding to actin (Chalovich 1987; Chalovich et al. 1987; Velaz et al. 1993). This correlation is more difficult to demonstrate with smooth muscle HMM, because caldesmon can bind to both HMM and to actin and thereby mask any detachment of HMM from actin. However, actin binding fragments of caldesmon that do not bind strongly to myosin do displace smooth muscle HMM from actin and inhibit ATP hydrolysis (Horiuchi and Chacko 1989; Velaz et al. 1990; Horiuchi et al. 1991). Even in this case, twice the fragment concentration was required for 50% inhibition of binding than for 50% inhibition of ATPase activity. In contrast to these results, inhibition of ATPase activity has been observed with virtually no reversal of S1–ATP binding to actin–tropomyosin (Marston and Redwood 1992, 1993). In these latter studies, the initial saturation of actin with S1 was much higher than in the studies where reversal of binding was observed. These experiments were recently repeated using ^{14}C-ρPDM labeled S1 (Sen and Chalovich 1995). This modification produces a stable S1–ATP-like state with a reduced rate of ATP hydrolysis so that ATP depletion at high S1 concentrations is not a problem. Furthermore, the use of a radioactive probe increases the sensitivity of binding measurements. Both the ATPase activity and the binding of ^{14}C-ρPDM labeled S1 to actin–tropomyosin decreased with increasing concentrations of caldesmon. However, the concentration of caldesmon required to inhibit 50% binding was often twice that required to inhibit 50% of the ATPase activity. Thus, although competition of binding is important, there do appear to be other factors that limit the ATPase rate.

The degree of inhibition of ATPase activity by caldesmon is greater than predicted from the fraction of actin containing bound caldesmon. This is particularly obvious with fragments from the COOH-terminal end of caldesmon such as 658C (Marston and Redwood 1992; Redwood and Marston 1993) and the 20-kDa chymotryptic fragment (Velaz et al. 1993). In the latter case, half-maximal inhibition occurs at 9–12% saturation of the actin in the presence of tropomyosin and at 20–33% saturation in the absence of tropomyosin. However, there is a good correlation between the inhibition of both ATPase activity and binding of S1–ATP to actin–tropomyosin (Velaz et al. 1990, 1993). This indicates that the fragments may be able to influence adjacent actin monomers, but it does not imply how this influence may occur.

The ability of caldesmon to compete with S1 depends on the relative affinity of caldesmon and S1 for actin. The affinity of S1 for actin is strongly nucleotide-dependent. In the presence of ATP, caldesmon binding to actin is favored by about 10^3; in the presence of ADP, S1 binding is favored. Direct binding competition has been demonstrated in solution between caldesmon and (i) S1–PPi in the presence and absence of tropomyosin (Chalovich et al. 1987; Chalovich et al. 1990), (ii) S1–AMP–PNP in the absence (Chalovich et al. 1990; Chen and Chalovich 1992; Velaz et al. 1993) and presence (Chen and Chalovich 1992) of tropomyosin, (iii) HMM–AMPPNP with tropomyosin (Hemric and Chalovich 1988), and (iv) S1 in the absence of nucleotides (Chalovich 1987; Chalovich et al. 1992). Several studies have confirmed the competition between caldesmon and myosin using fluorescence techniques (Dobrowolski et al. 1988; Nowak et al. 1989; Chalovich et al. 1995; Sen and Chalovich 1995). An important point is that when actin is totally saturated with S1, no caldesmon can bind to actin.

A detailed analysis of the competition of binding of S1–AMP–PNP and caldesmon to actin revealed that the binding is not simple competitive inhibition (Chen and Chalovich 1992). As the concentration of S1–AMP–PNP was increased, the amount of caldesmon bound to actin decreased in both the presence and absence of tropomyosin. The pattern of the competition could be simulated by a *mosaic multiple binding model*, shown in Figure 4. In this model, from one

Pure Competitive

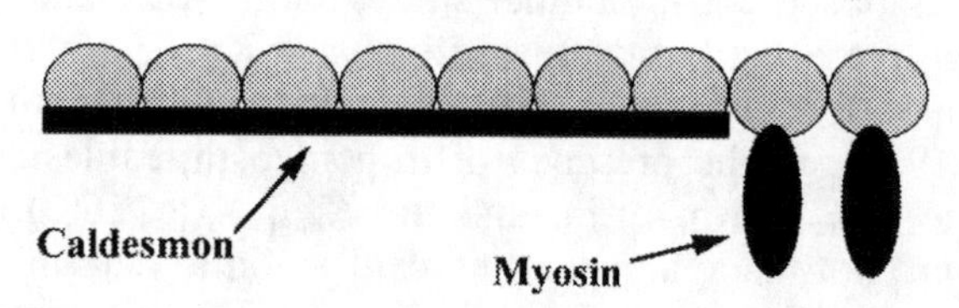

Mosaic Multiple-binding

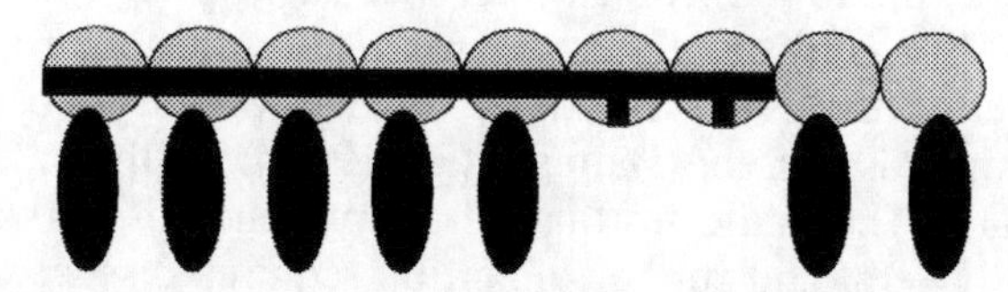

Figure 4. Diagrammatic representations of the pure competitive model and the mosaic multiple binding model, describing the competition of binding between caldesmon and myosin S1 to actin. In the mosaic model, myosin binding to some actin sites is blocked by caldesmon (shown by a projection from the caldesmon) while binding to other sites is allowed but at a lower affinity. Low-affinity binding is shown by lighter shading of the S1.

to three of the seven actin monomers that comprise a caldesmon binding site are prevented from binding to S1. The remaining actin monomers can bind to S1, but with a reduced affinity. An interesting aspect of the competition data was that tropomyosin increased the ability of caldesmon to displace S1 from actin. Therefore, one cause of the increased effectiveness of caldesmon in the presence of tropomyosin is an enhanced displacement of S1. The idea that there is a mosaic of binding sites is reasonable, since region 4 of caldesmon binds strongly to actin whereas region 3 binds weakly; various regions of caldesmon do appear to bind to actin differently. Those actin monomers that are bound to both caldesmon and to S1 may not promote ATP hydrolysis at the normal rapid rate. That is, both k_{cat} and K_{ATPase} effects may be required to simulate the inhibition of ATPase activity by this model. Direct evidence for an inactive C–A–M complex (Figure 3) has been difficult to find. There are indications of complexes between caldesmon, actin, and myosin (Marston 1989; Harricane et al. 1991; Szczesna et al. 1994), but it is unclear what type of ternary complex is formed and if caldesmon and S1 bind to the same actin monomer.

An analysis of the actin concentration dependence of the ATPase activity can provide helpful clues to the mechanism of regulation. A decrease in the K_{ATPase} by caldesmon supports a competitive binding mechanism, whereas a decrease in k_{cat} occurs when caldesmon can inhibit the rate of ATPase activity of S1 that remains bound to actin. Inhibition of the ATPase activity of thiophosphorylated smooth muscle (aorta) HMM by caldesmon was characterized by a decrease in the k_{cat} with no change in K_{ATPase} or $K_{binding}$ (Marston et al. 1988). Unlike other studies involving smooth muscle HMM and intact caldesmon, no tethering of HMM to actin was observed. In another study, the 35-kDa caldesmon fragment caused no change in the k_{cat} but decreased both the K_{ATPase} (to 26%) and $K_{binding}$ (to 21%) with smooth muscle actin and phosphorylated smooth muscle HMM (Horiuchi et al. 1991). In the presence of tropomyosin, caldesmon reduced the k_{cat} (to 42%), the K_{ATPase} (to 49%), and the $K_{binding}$ (to 26%). Unfortunately, measurements were not made at saturating caldesmon to determine which factor was most important under conditions where the ATPase rate was more completely inhibited. In experiments with skeletal S1, moderate concentrations of caldesmon had no effect on the k_{cat} but reduced the K_{ATPase} to 21%, whereas higher concentrations of caldesmon reduced both the k_{cat} (to 24%) and K_{ATPase} (to 8%) (Hemric et al. 1993a). In the presence of tropomyosin, caldesmon reduced the k_{cat} to 73% and the K_{ATPase} to 6%. In another approach, caldesmon was without effect on the ATPase activity of S1 covalently crosslinked to actin unless tropomyosin was present (Bartegi et al. 1990b). In the latter case, the rate was reduced to 41% This indicates that, in the presence of tropomyosin, the k_{cat} may be reduced by 60% and further inhibition requires dissociation of S1 from actin. In the absence of tropomyosin, virtually all of the inhibition may be attributable to inhibition of S1–ATP binding to actin. Taken together, these results suggest that both competitive and noncompetitive factors may contribute to the inhibition by caldesmon.

The added inhibition in the presence of tropomyosin may be due to the ability of caldesmon to reverse the potentiation of ATPase activity by tropomyosin. Smooth muscle tropomyosin normally enhances the k_{cat} of actin–activated ATP hydrolysis by a factor of two or three (Chacko and Eisenberg 1990). Thus, a reduction in the k_{cat} by caldesmon to 50–66% is consistent with reversal of the normal tropomyosin effect. An even greater effect of tropomyosin may occur with high concentrations of S1, where the presence of some strong binding crossbridges could potentiate the rate further. Caldesmon prevented the sixfold potentiation of ATPase activity caused by the binding of N-ethylmaleimide modified myosin to actin–tropomyosin (Horiuchi and Chacko 1989). The modified myosin binds strongly to actin but does not hydrolyze ATP appreciably. However, it is unclear to what extent the reversal of potentiation was due to displacement of NEM–S1 from actin and to what extent caldesmon caused an alteration in the tropomyosin–actin interaction. Taken as a whole, it appears that inhibition by caldesmon is due to a large change in S1–ATP binding and about a threefold reduction in the k_{cat}. The effect on the k_{cat} could be greater if there is further activation by strong binding crossbridges. The suggestion has been made that caldesmon functions like troponin, in that virtually all of the regulation of ATPase activity is due to an alteration of the tropomyosin position (Marston et al. 1994a). The observation was that caldesmon made the binding of S1–AMP–PNP to actin–tropomyosin cooperative. This cooperativity was interpreted as a transition of the tropomyosin molecule on actin as occurs with troponin–tropomyosin (Hill et al. 1980). However, unlike troponin, caldesmon is displaced from actin–tropomyosin when titrated with S1–AMP–PNP (Chen and Chalovich 1992). Because each caldesmon molecule can inhibit the binding of several molecules of S1, the displacement of caldesmon causes a cooperative increase in S1 binding.

Nevertheless, caldesmon does perturb tropomyosin on the actin filament. In three-dimensional reconstructions of native actin filaments, the presumed tropomyosin was seen as a continuous thin strand along the long pitch of actin near the inner domain of each actin monomer. Following removal of the caldesmon with Ca^{2+}–calmodulin, some filaments were found in which the long strands were closer to the outer domain of actin (Vibert et al. 1993). The position of tropomyosin stabilized by caldesmon is different from that stabilized by troponin in relaxed striated muscle (Milligan et al. 1990). It also appears that tropomyosin is not in a position where it can inhibit myosin binding in the presence of caldesmon (Popp and Holmes 1992). Therefore, although tropomyosin does facilitate the displacement of S1 from actin by caldesmon, it is caldesmon and not tropomyosin that competes with myosin binding. At the moment it is unclear what the relationship is between the change in tropomyosin binding on actin and the inhibition of actin-activated ATPase activity.

The mechanism of inhibition of contraction by caldesmon has been studied in chemically skinned single rabbit psoas fibers that naturally lack caldesmon. Both caldesmon and the 20-kDa actin binding fragment inhibited the stiffness (a measure of myosin binding to actin) in the presence of ATP but not in the presence of PP_i, ADP, or *in rigor* (Brenner et al. 1991; Kraft et al. 1995). Caldesmon and the 20-kDa fragment of caldesmon also inhibited active force in the psoas fibers. The decrease in active force occurred in parallel with the decrease in stiffness at 5°C and 20°C and ionic strengths up to 170 mM (Kraft et al. 1995). All of these effects of caldesmon in psoas fibers were reversible. Caldesmon did not affect the rate of force redevelopment, the maximum velocity of unloaded shortening, or the ratio of isometric ATPase activity to force in isometric conditions. Therefore, in this system, caldesmon inhibited force by inhibiting the binding of S1–ATP-like states to actin without much effect on the transition between actomyosin bound states.

Another paradigm that has been used to study the effects of caldesmon on contractility is the *in vitro* motility assay (Kron et al. 1991). Several laboratories have reported that caldesmon inhibits the velocity of actin filament movement in this assay (Ishikawa et al. 1991; Okagaki et al. 1991; Shirinsky et al. 1992). These studies support models in which caldesmon alters a transition between two bound actin–myosin states. In contrast, other reports state that caldesmon has no effect on the velocity. The COOH-terminal fragment of caldesmon was observed to decrease the number of attached actin filaments (Haeberle et al. 1992). However, this could be observed only in the absence of methyl cellulose, which retards diffusion of actin away from myosin. This effect was also more difficult to detect with intact caldesmon because of the tethering of actin to myosin with the intact molecule. These results (Haeberle et al. 1992; Haeberle and Hemric 1994; Horiuchi and Chacko 1995) support the competitive binding model. Two factors were suggested to have resulted in the apparent decrease in velocity observed by others: (1) the presence of caldesmon aggregates, which could be avoided by the use of very high dithiothreitol concentrations, and (2) the discontinuous movement caused by the increased probability of detachment of actin from myosin in the presence of caldesmon. Inclusion of these pauses during detachment into the velocity measurement reduces the apparent average velocity.

Advances have been made in localizing the regions of caldesmon involved in inhibition of ATPase activity. Region 4 appears to be very important in this regard. The 7.3-kDa chymotryptic fragment of caldesmon (Chalovich et al. 1992) constitutes the NH_2-terminal half of this region, while the 10-kDa CNBr fragment (Bartegi et al. 1990a) or the 658C recombinant fragment (Redwood and Marston 1993; Marston et al. 1994a) constitute the COOH-terminal region. Both fragments bind to actin and inhibit ATPase activity; moreover, their activity is reversed by Ca^{2+}–calmodulin. The 7.3-kDa fragment has also been shown to be competitive with S1 for binding to actin (Chalovich et al. 1992). Both fragments also inhibit stiffness (myosin binding) in the chemically skinned single psoas fiber model (Heubach et al. 1993). However, neither fragment is as effective as the parent 20-kDa fragment from which they are both derived. It appears likely that residues in both regions are very important for regulation. An interesting observation has been made that whereas the 7.3-kDa fragment binds to actin with a 1:1 stoichiometry (Chalovich et al. 1992), the 10-kDa fragment has been reported to bind to seven actin monomers (Bartegi et al. 1990a). It is unclear how cleavage of a 20-kDa fragment that binds with a stoichiometry of 1:2 yields fragments that bind 1:1 and 1:7. Redwood and Marston (1993) have reported that the 10-kDa fragment or 658C binds to two populations of sites, the first with a 1:37 stoichiometry. However, with increasing concentrations of 658C, the binding approaches a stoichiometry between 1:2 and 1:1. Thus there is considerable disagreement regarding the binding of caldesmon fragments.

6. Reversal of caldesmon function

The functions of caldesmon may be reversed by phosphorylation of caldesmon or by binding of one of several Ca^{2+}-binding proteins to caldesmon. It is uncertain whether one or all of these mechanisms are biologically functional.

In vitro, caldesmon is phosphorylated by Ca^{2+}–calmodulin-dependent protein kinase II (CaM kinase II), protein kinase A, protein kinase C (PKC), cdc2 kinase, and mitogen-activated protein kinase (MAP kinase) and is dephosphorylated by a type-2A protein phosphatase (Pato et al. 1993). However, in the intact smooth muscle, caldesmon appears to be phosphorylated only by MAP kinase (Adam and Hathaway 1993). In nonmuscle cells, caldesmon is phosphorylated during mitosis by cdc2 kinase, and in intact platelets by protein kinase C and protein kinase A (Litchfield and Ball 1987; Hettasch and Sellers 1991).

The N-terminal moiety of caldesmon is phosphorylated by CaM kinase II (up to 6 mol phosphate per 1 mol caldesmon, Ikebe and Reardon 1990) and casein kinase II (1 mol phosphate per 1 mol caldesmon, Bogatcheva et al. 1993a and Sutherland et al. 1994). CaM kinase II phosphorylates three sites; the preferential sites are Ser-73 and Ser-26 (Ikebe and Reardon 1990), which are also phosphorylated by casein kinase II (Bogatcheva et al. 1993b; Sutherland et al. 1994). Earlier studies demonstrated that caldesmon is phosphorylated with a stoichiometry of 1–2 mol phosphate per 1 mol caldesmon by a co-purifying protein kinase that requires Ca^{2+} and calmodulin for activity (Ngai and Walsh 1984, 1985). This was later identified to be the smooth muscle isoform of CaM kinase II (Abougou et al. 1989; Ikebe et al. 1990). Phosphorylation by the endogenous kinase (Sutherland and Walsh 1989; Hemric et al. 1993b) or by casein kinase II (Sutherland et al. 1994) weakens the binding to myosin.

CaM kinase II also phosphorylates the C-terminal actin–calmodulin-binding domain, albeit with a slower time course (Ikebe and Reardon 1990). Three phosphorylation sites were identified in the 20-kDa actin-binding fragment, Ser-587 and Ser-726 being preferentially phosphorylated (Ikebe and Reardon 1990). The phosphorylation of these sites may be responsible for reversal of the effects of caldesmon by the endogenous calcium–calmodulin-dependent protein kinase (Ngai and Walsh 1987). However, others found very little effect of phosphorylation by endogenous protein kinase(s) on the properties of caldesmon (Lash et al. 1986; Pinter and Marston 1992).

Phosphorylation of caldesmon by protein kinase C (Umekawa and Hidaka 1985) is restricted to the C terminus of the molecule (Vorotnikov et al. 1988; Tanaka et al. 1990; Ikebe and Hornick 1991; Vorotnikov et al. 1994). When corrected for molecular weight, the stoichiometry is about 2 mol phosphate per 1 mol caldesmon (Umekawa and Hidaka 1985; Ikebe and Hornick 1991), with phosphorylation occurring at Ser-587 as the preferred site and Ser-726 (Ikebe and Hornick 1991). Phosphorylation of caldesmon and a C-terminal 35-kDa fragment by PKC reduced the binding to both F actin and calmodulin (Tanaka et al. 1990), and reversed the inhibitory action on actomyosin ATPase (Tanaka et al. 1990; Ikebe and Hornick 1991). A proteolytic fragment of PKC purified from the myofibrillar fraction of sheep aorta phosphorylated up to six sites at the C terminus of caldesmon (Vorotnikov et al. 1994). These sites, which include Ser-587 and Ser-726, were phosphorylated even after incorporation of caldesmon into thin filaments. Despite this, there is no evidence to date that caldesmon is phosphorylated by PKC in intact smooth muscle (Adam et al. 1989).

From a physiological point of view, the two most interesting protein kinases that phosphorylate caldesmon are the proline-directed protein kinases, MAP kinase and cdc2 kinase. The latter, which is the catalytic subunit of maturation or M-phase promoting factor, is specifically active during mitosis and has been linked to the profound cellular changes associated with that process, including massive alteration in the organization of the microfilament cytoskeleton (Draetta 1990). Nonmuscle caldesmon is phosphorylated during mitosis by cdc2–kinase, and this cell cycle–dependent phosphorylation causes caldesmon to dissociate from actin (Yamashiro et al. 1990, 1991; Mak et al. 1991a,b). In the daughter cells, caldesmon reassociates with the stress fibers paralleled by its dephosphorylation (Hosoya et al. 1993). The cdc2 kinase phosphorylates mainly a 10-kDa cyanogen bromide fragment at the C-terminal end of caldesmon (Yamashiro et al. 1991; Mak et al. 1991a,b) that binds to actin and calmodulin (Bartegi et al. 1990a). Phosphorylation of chicken gizzard caldesmon *in vitro* by cdc2 kinase reached a maximum of about 1.5 mol of phosphate per 1 mol caldesmon; five of the potential six sites bearing the S(T)–P–X–Z motif were phosphorylated, although the preferred sites were Thr-673 and Ser-667 (40% and 20% of total phosphorylation, respectively). Thr-673 is the only site that has a basic residue at the fourth position required for recognition by cdc2 kinase (Mak et al. 1991a). *In vitro* phosphorylation decreased the actin-and calmodulin-binding affinities of nonmuscle and smooth muscle caldesmon, and reduced the inhibition of ATPase activity by caldesmon (Mak et al. 1991b; Yamakita et al. 1992).

MAP kinase incorporates 2 mol phosphate per 1 mol caldesmon and, in contrast to phosphorylation by cdc2 kinase, only 50% of the sites phosphorylated are located in the C-terminal cyanogen bromide fragment of caldesmon. MAP kinases belong to a family of 42–45-kDa Ser, Thr kinases that are activated by a number of growth factors (see Pelech and Sanghera 1992). Based on peptide mapping, MAP kinase and cdc2 kinase recognize similar sites for phosphorylation but have different preferred sites (Childs et al. 1992). Phosphorylation of caldesmon *in vitro* by MAP kinase does not affect its binding to calmodulin or tropomyosin; the decrease in actin-binding affinity is smaller than when caldesmon is phosphorylated by cdc2 kinase (Childs et al. 1992). *in vitro* phosphorylation was essentially blocked by F actin and Ca^{2+}–calmodulin, suggesting that the phosphorylation sites are inaccessible to MAP kinase when caldesmon is bound to either actin or calmodulin. This appears to be different in the intact tissue, since caldesmon appears to be phosphorylated by MAP kinase in the intact tissue. It could be speculated that, *in vivo*, the binding of caldesmon to the thin filaments is weakened at the beginning of a contraction when the concentration of Ca^{2+} is high, which then allows phosphorylation by MAP kinase to occur. The time course of caldesmon phosphorylation *in vivo* would be consistent with this possibility.

Ca^{2+}–calmodulin reverses many of the effects of caldesmon, including the binding of caldesmon to actin, the inhibition of ATPase activity, and the binding of caldesmon to myosin. Direct binding studies of caldesmon to Ca^{2+}–calmodulin are consistent with a stoichiometry of 1:1 (Shirinsky et al. 1988; Pritchard and Marston 1989) and with an affinity ranging between 3×10^5 M^{-1} and 1.3×10^7 M^{-1} (Shirinsky et al. 1988; Malencik et al. 1989; Pritchard and Marston 1989; Kasturi et al. 1993). The binding is weakened at high ionic strength and high temperature. The rate of association of caldesmon to Ca^{2+}–calmodulin is about 5.3×10^8 M^{-1} s^{-1}, and the dissociation rate is between 13 s^{-1} and 57 s^{-1} (Kasturi et al. 1993). The affinity is only 0.1% of that of other proteins which are regulated by calmodulin. It is unclear if the 30–40 μM calmodulin in smooth muscle is sufficient to bind to all of the caldesmon. Whether this can occur depends on the mechanism by which calmodulin reverses the activity of caldesmon. If Ca^{2+}–calmodulin must displace actin, then much higher concentrations of calmodulin are required than if calmodulin and actin can bind simultaneously to caldesmon. Ca^{2+}–calmodulin was first proposed to operate by a flip-flop mechanism in which caldesmon could bind either to actin or calmodulin at one time and Ca^{2+} favored binding to calmodulin (Sobue et al. 1982). However, several groups of investigators have reported that calmodulin can reverse the effects of caldesmon without displacing caldesmon from actin if the assays are done near physiological conditions (37°C and >70 mM ionic strength) (Horiuchi et al. 1986; Smith et al. 1987; Wang 1988; Pritchard and Marston 1989). Under these conditions, calmodulin was effective at much lower concentrations. Not all investigators have been able to observe these temperature and ionic strength effects on calmodulin effectiveness (Velaz et al. 1989). It is possible that the effectiveness of calmodulin depends on the source of myosin (Smith et al. 1987; Chacko et al. 1994). This is a question that requires additional study. An interesting but so far unexplained observation is that isolated thin filaments are more highly Ca^{2+}-sensitive than reconstituted proteins (Pritchard and Marston 1993).

The main calmodulin binding site is in the COOH-terminal region of caldesmon, from residues 658 through 725. These sites were identified using small polypeptide fragments of caldesmon (Marston et al. 1994b; Mezgueldi et al. 1994; Wang et al. 1994; Zhuang et al. 1995) and by NMR spectroscopy (Huber et al. 1996). The regions of importance appear to be Met-658–Ser-666 (site A), Ser-687–Lys-695 (site B) and Asp-714–Gln-725 (site B'). There is evidence for a second calmodulin binding site at the NH_2-terminal region of caldesmon (Wang 1988; Mabuchi and Wang 1991; Tsuruda et al. 1995). Calmodulin binding at the NH_2-terminal region of caldesmon is probably very weak, since NH_2-terminal caldesmon fragments do not bind significantly to calmodulin affinity columns. It is unclear if this calmodulin binding site is functional.

Several other Ca^{2+}-binding proteins have been found to bind to caldesmon and affect its interaction with other proteins. In general, they function best at near-physiological temperature and ionic strength. Caltropin from chicken gizzard muscle (Mani and Kay 1990) binds to caldesmon with an affinity of 0.5–1.3 $\times$ 10^7 M^{-1} and reverses the inhibitory effect of caldesmon in the presence of Ca^{2+} (Mani et al. 1992). Caltropin also reverses the binding of caldesmon to myosin (Mani and Kay 1993). S100 protein also binds to caldesmon in the presence of Ca^{2+} (Fujii et al. 1990), and reverses inhibition of ATPase activity by caldesmon (Pritchard and Marston 1991) even more effectively than calmodulin (Bogatcheva et al. 1993a). However, it is unclear if S100 is present in smooth muscles. Calcimedin binds to actin–tropomyosin and reverses the inhibitory activity of caldesmon (Bogatcheva et al. 1993a). Although calcimedin is more effective than S100, it is present only in low concentrations in smooth muscle.

7. Evidence that calponin affects contractility

Calponin, or p34K, is the most recently discovered candidate for an actin-associated regulatory protein in smooth muscles (Takahashi et al. 1986). The molecular mass of calponin is between 32,100 (Wills et al. 1993) and 31,400 (Stafford et al. 1995), and its dimensions have been estimated to be 16.2 nm by 2.6 nm. Calponin is present in smooth muscles at about the same concentration as tropomyosin and binds to actin, tropomyosin, calmodulin, and myosin. It is not expressed in striated muscle (Takahashi et al. 1987; Birukov et al. 1991; Takahashi and Nadal-Ginard 1991; Gimona et al. 1992). There is a differentiation-dependent increase in the expression of calponin (Gimona et al. 1992). When smooth muscle cells are taken into culture, calponin expression is down-regulated with increasing culture time and is associated with a switch of caldesmon from the high– to the low–molecular mass isoform (Birukov et al. 1991; Gimona et al. 1992).

The binding of calponin to actin, tropomyosin, myosin, and calmodulin – as well as its inhibitory effect on actomyosin ATPase – indicate a regulatory role for smooth muscle contraction (see Winder and Walsh 1993). Indirect evidence that calponin may be involved in the regulation of sustained contractions of arterial smooth muscle has been obtained by Bárány et al. (1992b). Experiments using skinned fibers and *in vitro* motility assays strengthen the evidence that calponin has the potential to modulate smooth muscle contractility.

Loading saponin skinned mesenteric arteries with calponin produced a 30% relaxation of maximal force and induced a small decrease in Ca^{2+} sensitivity (Itoh et al. 1994). Calponin also inhibited force in preparations activated by the constitutively active fragment of MLCK or by thiophosphorylation of MLC_{20} (Itoh et al. 1994;. Jaworowski et al. 1995; Obara et al. 1995). Tension development of permeabilized smooth muscle cells induced by the Ca^{2+}-independent isoform of PKC or by phenylephrine was also blocked by calponin (Horowitz et al. 1995). Calponin also decreased unloaded shortening velocity, and this effect was more pronounced than the effect on tension (Jaworowski et al. 1995; Obara et al. 1995). Extraction of calponin together with SM22 from Triton-skinned gizzard had only a minor effect on Ca^{2+} sensitivity. The relation between force and MLC_{20} phosphorylation was, however, significantly changed; in extracted fibers, less force per level of MLC_{20} phosphorylation was developed (Schmidt et al. 1995). The addition of calponin to single skinned skeletal muscle fibers, which normally lack calponin, also reduces the force (Heizmann et al. 1994).

Calponin inhibits actin filament sliding in *in vitro* motility assays. There is, however, some disagreement concerning whether this occurs in an "all or none" fashion (Shirinsky et al. 1992) or there is instead a gradual concentration-dependent decrease in the velocity of the filaments (Haeberle 1994; Jaworowski et al. 1995; Kolakowski et al. 1995). The simplest model that explains the effects of calponin on isometric force and shortening velocity in skinned fibers and the effects on actin filament sliding in *in vitro* motility systems is one in which calponin inhibits the dissociation of a high-affinity actomyosin complex. The idea that calponin affects a kinetic step in the crossbridge cycle is supported by biochemical studies of proteins in solution (Horiuchi and Chacko 1991; Miki et al. 1992).

There has been disagreement regarding the cellular location of calponin. Calponin was thought to be absent from native thin filaments isolated from smooth muscle (Lehman 1989). However, this was attributed to dissociation of calponin by ATP in the wash buffer (Nishida et al. 1990). Another preparation of native thin filaments did contain calponin (Marston 1991). Calponin was also observed by confocal immunofluorescence microscopy to be colocalized with actin and tropomyosin (Walsh et al. 1993). Other studies using dual-label immunocytochemical methods reported that calponin is localized primarily in the cytoskeleton together with β-cytoplasmic actin, filamin, and desmin, with only minor amounts associated with contractile proteins (North et al. 1994; Mabuchi et al. 1996). In cultured smooth muscle cells, calponin is colocalized with stress fibers but is absent from their ends where they approach the substrate anchorage sites (Birukov et al. 1991; Gimona et al. 1992). Finally, there is an indication that, in some smooth muscle cells, calponin is localized in the contractile domain in relaxed muscle but in the cytoskeletal domain in active muscle (Parker et al. 1994). Such a translocation may be part of a regulatory event. Even if calponin does not translocate to the contractile domain, calponin could have an indirect regulatory role by virtue of its ability to crosslink actin filaments and potentially provide a resistance to movement (Kolakowski et al. 1995). Thus, calponin could serve a regulatory function in either the contractile or the cytoskeletal domain of smooth muscle.

8. Calponin binding to actin

The reported stoichiometry of binding of calponin to actin ranges from 1:1 (Makuch et al. 1991; Kolakowski et al. 1995) to 1:3 (Takahashi et al. 1986; Winder et al. 1991; Nakamura et al. 1993). There is better agreement on the association constant which is between 0.3 and 3×10^7 M^{-1} and is independent of tropomyosin (Winder and Walsh, 1990; Makuch et al. 1991; Winder et al. 1991; Nakamura et al. 1993; Lu et al. 1995). Whereas tropomyosin does not enhance the binding of calponin to actin, calponin does bind directly to tropomyosin with an affinity near 10^6 M^{-1} (Nakamura et al. 1993). Calponin induces bundle formation in actin (Kolakowski et al. 1995). According to one report, these bundles form when 1 calponin is bound per 2 actin monomers while 1:1 complexes of calponin and actin are soluble (Kolakowski et al. 1995). In another study, the stoichiometry of binding of ^{14}C-iodoacetamide–labeled calponin to actin was found to be dependent on the ionic strength and was independent of actin bundling (Lu et al. 1995). Below 110 mM ionic strength, the stoichiometry was 1:1. At near-physiological ionic strength, calponin bound to actin with a 1:2 stoichiometry and an affinity of 6 x 10^6 M^{-1}. ATP decreased the affinity to 20%, in agreement with earlier observations (Nishida et al. 1990; Makuch et al. 1991). The rate constant for the association reaction was observed to be about 10^6 M^{-1} s^{-1} (Lu et al. 1995). Phosphorylation of the calponin reduced the affinity to actin to 10% of the original value (Nakamura et al. 1993).

Calponin can be cleaved by chymotrypsin into three fragments. The 22-kDa fragment (residues 7–182) binds to both calmodulin and actin, and inhibits the actin-activated ATPase activity of myosin as effectively as does intact calponin (Mezgueldi et al. 1992). This peptide inhibits 70% of the ATPase activity when bound to actin in a 1:1 complex (Mezgueldi et al. 1995). The 13-kDa peptide (residues 7–144) binds only to calmodulin (Mezgueldi et al. 1995). This means that the actin-binding and inhibitory region resides between residues 145 and 182. Experiments with synthetic calponin peptides have shown that the inhibitory region of calponin requires residues within the region 145–163; a larger peptide extending to residue 182 exhibits tighter binding to actin (Mezgueldi et al. 1995). The region including residues 153–163 is important for binding to both calmodulin and caltropin.

Caldesmon and calponin are competitive for binding to actin (Makuch et al. 1991; Lu et al. 1995). However, the binding sites for these proteins on actin are not identical. Thus, modification of the N-terminal acidic amino acids of actin weakens caldesmon binding (Adams et al. 1990) but has no effect on calponin binding (Miki et al. 1992). Calponin can be covalently crosslinked to actin in a 1:1 complex (Winder et al. 1992a) in which the COOH-terminal region of actin (residues 326–355) is attached to the NH_2-terminal region (residues 52–168) of calponin (Mezgueldi et al. 1992). There is additional evidence that the COOH-terminal region of actin is involved in calponin binding (Noda et al. 1992; Bonet-Kerrache and Mornet 1995).

Caldesmon and calponin do not appear to bind to each other (Vancompernolle et al. 1990). This, coupled with their competitive binding to actin, makes it unlikely that caldesmon and calponin function together as a regulatory complex.

9. Calponin binding to myosin

Calponin binds to myosin with an affinity near 10^6 M^{-1} and is antagonized by Ca^{2+}–calmodulin (Szymanski and Tao 1993). Large excesses of calponin increase the Mg^{2+} ATPase activity of myosin alone by a factor of 1.2 (Lin et al. 1993). In contrast to the binding to actin, calponin and caldesmon do not compete with each other for binding to myosin (Lin et al. 1993).

10. Calponin inhibition of actin-activated ATPase activity

Calponin inhibits the actin activated ATPase activity of skeletal and smooth muscle myosin and their subfragments (Winder and Walsh 1990; Horiuchi and Chacko 1991; Winder et al. 1991; Miki et al. 1992; Winder et al. 1992b). Although there are a few reports that tropomyosin alters the effectiveness of calponin, most reports state that tropomyosin has no effect on the inhibition of ATPase activity by calponin (Marston 1991; Mezgueldi et al. 1995; Shirinsky et al. 1992). To date, only one group has reported the effect of calponin on the steady-state kinetic parameters of actin-activated ATPase activity of smooth muscle HMM, and this has been done at low (20 mM) ionic strength (Horiuchi and Chacko 1991). At a ratio of calponin to actin of 0.6, calponin reduced the k_{cat} to 10% and decreased the K_{ATPase} (M^{-1}) to 58%. In the presence of tropomyosin, calponin reduced the k_{cat} to 39% and the K_{ATPase} to 19%; calponin was less effective in the presence of tropomyosin. The binding of HMM–ATP to actin was also inhibited by calponin, but 2.5 times as much calponin was required to give 50% inhibition of binding as for 50% inhibition of ATPase activity. Support for an effect of calponin on the k_{cat} comes from the observation that calponin inhibits the ATPase activity of covalently crosslinked acto-S1 (Miki et al. 1992). The extent of inhibition exceeds 40%, but the maximum amount of inhibition was not determined.

11. Reversal of calponin function

The functions of calponin, like those of caldesmon, may be attenuated either by phosphorylation or by association with one of several Ca^{2+}-binding proteins. Calponin is stoichiometrically phosphorylated *in vitro* by PKC and CaM kinase II (Winder and Walsh 1990; Nakamura et al. 1993). PKC phosphorylates Ser-175 and Ser-254 and some minor sites including Thr-184 (Winder et al. 1993a). The latter is the major site, according to Nakamura et al. (1993). In the case of CaM kinase II, approximately 75% of total phosphate incorporation was in Ser-175 (Winder et al. 1993a). Upon phosphorylation, calponin lost its ability to bind to actin and to inhibit actomyosin ATPase. Inhibitory function was restored when calponin was dephosphorylated by protein phosphatase 2A (reviewed in Winder and Walsh 1993). Phosphorylated calponin retained its ability to bind to immobilized tropomyosin (Winder and Walsh 1990) but possibly with a reduced affinity (Nakamura et al. 1993). Thus, calponin may be regulated by either protein kinase C and/or CaM kinase II during smooth muscle contraction. Reversal of the inhibitory action of calponin by phosphorylation by the Ca^{2+}-independent isoform of protein kinase C could account for Ca^{2+}-independent contractions elicited by norepinephrine (Collins et al. 1992).

In support of this hypothesis, several groups reported the phosphorylation of calponin in activated smooth muscle from toad stomach (Winder et al. 1993a), porcine carotid artery (Driska and Cummings 1993; Rokolya and Moreland 1993), and tracheal smooth muscle (Pohl et al. 1991). These findings have been challenged. Bárány and co-workers (Bárány et al. 1991; Bárány and Bárány 1993) could not detect any increase in phosphorylation in smooth muscles from porcine carotid artery, uterus, trachea, stomach, or bladder using different modes of stimulation. Similarly, chicken gizzard, guinea pig taenia coli, and lamb trachea were not phosphorylated during activation (Gimona et al. 1992; Steusloff and Pfitzer, unpublished observations). The reason for the discrepancy of these results is not clear at present, but there is an *in vitro* counterpart: it has been reported that F actin does (Nakamura et al. 1993) and does not (Winder et al. 1993a) prevent phosphorylation of calponin by PKC.

The inhibition of ATPase activity by calponin may be reversed by calmodulin, but there is no broad agreement that this represents a physiological mechanism. For example, the cellular calmodulin concentration is inadequate to bind to the 150-μM calponin in smooth muscle cells (Takahashi et al. 1986). Furthermore, in at least one report, calmodulin did not significantly reverse inhibition of ATPase activity by calponin (Marston 1991); in other reports, large excesses of calmodulin did reverse the effects of calponin (Makuch et al. 1991; Winder et al. 1993b). In another report, calmodulin was more effective but the activity of the calponin was also low (Abe et al. 1990). Calmodulin does reverse the inhibition of motility in the *in vitro* motility assay (Shirinsky et al. 1992).

Two molecules of calmodulin can bind per molecule of calponin with association constants near $5 \times 10^6\ M^{-1}$ and $4 \times 10^5\ M^{-1}$ (Wills et al. 1993). Other estimates of the affinity range from $1.6 \times 10^6\ M^{-1}$ to $2 \times 10^7\ M^{-1}$ (Winder et al. 1993b; Nakamura et al. 1993). Phosphorylation of calponin reduces calmodulin's affinity for calponin markedly (Winder et al. 1993b). The rate of association of calmodulin to calponin is $6 \times 10^6\ M^{-1}\ s^{-1}$; the rate of dissociation is about $0.3\ s^{-1}$ (Winder et al. 1993b). The Gln-153–Ile-163 region was identified as the high-affinity calmodulin binding site by the use of synthetic calponin polypeptides (Mezgueldi et al. 1995).

Each molecule of calponin may also bind to two molecules of caltropin or S100. The affinities of caltropin for calponin are $8 \times 10^6\ M^{-1}$ and $0.6–6 \times 10^6\ M^{-1}$ (Wills et al. 1993, 1994). Caltropin is roughly twice as effective as calmodulin in reversing the effects of calponin (Wills et al. 1994; Mezgueldi et al. 1995). The same region implicated in calmodulin binding (Gln-153–Ile-163) was identified as the high-affinity caltropin binding site (Mezgueldi et al. 1995). S100 is reportedly three times as effective as calmodulin in reversing the ATPase inhibition of calponin (Fujii et al. 1994). The affinity of S100 for calponin is $1.4 \times 10^6\ M^{-1}$. Two molecules of S100b (the β–β dimer of S100) were found to bind to calponin with affinities of $3 \times 10^7\ M^{-1}$ and $1.3–3 \times 10^6\ M^{-1}$ (Wills et al. 1993). It is not yet known which, if any, of the Ca^{2+}-binding proteins associates with calponin *in vivo*.

12. Summary

Caldesmon and calponin are important components of smooth muscle and non-muscle cells. Both proteins are abundant in these cells, and both inhibit actin-activated ATP hydrolysis of myosin. Both proteins are absent (or present in low levels) in skeletal muscle, although the skeletal muscle proteins nebulin and C protein do share some common features with caldesmon and calponin. There is evidence but not proof that caldesmon and calponin are involved in the regulation of contraction; there is disagreement regarding the mechanism by which these proteins might inhibit contraction. It is possible that additional proteins may also participate in the actin-based regulation of smooth muscle contraction. The ability of caldesmon and calponin to associate with multiple contractile elements is also consistent with a role in organizing the contractile apparatus. Regulation and organization are not mutually exclusive. Possible functions of caldesmon and calponin are summarized as follows.

1. *Competitive binding model of caldesmon.* There are many indications that myosin and caldesmon compete for binding to actin. In the skeletal muscle model, this competition is entirely responsible for inhibition of force production. However, solution studies indicate that there is some component of inhibition that does not require dissociation of myosin from actin. The ambiguity as to whether the competition of binding or inhibition of k_{cat} is most important results from differences in the correlation of myosin–ATP binding and ATPase activity with caldesmon concentration. It is likely that a kinetic model based on the mosaic multiple binding model (Figure 4) will be able to simulate the observed inhibition of ATPase activity and binding of myosin to actin.

2. *Allosteric model of caldesmon.* In this model, caldesmon is a troponin-like molecule that functions together with tropomyosin in much the same way that skeletal troponin functions with tropomyosin. The reversal of myosin binding to actin is thought to be unimportant at concentrations of caldesmon required for inhibition of ATPase activity.

3. *Dual regulation model of caldesmon.* Simply put, caldesmon alters ATPase activity by binding to actin and also by direct effects on myosin. The reported effects of caldesmon on the ATPase activity of myosin, in the absence of actin, range from 70% to 80% inhibition (Lim and Walsh 1986) to no effect (Chalovich et al. 1990; Mani and Kay 1993; Wang et al. 1994) to stimulation (Ishikawa et al. 1991). This issue remains unsettled.

4. *Caldesmon cross-linking model.* Caldesmon might crosslink actin to myosin and inhibit crossbridge dissociation, thus contributing to the latch state. Tethering actin to myosin by caldesmon can decrease the shortening velocity of actin in the *in vitro* motility system (Horiuchi and Chacko 1995), but only when few myosin molecules are active and the force is very low. These and other data make this an unlikely model.

5. *Nonregulatory model of caldesmon.* It remains possible that caldesmon is only a structural protein, and that the effects on ATPase, motility, and force production observed in solution do not occur in smooth muscle cells. Some who hold this view consider calponin to be the key protein in the actin-linked component of regulation of smooth muscle contraction.

6. *Integrated caldesmon–calponin model.* Haeberle and Hemric (1994) have

proposed that caldesmon, tropomyosin, and light chain phosphorylation control the binding of myosin to actin or the transition from nonforce- to force-producing crossbridges (f_{app}), while calponin controls the rate of the forward cycle in which the force-producing crossbridges return to nonforce-producing states (g_{app}). The activation of calponin (possibly by dephosphorylation) may be responsible for the maintenance of force during decreased levels of myosin light chain phosphorylation. This would occur as a result of the inhibition of crossbridge detachment by calponin.

7. *Calponin crosslinking model.* Caldesmon and calponin both contribute to regulation, but the effect of calponin is indirect. Calponin present in the cytoskeleton impedes myosin detachment from actin by crosslinking cytoskeletal actin and compressing the contractile domain (Dabrowska 1995). For such a mechanism to function, the rate of crosslinking must be fast relative to the decomposition of the crossbridge states that are to be trapped.

8. *Structural model.* Low concentrations of caldesmon enhance the motility of actin in the *in vitro* motility assay (Haeberle et al. 1992) and stimulate actin activation of skeletal and smooth muscle myosin ATPase activity (Lin et al. 1993) under conditions where the binding of myosin to actin is normally weak. Caldesmon has been observed to be essential for some types of motility (Walker et al. 1989; Hegmann et al. 1991; Hemric et al. 1994). All of these observations are consistent with caldesmon facilitating the interaction of actin and myosin as a result of the fact that the binding of caldesmon to both actin and to myosin is stronger than the binding of actin to myosin during ATP hydrolysis. Higher concentrations of caldesmon restrict the binding of myosin to actin and produce an inhibitory effect. Similarly, caldesmon might organize actin and myosin filaments under conditions where their interactions are weak. Caldesmon may also stabilize actin filaments (Ishikawa et al. 1989) and myosin filaments (Katayama et al. 1995). Stabilization of actin filaments may occur by inhibition of severing and capping by gelsolin and brevin, which are present in high concentrations in smooth muscle cells (Gailly et al. 1990).

Although the story is incomplete for now, these hypotheses are testable and the function and mechanism of both proteins can be unequivocally determined.

Acknowledgments

The authors thank Drs. A. Arner, S. Chacko, R. Dabrowska, A. Fattoum, J. R. Haeberle, A. Mak, S. B. Marston, R. Paul, W. F. Stafford, J. V. Small, and C. L. A. Wang for providing preprints and A. Sen, B. Leinweber, and Drs. A. M. Resetar and F. W. M. Lu for their helpful comments. Research was supported by grants from NIH (AR40540 and AR35216 to J.M.C.) and the Deutsche Forschungsgeminschaft (Pf226/4-1 to G.P.).

References

Abe, M., Takahashi, K., and Hiwada, K. (1990). Effect of calponin on actin-activated myosin ATPase activity. *J. Biochem. (Tokyo)* 108: 835–8.

Abe, Y., Kasuya, Y., Kudo, M., Yamashita, K., Goto, K., Masaki, T., and Takuwa, Y.

(1991). Endothelin-1-induced phosphorylation of the 20-kDa myosin light chain and caldesmon in porcine coronary artery smooth muscle. *Jpn. J. Pharmacol.* 57: 431–5.

Abougou, J.-C., Hagiwara, M., Hachiya, T., Terasawa, M., Hidaka, H., and Hartshorne, D. J. (1989). Phosphorylation of caldesmon. *FEBS Letts.* 257: 408–10.

Adam, L. P., Haeberle, J. R., and Hathaway, D. R. (1989). Phosphorylation of caldesmon in arterial smooth muscle. *J. Biol. Chem.* 264: 7698–7703.

Adam, L. P., and Hathaway, D. R. (1993). Identification of mitogen-activated protein kinase phosphorylation sequences in mammalian h-caldesmon. *FEBS Letts.* 322: 56–60.

Adam, L. P., Milio, L., Brengle, B., and Hathaway, D. R. (1990). Myosin light chain and caldesmon phosphorylation in arterial muscle stimulated with endothelin-1. *J. Mol. Cell. Cardiol.* 22: 1017–23.

Adams, S., DasGupta, G., Chalovich, J. M., and Reisler, E. (1990). Immunochemical evidence for the binding of caldesmon to the NH_2-terminal segment of actin. *J. Biol. Chem.* 265: 19652–7.

Bárány, M., and Bárány, K. (1993). Calponin phosphorylation does not accompany contraction of various smooth muscles. *Biochim. Biopyhys. Acta* 1179: 229–33.

Bárány, K., Poly, K. E., and Bárány, M. (1992a). Protein phosphorylation during the contraction-relaxation-contraction cycle of arterial smooth muscle. *Arch. Biochem. Biophys.* 294: 571–8.

Bárány, K., Poly, K. E., and Bárány, M. (1992b). Involvement of calponin and caldesmon in sustained contraction of arterial smooth muscle. *Biochem. Biophys. Res. Commun.* 187: 847–52.

Bárány M., Rokolya, A., and Bárány, K. (1991). Absence of calponin phosphorylation in contracting or resting arterial smooth muscle. *FEBS Letts.* 279: 65–8.

Bartegi, A., Fattoum, A., Derancourt, J., and Kassab, R. (1990a). Characterization of the carboxyl-terminal 10-kDa cyanogen bromide fragment of caldesmon as an actin-calmodulin-binding region. *J. Biol. Chem.* 265: 15231–8.

Bartegi, A., Fattoum, A., and Kassab, R. (1990b). Cross-linking of smooth muscle caldesmon to the NH_2-terminal region of skeletal F-actin. *J. Biol. Chem.* 265: 2231–7.

Birukov, K. G., Stepanova, O. V., Nanaev, A. K., and Shirinsky, V. P. (1991). Expression of calponin in rabbit and human aortic smooth muscle cells. *Cell Tissue Res.* 266: 579–84.

Bogatcheva, N. V., Panaiotov, M. P., Vorotnikov, A. V., and Gusev, N. B. (1993a). Effect of 67 kDa calcimedin on caldesmon functioning. *FEBS Letts.* 335: 193–7.

Bogatcheva, N. V., Vorotnikov, A. V., Birkukov, K. G., and Shirinsky, V. P. (1993b). Phosphorylation by casein kinase II affects the interaction of caldesmon with smooth muscle myosin and tropomyosin. *Biochem. J.* 290: 437–42.

Bonet-Kerrache, A., and Mornet, D. (1995). Importance of the C-terminal part of actin in interactions with calponin. *Biochem. Biophys. Res. Commun.* 206: 127–32.

Brenner, B., Yu, L. C., and Chalovich, J. M. (1991). Parallel inhibition of active force and relaxed fiber stiffness in skeletal muscle by caldesmon: Implications for the pathway to force generation. *Proc. Nat. Acad. Sci. USA* 88: 5739–43.

Bryan, J., Imai, M., Lee, R., Moore, P., Cook, R. G., and Lin, W.-G. (1989). Cloning and expression of a smooth muscle caldesmon. *J. Biol. Chem.* 264: 13873–9.

Chacko, S., and Eisenberg, E. (1990). Cooperativity of actin-activated ATPase of gizzard heavy meromyosin in the presence of gizzard tropomyosin. *J. Biol. Chem.* 265: 2105–110.

Chacko, S., Jacob, S. S., and Horiuchi, K. Y. (1994). Myosin I from mammalian smooth muscle is regulated by caldesmon-calmodulin. *J. Biol. Chem.* 269: 15803–7.

Chalovich, J. M. (1987). Caldesmon and thin filament regulation of muscle contraction. *Cell Biophys.* 12: 73–85.

Chalovich, J. M. (1992). Actin mediated regulation of muscle contraction. *Pharmac. Ther.* 55: 95–148.

Chalovich, J. M., Bryan, J., Benson, C. E., and Velaz, L. (1992). Localization and characterization of a 7.3-kDa region of caldesmon which reversibly inhibits actomyosin ATPase activity. *J. Biol. Chem.* 267: 16644–50.

Chalovich, J. M., Chen, Y., Dudek, R., and Luo, H. (1995). Kinetics of binding of caldesmon to actin. *J. Biol. Chem.* 270: 9911–16.

Chalovich, J. M., Cornelius, P., and Benson, C. E. (1987). Caldesmon inhibits skeletal actomyosin subfragment-1 ATPase activity and the binding of myosin subfragment-1 to actin. *J. Biol. Chem.* 262: 5711–16.

Chalovich, J. M., Hemric, M. E., and Velaz, L. (1990). Regulation of ATP hydrolysis by caldesmon: A novel change in the interaction of myosin with actin. *Ann. N. Y. Acad. Sci.* 599: 85–99.

Chen, Y., and Chalovich, J. M. (1992). A mosaic multiple-binding model for the binding of caldesmon and myosin subfragment-1 to actin. *Biophys. J.* 63: 1063–70.

Childs, T. J., Watson, M. H., Sanghera, J. S., Campbell, D. L., Pelech, S. L., and Mak, A. S. (1992). Phosphorylation of smooth muscle caldesmon by mitogen-activated protein (MAP) kinase and expression of MAP kinase in differentiated smooth muscle cells. *J. Biol. Chem.* 267: 22853–9.

Collins, E. M., Walsh, M. P., and Morgan, K. G. (1992). Contraction of single vascular smooth muscle cells by phenylephrine at constant $[Ca^{2+}]_i$. *Am. J. Physiol.* 262: H754–H762.

Crosbie, R., Adams, S., Chalovich, J. M., and Reisler, E. (1991). The interaction of caldesmon with the COOH terminus of actin. *J. Biol. Chem.* 266: 20001–6.

Crosbie, R. H., Chalovich, J. M., and Reisler, E. (1992). Interaction of caldesmon and myosin subfragment 1 with the C-terminus of actin. *Biochem. Biophys. Res. Commun.* 184: 239–45.

Crosbie, R. H., Chalovich, J. M., and Reisler, E. (1995). Flexation of caldesmon: Effect of conformation on the properties of caldesmon. *J. Muscle Res. Cell Motil.* 16: 509–18.

Crosbie, R. H., Miller, C., Chalovich, J. M., Rubenstein, P. A., and Reisler, E. (1994). Caldesmon, N-terminal yeast actin mutants, and the regulation of actomyosin interactions. *Biochemistry* 33: 3210–16.

Dabrowska, R. (1995). Actin and thin-filament-associated proteins in smooth muscle. In: *Airways Smooth Muscle: Biochemical Control of Contraction and Relaxation*, (D. Raeburn and M. A. Giembycz, eds.). Basel, Switzerland: Birkhaüser Verlag, pp. 31–59.

Dillon, P. F., Aksoy, M. O., Driska, S. P., and Murphy, R. A. (1981). Myosin phosphorylation and the cross-bridge cycle in arterial smooth muscle. *Science* 211: 495–7.

Dobrowolski, Z., Borovikov, Y. S., Nowak, E., Galazkiewicz, B., and Dabrowska, R. (1988). Comparison of Ca^{2+}-dependent effects of caldesmon-tropomyosin-calmodulin and troponin-tropomyosin complexes on the structure of F-actin in ghost fibers and its interaction with myosin heads. *Biochim. Biophys. Acta* 956: 140–50.

Draetta, G. (1990). Cell cycle control in eukaryotes: Molecular mechanisms of cdc2 activation. *Trends Biochem. Sci.* 15: 378–83.

Driska, S. P., and Cummings, J. J. (1993). Calponin is partially phosphorylated in smooth muscle strips. *Biophys. J.* 64: A31.

Fischer W., and Pfitzer, G. (1989). Rapid myosin phosphorylation transients in phasic contractions in chicken gizzard smooth muscle. *FEBS Letts.* 258: 59–62.

Fujii, T., Imai, M., Rosenfeld, G. C., and Bryan, J. (1987). Domain mapping of chicken gizzard caldesmon. *J. Biol. Chem.* 262: 2757–63.

Fujii, T., Machino, K., Andoh, H., Satoh, T., and Kondo, Y. (1990). Calcium-dependent control of caldesmon-actin interaction by S100 protein. *J. Biochem. (Tokyo)* 107: 133–7.

Fujii, T., Oomatsuzawa, A., Kuzumaki, N., and Kondo, Y. (1994). Calcium-dependent regulation of smooth muscle calponin by S100. *J. Biochem. (Tokyo)* 116: 121–7.

Furst, D. O., Cross, R. A., DeMey, J., and Small, J. V. (1986). Caldesmon is an elongated, flexible molecule localized in the actomyosin domains of smooth muscle. *EMBO J.* 5: 251–7.

Gailly, Ph., Lejeune, Th., Capony, J. P., and Gillis, J. M. (1990). The action of brevin, an F-actin severing protein, on the mechanical properties and ATPase activity of skinned smooth muscle. *J. Muscle Res. Cell Motil.* 11: 293–301.

Gerthoffer, W. T. (1987). Dissociation of myosin phosphorylation and active tension during muscarinic stimulation of tracheal smooth muscle. *J. Pharmacol. Exp. Ther.* 240: 8–15.

Gerthoffer, W. T., and Murphy, R. A. (1983). Ca^{2+}, myosin phosphorylation, and relaxation of arterial smooth muscle. *Am. J. Physiol.* 245: C271–C277.

Gimona, M., Sparrow, M. P., Strasser, P., Herzog, M., and Small, V. J. (1992). Calponin and SM 22 isoforms in avian and mammalian smooth muscle. *Eur. J. Biochem.* 205: 1067–75.

Graceffa, P., Adam, L. P., and Lehman, W. (1993). Disulphide cross-linking of smooth-muscle and non-muscle caldesmon to the C-terminus of actin in reconstituted and native thin filaments. *Biochem. J.* 294: 63–7.

Graceffa, P., Wang, C. L. A., and Stafford, W. F. (1988). Caldesmon: Molecular weight and subunit composition by analytical ultracentrifugation. *J. Biol. Chem.* 263: 14196–202.

Haeberle, J. R. (1994). Calponin decreases the rate of cross-bridge cycling and increases maximum force production by smooth muscle myosin in an *in vitro* motility assay. *J. Biol. Chem.* 269: 12424–31.

Haeberle, J. R., Hathaway, D. R., and Smith, C. L. (1992). Caldesmon content of mammalian smooth muscles. *J. Muscle Res. Cell Motil.* 13: 582–5.

Haeberle, J. R., and Hemric, M. E. (1994). A model for the coregulation of smooth muscle actomyosin by caldesmon, calponin, tropomyosin, and the myosin regulatory light chain. *Can. J. Physiol. Pharmacol.* 72: 1400–09.

Haeberle, J. R., Trybus, K. M., Hemric, M. E., and Warshaw, D. M. (1992). The effects of smooth muscle caldesmon on actin filament motility. *J. Biol. Chem.* 267: 23001–6.

Harricane, M-C., Bonet-Kerrache, A., Cavadore, C., and Mornet, D. (1991). Actin-caldesmon-myosin-subfragment-1 ternary complex viewed by electron microscopy – Competitive actin binding region for caldesmon and myosin subfragment-1. *Eur. J. Biochem.* 196: 219–24.

Hayashi, K., Fujio, Y., Kato, I., and Sobue, K. (1991). Structural and functional relationships between *h*- and *l*-caldesmons. *J. Biol. Chem.* 266: 355–61.

Hegmann, T. E., Schulte, D. L., Lin, J. L-C., and Lin, J. J-C. (1991). Inhibition of intracellular granule movement by microinjection of monoclonal antibodies against caldesmon. *Cell Motil. Cytoskeleton* 20: 109–20.

Heizmann, S. I., Lu, F. W. M., Chalovich, J. M., and Brenner, B. (1994). Effect of calponin on skinned rabbit psoas muscle. *Biophys. J.* 66: A188. (abstract).

Hemric, M. E., and Chalovich, J. M. (1988). Effect of caldesmon on the ATPase activity and the binding of smooth and skeletal myosin subfragments to actin. *J. Biol. Chem.* 263: 1878–85.

Hemric, M. E., and Chalovich, J. M. (1990). Characterization of caldesmon binding to myosin. *J. Biol. Chem.* 265: 19672–8.

Hemric, M. E., Freedman, M. V., and Chalovich, J. M. (1993a). Inhibition of actin stimulation of skeletal muscle (A1)S-1 ATPase activity by caldesmon. *Arch. Biochem. Biophys.* 306: 39–43.

Hemric, M. E., Lu, F. W. M., Shrager, R., Carey, J., and Chalovich, J. M. (1993b). Reversal of caldesmon binding to myosin with calcium-calmodulin or by phosphorylating caldesmon. *J. Biol. Chem.* 268: 15305–11.

Hemric, M. E., Tracy, P. B., and Haeberle, J. R. (1994). Caldesmon enhances the binding of myosin to the cytoskeleton during platelet activation. *J. Biol. Chem.* 269: 4125–8.

Hettasch, J. M., and Sellers, J. R. (1991). Caldesmon phosphorylation in intact human platelets by cAMP-dependent protein kinase and protein kinase C. *J. Biol. Chem.* 266: 11876–81.

Heubach, J., Brenner, B., and Chalovich, J. M. (1993). Mapping the caldesmon molecule: Fragment effects on muscle mechanics. *Biophys. J.* 64: A1348.

Hill, T. L., Eisenberg, E., and Greene, L. E. (1980). Theoretical model for the cooperative equilibrium binding of myosin subfragment 1 to the actin-troponin-tropomyosin complex. *Proc. Nat. Acad. Sci. USA* 77: 3186–90.

Horiuchi, K. Y., and Chacko, S. (1988). Interaction between caldesmon and tropomyosin in the presence and absence of smooth muscle actin. *Biochemistry* 27: 8388–93.

Horiuchi, K. Y., and Chacko, S. (1989). Caldesmon inhibits the cooperative turning-on of the smooth muscle heavy meromyosin by tropomyosin-actin. *Biochemistry* 28: 9111–16.

Horiuchi, K. Y., and Chacko, S. (1991). The mechanism for the inhibition of actin-activated ATPase of smooth muscle heavy meromyosin by calponin. *Biochem. Biophys. Res. Commun.* 176: 1487–93.

Horiuchi, K. Y., and Chacko, S. (1995). Effect of unphosphorylated smooth muscle myosin on caldesmon-mediated regulation of actin filament velocity. *J. Muscle Res. Cell Motil.* 16: 11–19.

Horiuchi, K. Y., Miyata, H., and Chacko, S. (1986). Modulation of smooth muscle actomyosin ATPase by thin filament associated proteins. *Biochem. Biophys. Res. Commun.* 136: 962–8.

Horiuchi, K. Y., Samuel, M., and Chacko, S. (1991). Mechanism for the inhibition of acto-heavy meromyosin ATPase by the actin/calmodulin binding domain of caldesmon. *Biochemistry* 30: 712–17.

Horowitz, A., Chomienne, O., Walsh, M. P., Tao, T., Katsuyama, H., and Morgan, K. G. (1995). PKCε but not ζ causes a Ca^{2+}-independent contraction of smooth muscle cells that is reversed by calponin. *Biophys. J.* 68: A74.

Hosoya, N., Hosoya, H., Yamashiro, S., Mohri, H., and Matsumura, F. (1993). Localization of caldesmon and its dephosphorylation during cell division. *J. Cell Biol.* 121: 1075–82.

Huber, P. A., El-Mezgueldi, M., Grabarek, Z., Slatter, D. A., Levine, B. A., and Marston, S. B. (1996). Multiple-sited interaction of caldesmon with Ca^{2+}-calmodulin. *Biochem. J.* 316: 413–20.

Huber, P. A. J., Redwood, C. S., Avent, N. D., Tanner, M. J. A., and Marston, S. B.

(1993). Identification of functioning regulatory sites and a new myosin binding site in the C-terminal 288 amino acids of caldesmon expressed from a human clone. *J. Muscle Res. Cell Motil.* 14: 385–91.

Ikebe, M., and Hornick, T. (1991). Determination of the phosphorylation sites of smooth muscle caldesmon by protein kinase C. *Arch. Biochem. Biophys.* 288: 538–42.

Ikebe, M., and Reardon, S. (1988). Binding of caldesmon to smooth muscle myosin. *J. Biol. Chem.* 263: 3055–8.

Ikebe, M., and Reardon, S. (1990). Phosphorylation of smooth muscle caldesmon by calmodulin-dependent protein kinase II. *J. Biol. Chem.* 265: 17607–12.

Ikebe, M., Reardon, S., Scott-Woo, G. C., Zhou, Z., and Koda, Y. (1990). Purification and characterization of calmodulin-dependent multifunctional protein kinase from smooth muscle: Isolation of caldesmon kinase. *Biochemistry* 29: 11242–8.

Ishikawa, R., Okagaki, T., Higashi-Fujime, S., and Kohama, K. (1991). Stimulation of the interaction between actin and myosin by *Physarum* caldesmon-like protein and smooth muscle caldesmon. *J. Biol. Chem.* 266: 21784–90.

Ishikawa, R., Yamashiro, S., and Matsumura, F. (1989). Annealing of gelsolin-severed actin fragments by tropomyosin in the presence of Ca^{2+}. *J. Biol. Chem.* 264: 16764–70.

Itoh, T., Ikebe, M., Kargacin, G. J., Hartshorne, D. J., Kemp, B. E., and Fay, F. S. (1989). Effects of modulators of myosin light-chain kinase activity in single smooth muscle cells. *Nature* 338: 164–7.

Itoh, T., Suzuki, S., Suzuki, A., Nakamura, F., Naka, M., and Tanaka, T. (1994). Effects of exogenously applied calponin on Ca^{2+}-regulated force in skinned smooth muscle of the rabbit mesenteric artery. *Pflügers Arch.* 427: 301–8.

Jaworowski, Å., Anderson, K. I., Arner, A., Enström, M., Gimona, M., Strasser, P., and Small J. V. (1995). Calponin reduces shortening velocity in skinned taenia coli smooth muscle fibers. *FEBS Letts.* 365: 167–71.

Kamm, K. E., and Stull, J. T. (1985). The function of myosin and myosin light chain kinase phosphorylation in smooth muscle. *Annu. Rev. Pharmacol. Toxicol.* 25: 593–620.

Kasturi, R., Vasulka, C., and Johnson, J. D. (1993). Ca^{2+}, caldesmon, and myosin light chain kinase exchange with calmodulin. *J. Biol. Chem.* 268: 7958–64.

Katayama, E., and Ikebe, M. (1995). Mode of caldesmon binding to smooth muscle thin filament: Possible projection of the amino-terminal domain of caldesmon from native thin filament. *Biophys. J.* 68: 2419–28.

Katayama, E., Scott-Woo, G. C., and Ikebe, M. (1995) Effect of caldesmon on the assembly of smooth muscle myosin. *J. Biol. Chem.* 270: 3919–25.

Katsuyama, H., Wang, C.-L. A., and Morgan, K. G. (1992). Regulation of vascular smooth muscle tone by caldesmon. *J. Biol. Chem.* 267: 14555–8.

Kolakowski, J., Makuch, R., and Dabrowska, R. (1992). Lys-373 of actin is involved in binding to caldesmon. *FEBS Letts.* 309: 65–7.

Kolakowski, J., Makuch, R., Stepkowski, D., and Dabrowska, R. (1995). Interaction of calponin with actin and its functional implications. *Biochem. J.* 306: 199–204.

Kraft, T., Chalovich, J. M., Yu, L. C., and Brenner, B. (1995). Parallel inhibition of active force and relaxed fiber stiffness by caldesmon at near physiological conditions. Further evidence that weak cross-bridge binding to actin is an essential intermediate for force generation. *Biophys. J.* 68: 2404–18.

Lash, J. A., Sellers, J. R., and Hathaway, D. R. (1986). The effects of caldesmon on smooth muscle heavy actomeromyosin ATPase activity and binding of heavy meromyosin to actin. *J. Biol. Chem.* 261: 16155–60.

Lehman, W. (1989). 35 kDa proteins are not components of vertebrate smooth muscle thin filaments. *Biochim. Biophys. Acta* 996: 57–61.

Lehman, W., Craig, R., Lui, J., and Moody, C. (1989). Caldesmon and the structure of smooth muscle thin filaments: Immunolocalization of caldesmon on thin filaments. *J. Muscle Res. Cell Motil.* 10: 101–12.

Lehman, W., Denault, D., and Marston, S. (1992). Letter to the editor. *J. Muscle Res. Cell Motility* 13: 582–3.

Leszyk, J., Mornet, D., Audemard, E., and Collins, J. H. (1989). Amino acid sequence of a 15 kilodalton actin-binding fragment of turkey gizzard caldesmon: Similarity with dystrophin, tropomyosin and the tropomyosin-binding region of troponin T. *Biochem. Biophys. Res. Commun.* 160: 210–16.

Levine, B. A., Moir, A. J. G., Audemard, E., Mornet, D., Patchell, V. B., and Perry, S. V. (1990). Structural study of gizzard caldesmon and its interaction with actin – Binding involves residues of actin also recognized by myosin subfragment 1. *Eur. J. Biochem.* 193: 687–96.

Lim, M. S., and Walsh, M. P. (1986). The effects of caldesmon on the ATPase activities of rabbit skeletal-muscle myosin. *Biochem. J.* 238: 523–30.

Lin, Y., Ye, L.-H., Ishikawa, R., Fujita, K., and Kohama, K. (1993). Stimulatory effect of calponin on myosin ATPase activity. *J. Biochem. (Tokyo)* 113: 643–5.

Litchfield, D. W., and Ball, E. H. (1987). Phosphorylation of caldesmon$_{77}$ by protein kinase C *in vitro* and in intact human platelets. *J. Biol. Chem.* 262: 8056–60.

Lu, F. W. M., and Chalovich, J. M. (1995). Role of ATP in the binding of caldesmon to smooth muscle myosin. *Biochemistry* 34: 6359–65.

Lu, F. W. M., Freedman, M. V., and Chalovich, J. M. (1995). Characterizaton of calponin binding to actin. *Biochemistry* 34: 11864–71.

Lynch, W. P., Riseman, V. M., and Bretscher, A. (1987). Smooth muscle caldesmon is an extended flexible monomeric protein in solution that can readily undergo reversible intra- and intermolecular sulfhydryl cross-linking. *J. Biol. Chem.* 262: 7429–37.

Mabuchi, K., Li, Y., Tao, T., and Wang, C.-L. A. (1996). Immunocytochemical localization of caldesmon and calponin in chicken gizzard smooth muscle. *J. Muscle Res. Cell Motil.* 17: 243–60.

Mabuchi, K., Lin, J. J-C., and Wang, C-L. A. (1993). Electron microscopic images suggest both ends of caldesmon interact with actin filaments. *J. Muscle Res. Cell Motil.* 14: 54–64.

Mabuchi, K., and Wang, C-L. A. (1991). Electron microscopic studies of chicken gizzard caldesmon and its complex with calmodulin. *J. Muscle Res. Cell Motil.* 12: 145–51.

Mak, A. S., Carpenter, M., Smillie, L. B., and Wang, J. H. (1991a). Phosphorylation of caldesmon by p34^{cdc2} kinase. Identification of phosphorylation sites. *J. Biol. Chem.* 266: 19971–5.

Mak, A. S., Watson, M. H., Litwin, C. M. E., and Wang, J. H. (1991b). Phosphorylation of caldesmon by p34^{cdc2} kinase. *J. Biol. Chem.* 266: 6678–81.

Makuch, R., Birukov, K., Shirinsky, V., and Dabrowska, R. (1991). Functional interrelationship between calponin and caldesmon. *Biochem. J.* 280: 33–8.

Makuch, R., Kolakowski, J., and Dabrowska, R. (1992). The importance of C-terminal amino acid residues of actin to the inhibition of actomyosin ATPase activity by caldesmon and troponin I. *FEBS Letts.* 297: 237–40.

Makuch, R., Walsh, M. P., and Dabrowska, R. (1989). Location of the calmodulin- and actin-binding domains at the C-terminus of caldesmon. *FEBS Letts.* 247: 411–14.

Malencik, D. A., Ausio, J., Byles, C. E., Modrell, B., and Anderson, S. R. (1989).

Turkey gizzard caldesmon: Molecular weight determination and calmodulin binding studies. *Biochemistry* 28: 8227–33.

Mani, R. S., and Kay, C. M. (1990). Isolation and characterization of a novel molecular wt 11,000 Ca^{2+}-binding protein from smooth muscle. *Biochemistry* 29: 1398–1404.

Mani, R. S., and Kay, C. M. (1993). Calcium-dependent regulation of the caldesmon-heavy meromyosin interaction by caltropin. *Biochemistry* 32: 11217–23.

Mani, R. S., McCubbin, W. D., and Kay, C. M. (1992). Calcium-dependent regulation of caldesmon by an 11-kDa smooth muscle calcium-binding protein, caltropin. *Biochemistry* 31: 11896–11901.

Marston, S. B. (1989). A tight-binding interaction between smooth-muscle native thin filaments and heavy meromyosin in the presence of MgATP. *Biochem. J.* 259: 303–6.

Marston, S. B. (1991). Properties of calponin isolated from sheep aorta thin filaments. *FEBS Letts.* 292: 179–82.

Marston, S. B., Fraser, I. D. C., and Huber, P. A. J. (1994a). Smooth muscle caldesmon controls the strong binding interaction between actin-tropomyosin and myosin. *J. Biol. Chem.* 269: 32104–9.

Marston, S. B., Fraser, I. D. C., Huber, P. A. J., Pritchard, K., Gusev, N. B., and Torok, K. (1994b). Location of two contact sites between human smooth muscle caldesmon and Ca^{2+}-calmodulin. *J. Biol. Chem.* 269: 8134–9.

Marston, S., Lehman, W., Moody, C., Pritchard, K., and Smith, C. (1988). Caldesmon and Ca^{2+} regulation in smooth muscles. In: *Calcium and Calcium Binding Proteins,* Ch. Gerday, R. Gilles, and L. Bolis, eds. Berlin: Springer Verlag pp. 69–81.

Marston, S., Pinter, K., and Bennett, P. (1992). Caldesmon binds to smooth muscle myosin and myosin rod and crosslinks thick filaments to actin filaments. *J. Muscle Res. Cell Motil.* 13: 206–18.

Marston, S. B., and Redwood, C. S. (1992). Inhibition of actin-tropomyosin activation of myosin MgATPase activity by the smooth muscle regulatory protein caldesmon. *J. Biol. Chem.* 267: 16796–16800.

Marston, S. B., and Redwood, C. S. (1993). The essential role of tropomyosin in cooperative regulation of smooth muscle thin filament activity by caldesmon. *J. Biol. Chem.* 268: 12317–20.

Marston, S. B., and Smith, C. W. J. (1985). The thin filaments of smooth muscles. *J. Muscle Res. Cell Motil.* 6: 669–708.

Martin, F., Harricane, M-C., Audemard, E., Pons, F., and Mornet, D. (1991). Conformational change of turkey-gizzard caldesmon induced by specific chemical modification with carbodiimide. *Eur. J. Biochem.* 195: 335–42.

McGhee, J. D., and vonHippel, P. H. (1974). Theoretical aspects of DNA-protein interactions: Co-operative and non-co-operative binding of large ligands to a one-dimensional homogeneous lattice. *J. Mol. Biol.* 86: 469–89.

Mezgueldi, M., Derancourt, J., Calas, B., Kassab, R., and Fattoum, A. (1994). Precise identification of the regulatory F-actin- and calmodulin-binding sequences in the 10-kDa carboxyl-terminal domain of caldesmon. *J. Biol. Chem.* 269: 12824–32.

Mezgueldi, M., Fattoum, A., Derancourt, J., and Kassab, R. (1992). Mapping of the functional domains in the amino-terminal region of calponin. *J. Biol. Chem.* 267: 15943–51.

Mezgueldi, M., Mendre, C., Calas, B., Kassab, R., and Fattoum, A. (1995). Characterization of the regulatory domain of gizzard calponin. Interactions of the 145–163 region with F-actin, calcium-binding proteins, and tropomyosin. *J. Biol. Chem.* 270: 8867–76.

Miki, M., Walsh, M. P., and Hartshorne, D. J. (1992). The mechanism of inhibition of the actin-activated myosin MgATPase by calponin. *Biochem. Biophys. Res. Commun.* 187: 867–71.

Milligan, R. A., Whittaker, M., and Safer, D. (1990). Molecular structure of F-actin and location of surface binding sites. *Nature* 348: 217–21.

Moody, C., Lehmann, W., and Craig, R. (1990). Caldesmon and the structure of smooth muscle thin filaments: Electron microscopy of isolated thin filaments. *J. Muscle Res. Cell Motil.* 11: 176–85.

Mornet, D., Bonet-Kerrache, A., Strasburg, G. M., Patchell, V. B., Perry, S. V., Huber, P. A. J., Marston, S. B., Slatter, D. A., Evans, J. S., and Levine, B. A. (1995). The binding of distinct segments of actin to multiple sites in the C-terminus of caldesmon: Comparative aspects of actin interaction with troponin-I and caldesmon. *Biochemistry* 34: 1893–1901.

Mornet, D., Harricane, M. C., and Audemard, E. (1988). A 35-kilodalton fragment from gizzard smooth muscle caldesmon that induces F-actin bundles. *Biochem. Biophys. Res. Commun.* 155: 808–15.

Nakamura, F., Mino, T., Yamamoto, J., Naka, M., and Tanaka, T. (1993). Identification of the regulatory site in smooth muscle calponin that is phosphorylated by protein kinase C. *J. Biol. Chem.* 268: 6194–6201.

Ngai, P. K., and Walsh, M. P. (1984). Inhibition of smooth muscle actin-activated myosin Mg^{2+}-ATPase activity by caldesmon. *J. Biol. Chem.* 259: 13656–9.

Ngai, P. K., and Walsh, M. P. (1985). Properties of caldesmon isolated from chicken gizzard. *Biochem. J.* 230: 695–707.

Ngai, P. K., and Walsh, M. P. (1987). The effects of phosphorylation of smooth-muscle caldesmon. *Biochem. J.* 244: 417–25.

Nishida, W., Abe, M., Takahashi, K., and Hiwada, K. (1990). Do thin filaments of smooth muscle contain calponin? A new method for the preparation. *FEBS Letts.* 268: 165–8.

Noda, S., Ito, M., Watanabe, S., Takahashi, K., and Maruyama, K. (1992). Conformational changes of actin induced by calponin. *Biochem. Biophys. Res. Commun.* 185: 481–7.

North, A. J., Gimona, M., Cross, R. A., and Small, J. V. (1994). Calponin is localized in both the contractile apparatus and the cytoskeleton of smooth muscle cells. *J. Cell Sci.* 107: 437–44.

Nowak, E., Borovikov, Y. S., and Dabrowska, R. (1989). Caldesmon weakens the binding between myosin heads and actin in ghost fibers. *Biochim. Biophys. Acta* 999: 289–92.

Obara, K., Szymanski, P. T., Tao, T., and Paul, R. J. (1995). Calponin inhibits shortening velocity and isometric force in skinned taenia coli smooth muscle. *Biophys. J.* 68: A75.

Okagaki, T., Higashi-Fujime, S., Ishikawa, R., Takano-Ohmuro, H., and Kohama, K. (1991). *In vitro* movement of actin filaments on gizzard smooth muscle myosin: Requirement of phosphorylation of myosin light chain and effects of tropomyosin and caldesmon. *J. Biochem. (Tokyo)* 109: 858–66.

Park, S., and Rasmussen, H. (1986). Carbachol-induced protein phosphorylation changes in bovine tracheal smooth muscle. *J. Biol. Chem.* 261: 15734–9.

Parker, C. A., Takahashi, K., Tao, T., and Morgan, K. G. (1994). Agonist-induced redistribution of calponin in contractile vascular smooth muscle cells. *Am. J. Physiol. (Cell Physiol.)* 267: C1262–C1270.

Pato, M. D., Sutherland, C., Winder, S. J., and Walsh, M. P. (1993). Smooth-muscle caldesmon phosphatase is SMP-1, a type 2A protein phosphatase. *Biochem. J.* 293: 35–41.

Pelech, S. L., and Sanghera, J. S. (1992). Mitogen-activated protein kinases: Versatile transducers for cell signalling. *Trends Biochem. Sci.* 17: 233–8.

Pfitzer, G., Zeugner, C., Troschka, M., and Chalovich, J. M. (1993). Caldesmon and a 20-kDa actin-binding fragment of caldesmon inhibit tension development in skinned gizzard muscle fiber bundles. *Proc. Nat. Acad. Sci. USA* 90: 5904–8.

Pinter, K., and Marston, S. B. (1992). Phosphorylation of vascular smooth muscle caldesmon by endogenous kinase. *FEBS Letts.* 305: 192–6.

Pohl, J., Walsh, M. P., and Gerthoffer, W. T. (1991). Calponin and caldesmon phosphorylation in canine tracheal smooth muscle. *Biophys. J.* 59: A58.

Popp, D., and Holmes, K. C. (1992). X-ray diffraction studies on oriented gels of vertebrate smooth muscle thin filaments. *J. Mol. Biol.* 224: 65–76.

Pritchard, K., and Marston, S. B. (1989). Ca^{2+}-calmodulin binding to caldesmon and the caldesmon-actin-tropomyosin complex. Its role in Ca^{2+} regulation of the activity of synthetic smooth-muscle thin filaments. *Biochem. J.* 257: 839–43.

Pritchard, K., and Marston, S. B. (1991). Ca^{2+}-dependent regulation of vascular smooth-muscle caldesmon by S. 100 and related smooth-muscle proteins. *Biochem. J.* 277: 819–24.

Pritchard, K., and Marston, S. B. (1993). The Ca^{2+}-sensitizing component of smooth muscle thin filaments: Properties of regulatory factors that interact with caldesmon. *Biochem. Biophys. Res. Commun.* 190: 668–73.

Redwood, C. S., and Marston, S. B. (1993). Binding and regulatory properties of expressed functional domains of chicken gizzard smooth muscle caldesmon. *J. Biol. Chem.* 268: 10969–76.

Remboldt, C. M., and Murphy, R. A. (1993). Models of the mechanism for cross-bridge attachment in smooth muscle. *J. Muscle Res. Cell Motil.* 14: 325–33.

Riseman, V. M., Lynch, W. P., Nefsky, B., and Bretscher, A. (1989). The calmodulin and F-actin binding sites of smooth muscle caldesmon lie in the carboxyl-terminal domain whereas the molecular weight heterogeneity lies in the middle of the molecule. *J. Biol. Chem.* 264: 2869–75.

Rokolya, A., and Moreland, R. S. (1993). Calponin phosphorylation during endothelin-1 induced contraction of intact swine carotid artery. *Biophys. J.* 64: A31.

Rüegg, J. C., Zeugner, C., Strauss, J. D., Paul, R. J., Kemp, B., Chem, M., Li, A-Y., and Hartshorne, D. (1989). A calmodulin-binding peptide relaxes skinned muscle from guinea-pig taenia coli. *Pflügers Arch.* 414: 282–5.

Schmidt, U. S., Troschka, M., and Pfitzer, G. (1995). The variable coupling between force and myosin light chain phosphorylation in triton-skinned chicken gizzard fibre bundles: Role of myosin light chain phosphatase. *Pflügers Arch.* 429: 708–15.

Sen, A., and Chalovich, J. M. (1995). An evaluation of the competition of binding of caldesmon and S1 to actin. *Biophys. J.* 68: A163 (abstract).

Shirinsky, V. P., Biryukov, K. G., Hettasch, J. M., and Sellers, J. R. (1992). Inhibition of the relative movement of actin and myosin by caldesmon and calponin. *J. Biol. Chem.* 267: 15886–92.

Shirinsky, V. P., Bushueva, T. L., and Frolova, S. I. (1988). Caldesmon-calmodulin interaction. Study by the method of protein intrinsic tryptophan fluorescence. *Biochem. J.* 255: 203–9.

Smith, C. W. J., Pritchard, K., and Marston, S. B. (1987). The mechanism of Ca^{2+} regulation of vascular smooth muscle thin filaments by caldesmon and calmodulin. *J. Biol. Chem.* 262: 116–22.

Sobue, K., Morimoto, K., Inui, M., Kanda, K., and Kakiuchi, S. (1982). Control of

actin–myosin interaction of gizzard smooth muscle by calmodulin- and caldesmon-linked flip-flop mechanism. *Biomed. Res.* 3: 188–96.
Sobue, K., Muramoto, Y., Fujita, M., and Kakiuchi, S. (1981). Purification of a calmodulin-binding protein from chicken gizzard that interacts with F-actin. *Proc. Nat. Acad. Sci. USA.* 78: 5652–5.
Sobue, K., and Sellers, J. R. (1991). Caldesmon, a novel regulatory protein in smooth muscle and nonmuscle actomyosin systems. *J. Biol. Chem.* 266: 12115–18.
Somlyo, A. P., and Somlyo, A. V. (1994). Signal transduction and regulation in smooth muscle. *Nature* 372: 231–6.
Stafford, W. F., Chalovich, J. M., and Graceffa, P. (1994). Turkey gizzard caldesmon molecular weight and shape. *Arch. Biochem. Biophys.* 313: 47–9.
Stafford, W. F. I., Mabuchi, K., Takahashi, K., and Tao, T. (1995). Physical characteristics of calponin. A circular dichroism, analytical ultracentrifuge, and electron microscopy study, *J. Biol. Chem.* 270: 10576–9.
Steusloff, A., Paul, E., Semenchuk, L. A., Di Salvo, J., and Pfitzer, G. (1995). Modulation of Ca^{2+}-sensitivity in smooth muscle by genistein and protein tyrosine-phosphorylation. *Arch. Biochem. Biophys.* 320: 236–42.
Suematsu, E., Resnick, M., and Morgan, K. G. (1991). Change of Ca^{2+} requirement for myosin phosphorylation by prostaglandin $F_{2\alpha}$. *Am. J. Physiol.* 261: C253–C258.
Sutherland, C., Renaux, B. S., McKay, D. J., and Walsh, M. P. (1994). Phosphorylation of caldesmon by smooth-muscle casein kinase II. *J. Muscle Res. Cell Motil.* 15: 440–56.
Sutherland, C., and Walsh, M. P. (1989). Phosphorylation of caldesmon prevents its interaction with smooth muscle myosin. *J. Biol. Chem.* 264: 578–83.
Szczesna, D., Graceffa, P., Wang, C.-L. A., and Lehrer, S. S. (1994). Myosin S1 changes the orientation of caldesmon on actin. *Biochemistry* 33: 6716–20.
Szpacenko, A. and Dabrowska, R. (1986). Functional domain of caldesmon. *FEBS Letts.* 202: 182–6.
Szpacenko, A., Wagner, J., Dabrowska, R., and Rüegg, J. C. (1985). Caldesmon-induced inhibition of ATPase activity of actomyosin and contraction of skinned fibers of chicken gizzard smooth muscle. *FEBS Letts.* 192: 9–12.
Szymanski, P. T., and Tao, T. (1993). Interaction between calponin and smooth muscle myosin. *FEBS Letts.* 331: 256–9.
Takahashi, K., Hiwada, K., and Kokubu, T. (1986). Isolation and characterization of a 34000-dalton calmodulin- and F-actin-binding protiein from chicken gizzard smooth muscle. *Biochem. Biophys. Res. Commun.* 141: 20–6.
Takahashi, K., Hiwada, K., and Kokubu, T. (1987). Occurence of anti-gizzard P34K antibody cross-reactive components in bovine smooth muscles and non-muscle tissues. *Life Sci.* 41: 291–6.
Takahashi, K., and Nadal-Ginard, B. (1991). Molecular cloning and sequence analysis of smooth muscle calponin. *J. Biol. Chem.* 266: 13284–8.
Tanaka, T., Ohta, H., Kanda, K., Hidaka, H., and Sobue K. (1990). Phosphorylation of high-M_r caldesmon by protein kinase C modulates the regulatory function of this protein on the interaction between actin and myosin. *Eur. J. Biochem.* 188: 495–500.
Tansey, M. G., Hori, M., Karaki, H., Kamm, K. E., and Stull, J. T. (1990). Okadaic acid uncouples myosin light chain phosphorylation and tension in smooth muscle. *FEBS Letts.* 270: 219–21.
Tsuruda, T. S., Watson, M. H., Foster, D. B., Lin, J. J. J.-C., and Mak, A. S. (1995). Alignment of caldesmon on the actin-tropomyosin filaments. *Biochem J.* 309: 951–7.

Umekawa, H., and Hidaka, H. (1985). Phosphorylation of caldesmon by protein kinase C. *Biochem. Biophys. Res. Commun.* 132: 56–62.

Vancompernolle, K., Gimona, M., Herzog, M., Van Damme, J., Vandekerckhove, J., and Small, V. (1990). Isolation and sequence of a tropomyosin-binding fragment of turkey gizzard calponin. *FEBS Letts.* 274: 146–50.

Velaz, L., Chen, Y., and Chalovich, J. M. (1993). Characterization of a caldesmon fragment that competes with myosin-ATP binding to actin. *Biophys. J.* 65: 892–8.

Velaz, L., Hemric, M. E., Benson, C. E., and Chalovich, J. M. (1989). The binding of caldesmon to actin and its effect on the ATPase activity of soluble myosin subfragments in the presence and absence of tropomyosin. *J. Biol. Chem.* 264: 9602–10.

Velaz, L., Ingraham, R. H., and Chalovich, J. M. (1990). Dissociation of the effect of caldesmon on the ATPase activity and on the binding of smooth heavy meromyosin to actin by partial digestion of caldesmon. *J. Biol. Chem.* 265: 2929–34.

Vibert, P., Craig, R., and Lehman, W. (1993). Three-dimensional reconstruction of caldesmon-containing smooth muscle thin filaments. *J. Cell Biol.* 123: 313–21.

Vorotnikov, A. V., Gusev, N. B., Hua, S., Collins, J. H., Redwood, C. S., and Marston, S. B. (1994). Phosphorylation of aorta caldesmon by endogenous proteolytic fragments of protein kinase C. *J. Muscle Res. Cell Motil.* 15: 37–48.

Vorotnikov, A. V., Shirinsky, V. P., and Gusev, N. B. (1988). Phosphorylation of smooth muscle caldesmon by three protein kinases: Implication for domain mapping. *FEBS Letts.* 236: 321–4.

Walker, G., Kerrick, W. G. L., and Bourguignon, Y. W. (1989). The role of caldesmon in the regulation of receptor capping in mouse T-lymphoma cell. *J. Biol. Chem.* 264: 496–500.

Walsh, M. P., Carmichael, J. D., and Kargacin, G. J. (1993). Characterization and confocal imaging of calponin in gastrointestinal smooth muscle. *Am. J. Physiol.* 265: C1371–C1378.

Wang, C.-L. A. (1988). Photocrosslinking of calmodulin and/or actin to chicken gizzard caldesmon. *Biochem. Biophys. Res. Commun.* 156: 1033–8.

Wang, Z., Horiuchi, K. Y., Jacob, S. S., Gopalakurup, S., and Chacko, S. (1994). Overexpression, purification, and characterization of full-length and mutant caldesmons using a baculovirus expression system. *J. Musc. Res. Cell Motil.* 15: 646–58.

Wills, F. L., McCubbin, W. D., and Kay, C. M. (1993). Characterization of the smooth muscle calponin and calmodulin complex. *Biochemistry* 32: 2321–8.

Wills, F. L., McCubbin, W. D., and Kay, C. M. (1994). Smooth muscle calponin-caltropon interaction: Effect on biological activity and stability of calponin. *Biochemistry* 33: 5562–9.

Winder, S. J., Allen, B. G., Fraser, E. D., Kang, H-M., Kargacin, G. J., and Walsh, M. P. (1993a). Calponin phosphorylation *in vitro* and in intact muscle. *Biochem J.* 296: 827–36.

Winder, S. J., Kargacin, G. J., Bonet-Kerrache, A. A., Pato, M. D., and Walsh, M. P. (1992a). Calponin: Localization and regulation of smooth muscle actomyosin MgATPase. *Jpn. J. Pharmacol.* 58 (suppl.) 2: 29P–34P.

Winder, S. J., Sutherland, C., and Walsh, M. P. (1991). Biochemical and functional characterization of smooth muscle calponin. *Adv. Exp. Med. Biol.* 304: 37–51.

Winder, S. J., Sutherland, C., and Walsh, M. P. (1992b). A comparison of the effects of calponin on smooth and skeletal muscle actomyosin systems in the presence and absence of caldesmon. *Biochem. J.* 288: 733–9.

Winder, S. J., and Walsh, M. P. (1990). Smooth muscle calponin. Inhibition of actomyosin MgATPase and regulation by phosphorylation. *J. Biol. Chem.* 265: 10148–55.

Winder, S. T., and Walsh, M. P. (1993). Calponin: Thin filament-linked regulation of smooth muscle contraction. *Cell. Signal.* 5: 677–86.

Winder, S. J., Walsh, M. P., Vasulka, C., and Johnson, J. D. (1993b). Calponin-calmodulin interaction: Properties and effects on smooth and skeletal muscle actin binding and actomyosin ATPases. *Biochemistry* 32: 13327–33.

Yamakita, Y., Yamashiro, S., and Matsumura, F. (1992). Characterization of mitotically phosphorylated caldesmon. *J. Biol. Chem.* 267: 12022–9.

Yamashiro, S., Yamakita, Y., Hosoya, H., and Matsumura, F. (1991). Phosphorylation of non-muscle caldesmon by $p34^{cdc2}$ kinase during mitosis. *Nature* 349: 169–72.

Yamashiro, S., Yamakita, Y., Ishikawa, R., and Matsumura, F. (1990). Mitosis-specific phosphorylation causes 83k non-muscle caldesmon to dissociate from microfilaments. *Nature* 344: 675–8.

Zhan, Q., Wong, S. S., and Wang, C.-L. A. (1991). A calmodulin-binding peptide of caldesmon. *J. Biol. Chem.* 266: 21810–14.

Zhuang, S. B., Wang, E. Z., and Wang, C. L. A. (1995). Identification of the functionally relevant calmodulin binding site in smooth muscle caldesmon. *J. Biol. Chem.* 270: 19964–8.

Index